W9-CRL-657

ACS SYMPOSIUM SERIES **637**

Biotechnology for Improved Foods and Flavors

Gary R. Takeoka, EDITOR
Agricultural Research Service

Roy Teranishi, EDITOR
Agricultural Research Service

Patrick J. Williams, EDITOR
Australian Wine Research Institute

Akio Kobayashi, EDITOR
Ochanomizu University

Developed from a symposium sponsored by
the ACS Division of Agricultural and Food Chemistry at the
1995 International Chemical Congress of Pacific Basin Societies

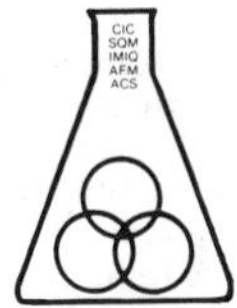

American Chemical Society, Washington, DC

Library of Congress Cataloging-in-Publication Data

Biotechnology for improved foods and flavors / Gary R. Takeoka, editor . . . [et al.].

p. cm.—(ACS symposium series; ISSN 0097–6156; 637)

"Developed from a symposium sponsored by the ACS Division of Agricultural and Food Chemistry at the 1995 International Chemical Congress of Pacific Basin Societies."

Includes bibliographical references and indexes.

ISBN 0–8412–3421–3

1. Food—Biotechnology—Congresses. 2. Agricultural biotechnology—Congresses.

I. Takeoka, Gary R. II. American Chemical Society. Division of Agricultural and Food Chemistry. III. International Chemical Congress of Pacific Basin Societies (1995: Honolulu, Hawaii) IV. Series.

TP248.65.F66B635 1996
664′.024—dc20 96–24783
CIP

This book is printed on acid-free, recycled paper.

PRINTED IN THE UNITED STATES OF AMERICA

Advisory Board

ACS Symposium Series

Foreword

THE ACS SYMPOSIUM SERIES was first published in 1974 to provide a mechanism for publishing symposia quickly in book form. The purpose of this series is to publish comprehensive books developed from symposia, which are usually "snapshots in time" of the current research being done on a topic, plus some review material on the topic. For this reason, it is necessary that the papers be published as quickly as possible.

Before a symposium-based book is put under contract, the proposed table of contents is reviewed for appropriateness to the topic and for comprehensiveness of the collection. Some papers are excluded at this point, and others are added to round out the scope of the volume. In addition, a draft of each paper is peer-reviewed prior to final acceptance or rejection. This anonymous review process is supervised by the organizer(s) of the symposium, who become the editor(s) of the book. The authors then revise their papers according to the recommendations of both the reviewers and the editors, prepare camera-ready copy, and submit the final papers to the editors, who check that all necessary revisions have been made.

As a rule, only original research papers and original review papers are included in the volumes. Verbatim reproductions of previously published papers are not accepted.

ACS BOOKS DEPARTMENT

Contents

INDEXES

Preface

"Biotechnology is the application of scientific and engineering principles to the processing of materials by biological agents to provide goods and services" (Organization for Economic Co-operation and Development).

In the Western World, biotechnology's ancient roots lie in the production of a variety of items, such as brews, leavened bread and other cereal products, and cheeses. In the East, organisms have been used for centuries to produce soy sauce, miso, sake, as well as many other foods and drinks. Today, the production of these foods and drinks has improved because of the increase in scientific knowledge of the organisms used. Although fear exists with respect to the use of genetic engineering in the biotechnological production of foods and drinks, there is no fear in the old definition of biotechnology: the use of biological agents to provide foods and drinks. Recently, remarkable advances have been made in the knowledge and applications of biological agents and the processes utilizing them.

The symposium on which this volume is based was sponsored by the ACS Division of Agricultural and Food Chemistry at the 1995 International Chemical Congress of Pacific Basin Societies (Pacifichem '95) from December 17–22, 1995, in Honolulu, Hawaii. The purpose of this symposium was to discuss the various biotechnological projects that are directly and indirectly involved in food and flavor production and the actual and proposed benefits evolving from such efforts. Topics covered included industrial applications and products and how some biotechnological methods are used specifically to improve quality; enzyme and microbial transformations and how detailed knowledge is helping to control their effects; and methodology developed in recent years and how it helps in following various biotechnological transformations.

Internationally known experts in governmental, industrial, and academic organizations have contributed to this book. Biotechnology is becoming more and more important for producing foods and flavors by "natural" methods. Those engaged in or interested in such endeavors will find the topics discussed in this book interesting and useful.

Acknowledgments

The symposium and book would not have been possible without the financial assistance of: Ajinomoto Company, Inc.; Hasegawa Company, Ltd.; Kirin Brewery Company, Ltd.; Ogawa & Company, Ltd.; and Takasago International Corporation. We are grateful for their generous contributions.

GARY R. TAKEOKA
ROY TERANISHI
Western Regional Research Center
Agricultural Research Service
U.S. Department of Agriculture
800 Buchanan Street
Albany, CA 94710

PATRICK J. WILLIAMS
Australian Wine Research Institute
P.O. Box 197
Glen Osmond
South Australia 5064
Australia

AKIO KOBAYASHI
Laboratory of Food Chemistry
Ochanomizu University
2–1–1, Ohtsuka, Bunkyo-ku,
Tokyo 112
Japan

March 21, 1996

Chapter 1

Food and Agricultural Biotechnology: An Overview

Daniel D. Jones and Alvin L. Young[1]

Office of Agricultural Biotechnology, Cooperative State Research, Education, and Extension Service, U.S. Department of Agriculture, 14th Street and Independence Avenue, Southwest, Washington, DC 20250–0904

Biotechnology is an invaluable process for the quick, safe, and precise transfer of specific genetic information from one organism to another, in order to create predictable end results. As such, biotechnology represents a tool that is an important component of a balanced, efficient, well-managed, and environmentally responsible agricultural system, which uses the very best of technology and science. Recent discoveries in the field of biotechnology have made the life sciences some of the most exciting fields of scientific endeavor--especially for those with creativity and vision. This overview provides a "sampling" of how biotechnology will impact global agriculture, and hence, the world population of the third millennium.

For centuries, people have sought to improve plants and animals and food products derived from them by selecting and breeding individuals that excel in some desirable property or characteristic. The usual goals of traditional breeding have been to develop plants and animals that grow faster, produce more, provide better quality products, use resources more efficiently, and show increased fertility or resistance to disease and stress. The tools and techniques provided by modern biotechnology do not change these traditional goals of agriculture. Instead, they offer new opportunities for changing biological traits in a much more direct, predictable, and timely manner than is possible with conventional plant and animal breeding.

Several products of modern biotechnology are now on the market (*1*). They include recombinant chymosin for use in cheese-making, recombinant bovine somatotropin for increasing feed efficiency in dairy cattle, and antisense tomatoes that can be picked ripe and have a longer shelf-life. Many more products of food and agricultural biotechnology are in the development pipeline and they should begin appearing on supermarket shelves in the coming months and years (*1,2*).

The Federal government recently released a report entitled *Biotechnology for the 21st Century: New Horizons* (*3*). In it are described emerging biotechnology

[1]Corresponding author

research opportunity areas in agriculture, the environment, manufacturing/bioprocessing, and marine biotechnology and aquaculture. The report outlines agricultural research opportunities in five areas: 1) genome mapping in plants, animals, and microbes; 2) biochemical and genetic control of metabolic pathways; 3) the molecular basis of growth and development in plants and animals; 4) the molecular basis of interactions of plants and animals with their physical and biological environments; and 5) enhancement of food safety through development of molecular probes, biosensors, and microbial ecological control methods (3).

The purpose of this review is to survey recent developments in several areas of food and agricultural biotechnology. These areas include plant biotechnology, animal biotechnology, aquaculture, genome mapping, food safety, diet and nutrition, functional foods, and patenting. Several of these issues have also received extensive coverage in the science press (*4*).

Plant Biotechnology

USDA has issued permits and ackowledgements for testing genetically engineered plants in thousands of field test sites (*5*). Six plant species accounted for about 85 percent of the approved field tests; tobacco, tomato, potato, corn, soybean, and cotton. The early genetic changes were largely single-gene changes such as herbicide tolerance, insect resistance, virus resistance, and fungus resistance. Now a wider variety of genetic changes in product quality is being tested including: delayed ripening in tomatoes, peas, and peppers; modified oils, enzymes, and storage proteins; higher solids content in tomato and potato; improved nutrition in corn, sunflower, soybean, and oilseeds; and freeze tolerance in tomatoes and other fruits and vegetables (*2*).

Many transgenic plants field tested heretofore have involved only one or a few gene changes. Current research on genome mapping and gene regulation will facilitate the future manipulation of whole clusters of genes that influence agronomic and other economically important traits. Multi-gene traits in plants that may be amenable to the methods of biotechnology include improvements in photosynthesis, nitrogen fixation, heat tolerance, drought tolerance, cold tolerance, and salt tolerance. For example, scientists have transferred a gene for salt tolerance from an Old World plant called "ice plant" to tomato, tobacco, and *Arabidopsis thaliana* (*3,6,7*). The ability to clone ion transport genes and their genetic regulatory sequences from salt-tolerant plants such as ice plant and mangrove could lead to more crop species better able to withstand salt build-up from irrigated soils with poor drainage.

Another plant biotechnology development with implications for agricultural production is the use of plants as bioreactors to produce valuable substances which are then isolated from the plant and marketed. A recent example is transgenic tobacco that produces human glucocerebrosidase (hGC), an enzyme used in the treatment of Gaucher's disease (*8*). A single clinically useful dose of hGC has previously taken thousands of human placentae or Chinese hamster ovary cells, thus contributing to the high cost of the drug. Other examples of exogenous substances from transgenic plants include engineered antibodies in transgenic tobacco (*9*) foot-and-mouth disease antigen in transgenic cowpea (*10,11*), and rabies virus glycoprotein in transgenic tomatoes (*12*). A number of companies appear to be working on the production of enzymes, therapeutics, and vaccines from transgenic tobacco (*12*), but there is uncertainty that tobacco "pharming" will be a profitable endeavor for traditional tobacco farmers (*13*).

Animal Biotechnology

Animal biotechnology can lead to improvements in growth and feed efficiency, animal health, growth and development, reproductive efficiency, food product quality, and lactational production of novel or valuable proteins. One of the greatest advantages of transgenic animal technology is that it can produce specific, significant genetic changes that are not constrained by existing genetic variation in the host population (*14*).

For technical and economic reasons, animal biotechnology is proceeding at a somewhat slower pace than plant biotechnology. Early experiments involved the use of animal growth hormones, administered either by repeated injection or by genetic incorporation, to improve feed efficiency and growth rate (*15*). Repeated injection is very labor-intensive and the success rate of most current methods of genetic incorporation is disappointingly low (*15*). Consequently, animal scientists are looking for better methods of transferring genetic material and for controlling the expression of genetic material once it is incorporated into a host animal.

Growth Hormones. One of the more controversial areas of animal biotechnology has been the administration of bovine somatotropin (BST) to cows to increase the efficiency of milk production. This development has posed issues of both food safety and food labeling. The Food and Drug Administration (FDA) concluded in 1993 that milk from cows treated with recombinant BST (rBST) is safe for human consumption (*16*). The FDA approval was based on over 100 studies of the human food safety of BST (17).

The use of rBST posed difficult food labeling issues for dairy products. The FDA concluded that it did not have a statutory or scientific basis for requiring special labeling of dairy products from rBST-treated cows. Food companies, however, may voluntarily label dairy products from rBST-treated cows provided the information is truthful and not misleading (*16*). Some companies have seen more market value in milk from cows not treated with rBST. FDA also permitted special labeling for milk from untreated cows as long as the labeling was truthful and not misleading (*18*).

Some authors have focused on opportunities for genetic manipulation of dairy cattle beyond BST (*19*). Opportunities include production of heterologous proteins in milk (*20,21*), modification of milk and milk protein composition (*22,23*), and genetic modification of the rumen or intestinal microflora of cows to improve their digestive capacity, feed efficiency, or ability to degrade plant toxicants (*24,25*).

Scientists have also begun to think about transgenic arthropods as a component of pest control programs (*26*). Contained research on transgenic insects is in progress in many laboratories around the world (*27*) and it may lead to advances in the control of insect pests as well as to help for beneficial insects such as bees.

Aquaculture

Aquaculture, the cultivation of certain aquatic animals and plants in farms on land and at sea, offers a partial solution to growing global consumption of fish and declining natural fishery populations. The World Bank estimates that aquaculture could meet some 40 percent of the demand for fish within 15 years if the proper investments in research and technology are made by governments today (*28*). Those practicing aquaculture have to contend with the same problems that any farmer faces: disease, pollution, space limitations, volatile markets, and unintended effects on the environment. Genetic engineering offers a promising approach to managing many of these problems (3). China, for example, has 10 million farmers engaged in the annual production of 21 million metric tons of fish. The government of China has announced that their goal for the next five years (1996-2000) is to increase production to 30 million metric tons. The application of biotechnology to enhance growth, feed efficiency, and reproduction will be essential in meeting this goal (*29*). International organizations have also begun the process of developing criteria for evaluating the food safety of products of aquaculture (30).

Some people are also very concerned about the potential environmental effects of aquaculture operations. A USDA advisory committee on aquatic biotechnology and environmental safety recently drew up a set of performance standards and logic flowcharts for safely conducting research with genetically modified fish and shellfish (ABRAC). This effort was especially noteworthy because it enjoyed the collective support of the aquatic research community, private industry, environmental interest groups, and state natural resource officials.

Genome Mapping

Genome mapping is an ongoing scientific effort that will provide useful background information for genetic modification of plants and animals. Just as improvements in mapmaking contributed to the age of world exploration in the 15th and 16th centuries, so will genome mapping make significant contributions to the breeding of plants and animals and the improvement of foods derived from them.

Since the mid-1980's, scientists have been able to construct maps of the genetic makeup of various organisms. Programs of genome mapping for economically important traits have been initiated. Genome maps for several agronomically important species are near completion. These maps offer tremendous opportunities for researchers to pinpoint the genetic sequences that control specific traits (*31*). USDA supports genome research programs in both plants and animals. In the plant area for FY 1994, for example, USDA awarded about $12 M in grants to over 100 investigators to map all or part of the genomes of over 50 species of plants (*32*). USDA also supports an animal genome mapping project although not as many species are involved as in the plant area. Agricultural Experiment Stations associated with many of the Land Grant Universities have also recognized the importance of investing in genome programs that involve species to state economies (*31*).

Genetically Engineered Foods

Biotechnology permits the transfer of the genetic material that codes for desirable traits from one organism, variety, or species to another to produce transgenic plants and animals. These technologies have the potential to improve the productivity and quality of many foods.

In May, 1994, the Food and Drug Administration concluded that FLAVR SAVR, a new variety of tomato developed through biotechnology, is as safe as tomatoes bred by conventional means (*33*). The FLAVR SAVR tomato was developed by a technique called antisense in which an inverted polynucleotide segment which cannot be interpreted correctly by a cell's biosynthetic machinery is added to the cell thereby interfering with the function of the target gene. The antisense technique eliminates a tissue softening enzyme from the tomato thus slowing the biological process by which a tomato ripens and eventually rots. The tomato can thus remain on the vine longer before being harvested, yet remain firm enough to ship in a ripe state most times of the year.

Genetically engineered foods were the subject of controversy since before the antisense tomato first appeared on supermarket shelves (*34,35,36,37*). The prospect of commercial availability of genetically engineered foods has raised questions of safety, ethics, and social acceptability in the minds of many consumers, advocacy groups, and public officials (*38*).

Supporters of genetic engineering say these techniques will provide healthier, cheaper, better-tasting foods; reduce farmers' dependence on toxic chemicals to control weeds and pests; and increase the world's food supply to meet the needs of a growing population. In some cases, genetic engineering will provide entirely new approaches to controlling plant pathogens such as viruses.

Critics say scientists do not fully understand the impact that genetic changes can have on the nutrition, toxicity, or other properties of foods. Critics fear that genetic manipulations may permit the wider spread of allergy-producing proteins in the food supply. They also fear that the artificial splicing of novel genes into agricultural plants and animals could have unintended ecological consequences such as more aggressive weeds, voracious, oversized fish, and rapidly evolving plant viruses.

Federal agencies have addressed many of these concerns under various statutory authorities and other ongoing programs. For example, for early field testing of genetically engineered plants under plant pest statutes, USDA prepared detailed environmental assessments before issuing permits and these assessments were critically analyzed by outside parties (*39*). As the Department of Agriculture gained experience and biosafety data on field tests of engineered plants. The permit application system was largely replaced in 1993 by a notification system (*40,49*). Similarly, for pioneering outdoor research on transgenic fish, USDA prepared an environmental assessment that was open to public comment (*42*). Subsequently, a USDA advisory committee developed a set of *Performance Standards for Safely Conducting Research with Genetically Modified Fish and Shellfish* (*43*).

Federal agencies have also addressed the safety of foods derived from new varieties of plants and animals. For example, the Food and Drug Administration published a statement of policy on foods derived from new plant varieties that addressed issues of potential allergenicity and changes in the content of nutrients and toxicants (*44*). Similarly, USDA, with the assistance of a Federal advisory committee, announced criteria for evaluating the food safety of transgenic (*45*) and non-transgenic (*46*) animals from transgenic experiments.

Biotechnology and Food Safety

Food-borne illness can become a significant threat in a large, complex food processing and distribution system. In 1993, for example, more than 500 persons in the Pacific Northwest became ill from eating undercooked hamburger and four persons died. This outbreak of food-borne illness was attributed to a pathogenic strain of the common intestinal bacterium *E. coli* 0157:H7 which appears to be harmless to cattle, but which can cause serious disease in humans (*3*).

There are several ways in which the tools of biotechnology can be used to improve the safety of food products. Biotechnology can help to improve both the detection and control of food-borne microorganisms. It can do this, for example, through improved methods for detecting microbial contaminants in food and by the design of food processing enzymes that can withstand higher processing temperatures.

Biotechnology offers several different technologies for improved detection of microbial agents in food products. Two of these are monoclonal antibody technology and DNA probe technology. Antibodies are protein molecules that can recognize foreign antigens to mark them for identification, removal or destruction. Monoclonal antibodies are antibodies that are produced by specialized cells in large amounts and high purities that make them very useful for a wide variety of detection applications including the analysis of food products for microbial contaminants and pesticide residues (*47*).

A second technology for improving microbial detection in food products is DNA probes. These are carefully designed fragments of deoxyribonucleic acid (DNA) that can bind to the genetic material of viruses, bacteria, or parasites for purposes of identification, detection, or, in some cases, inactivation. DNA probe kits are available for food-borne pathogens such as *Salmonella, Listeria, E. coli,* and *Staphylococcus aureus*. These diagnostic kits have several advantages over traditional plating methods including greater precision, shorter turnaround times, and reduced need for highly trained personnel (*3*).

An outstanding example of DNA technology is the polymerase chain reaction (PCR). This is a technique for producing millions of copies of a single DNA molecule so that it can be analyzed almost as easily as a purified chemical substance. The Food and Drug Administration has developed and deployed in its field laboratories a PCR method for the detection of *Vibrio cholerae* in imported food (*3*). Practical experience with this and other methods will no doubt highlight areas where further technical improvements can be made for purposes of food analysis.

Biotechnology can also provide methods for controlling and discouraging the growth of microorganisms during food processing. Enzymes from a number of naturally occurring organisms, such as those living around ocean-floor hydrothermal vents, are stable under conditions of high temperature and high acidity that kill or inactivate many other microorganisms. Some of these enzymes show great promise for food processing applications, and several companies have started to market thermostable enzymes for that purpose (*48*).

Biosensors. Another exciting area for food biotechnology is the development of biosensors. These are devices with special surfaces or membranes that can respond to the presence of a specific substance or cell type in a food or other substance. Depending on how the device is designed, its use could be as simple as placing the device in contact with the food product and reading the amount of the specific substance from a meter or dial. Biosensors are commercially available to detect a variety of sugars, alcohols, esters, peptides, amino acids, cell types, and antibiotics (*49,50*).

Miniaturization and mass production of biosensors could increase their availability and decrease their unit cost. Technologies such as microlithography, ultrathin membranes, and molecular self-assembly have the potential to facilitate the development and diversification a wide variety of biosensors. Miniature biosensors could be incorporated into food packages to monitor temperature stress, microbial contamination, or remaining shelf life, and to provide a visual indicator to consumers of product state at the time of purchase (*3,51,52*).

Biotechnology, Diet, and Nutrition

The link between diet, the maintenance of health, and the development of chronic disease has become increasingly evident in recent years. Many consumers are looking for inexpensive and readily available food products that meet their requirements for less fat, more nutrients, and fewer additives. The technologies of the new biotechnology hold great promise for helping nutritionists and physicians to explore the design of new foods to meet dietary and health goals, particularly for subpopulations such as children (*53*). An example is the improvement of the nutritional attributes of animal products through decreases in fat, saturated fatty acids, and cholesterol.

Porcine somatotropin (PST), for example, may help to improve human health, while at the same time lowering the farmer's cost of production. PST not only improves the feed efficiency in hogs by 25 to 30 percent -- but, perhaps more importantly in this age of health consciousness, it reduces fat deposition, allowing PST-treated hogs to provide consumers with leaner cuts of pork (*54*). This ability to produce lean pork has important implications for improving human health by reducing dietary fat and cholesterol. As producers of animal products begin to understand more about how diet relates to health and to implement appropriate feeding, breeding, and selection programs, biotechnology offers them a valuable set of tools.

Functional Foods. "Functional food" may be broadly defined as any modified food or food ingredient that may provide a health benefit beyond the traditional nutrients it contains (*55*). Other terms sometimes used include "nutraceutical" and "designer food" which usually denote a processed food that is supplemented with food ingredients naturally rich in disease-preventing substances (*55*). These disease-preventing substances include phytochemicals, substances found in fruits and vegetables that exhibit a potential for modulating human metabolism in a manner favorable for disease prevention.

The National Cancer Institute, through its Experimental Food Program, has identified many phytochemicals that can interfere with and potentially block the biochemical pathways that lead to malignancy in animals. These phytochemicals include sulfides, phytates, flavonoids, glucarates, carotenoids, coumarins, terpenes, lignans, phenolic acids, indoles, isothiocyanates, and polyacetylenes. Foods that appear to contain significant amounts of these phytochemicals include garlic, onions, broccoli, cabbage, soybeans, citrus fruits, cereal grains, and green tea (*56*).

Biotechnology is only one method of producing functional foods (*57*), but it is one that may find favor among producers in the future because of its molecular precision, genetic specificity, and relative speed.

Biotechnology and the Environment

The application of biotechnology to environmental problems is one of its areas of greatest potential. Biotechnology has already contributed to a safer environment, with many anticipated breakthroughs on the horizon. A few examples of areas where biotechnology has already had an impact include the replacement of environmentally hazardous pesticides with safer biotechnologically-produced pesticides, the development of new microbial techniques for cleaning up pollution, the creation of alternative fuels that are less environmentally damaging, and the formulation of new biodegradable materials (*3*).

Biosystems developed through biotechnology research can be used to treat contaminated wastes, oil spills, acid wastes, municipal wastes, and pesticides. Bioremediation is a biological conversion process in which living organisms assimilate and store waste byproducts such as toxic materials, heavy metals, uncollected residues from oil spills, and other pollutants that endanger the environment. Genetic engineering promises to dramatically expand our ability to create microbes that will break down a wider range of wastes (*3*). Plants can also be used to absorb heavy metals through their roots thus helping to cleanse contaminated soils (*58*).

Biotechnology promises to expand the set of tools at our disposal for control of pests in agricultural production systems. Natural and engineered microbial pesticides are being developed to control a variety of lepidopteran pests (*59*). Some of these biological pesticides may act as replacements for environmentally persistent chemicals, quickly breaking down into harmless components and thus reducing residue problems.

Biotechnology research is seeking to develop new foods, feeds, fiber, and biomass energy production processes that are environmentally safe. Researchers are developing new uses for agricultural products to replace non-renewable sources of raw materials. Their work promises to have broad commercial applications and has already led to the creation of new industries. These discoveries have led to environmentally compatible commercial products such as biodegradable plastics (*60*), soybean oil printing inks, and super absorbent polymers (*61*).

Biotechnology techniques such as gene amplification, DNA probes, bioluminescence, and immunological assays also are being used to increase our understanding of the complexities of agricultural production and the environment.

It is clear that biotechnology research will become increasingly significant in global efforts to improve agricultural production and protect the environment.

Patenting

A patent is a limited monopoly that protects the interests of an inventor by excluding others from practicing an invention for a specified period of time. Patenting of biotechnology inventions is necessary, in the view of some, to provide incentive for investing time, money, and resources in product research and development (*62*). In the view of others, patenting biotechnology inventions, particularly living organisms, raises serious ethical and religious concerns (*63*). In addition, many scientists are concerned about the chilling effect of overly-broad patents on the progress of scientific research (*64*). And farmers are concerned about exempting on-farm uses of a patented product from royalty payments (*65*). Finally, there are difficult international questions involving developing countries (*66*) and the effects of patents on international trade (*67*).

Conclusion

Our ability to manipulate genetic material for biomedical, agricultural, and environmental purposes represents a major technological advance. Biotechnology can have a dramatic impact on the agriculture and food producing sector. It has the potential to reduce the need for agricultural chemicals; improve the productivity, efficiency, and profitability of food production and processing; open new markets for improved or unique processed food products; and, improve the nutritional quality, safety, cost, and convenience of consumer food products.

Literature Cited

1. Niebling, K. *Genet. Engineering News* **1995,** *15*(13), 1, 20-21.
2. Beck, C.I.; Ulrich, T. *Bio/Technology* **1993,** *11*, 895.
3. Biotechnology Research Subcommittee, Committee on Fundamental Science, National Science and Technology Council, *Biotechnology for the 21st Century: New Horizons*, U.S. Government Printing Office, Washington, D.C., 1995.
4. Hileman, B. *Chem. & Engin. News*, Aug. 21, 1995, 8.
5. Animal and Plant Health Inspection Service, United States Department of Agriculture. Updated information on field test notifications and modifications of genetically modified plants can be obtained from several Internet addresses via gopher, ftp, telnet, or WWW (http://) including "ftp.nbiap.vt.edu, ""ftp.aphis.ag.gov/pub/bbep,"and"ftp://ftp.aphis.ag.gov/pub/ bbep/home.htm."
6. Bohnert, H.J.; Thomas, J.C.; DeRocher, E.J.; Michalowski, C.B.; Breiteneder, H.; Vernon, D.M.; Deng, W.; Yamada, S.; Jensen, R.G. in *Proceedings of a NATO Advanced Research Workshop on Biochemical and Cellular Mechanisms of Stress Tolerance in Plants*," J.H. Cherry, Ed., Springer-Verlag, Berlin, New York, 1994, p. 415.

7. Vernon, D.M.; Tarczynski, M.C.; Jensen, R.G.; Bohnert, H.J. *Plant J.* **1993,** *4*(1), 199.
8. Cramer, C.L.; Weissenborn, D.L.; Oishi, K.K.; Grabau, E.A.; Bennett, S.; Ponce, E.; Grabowski, G.A.; Radin, D.A. presentation abstract, International Symposium on Engineering Plants for Commercial Products and Apppications, Oct. 1-4, 1995, Lexington, Kentucky.
9. Fiedler, U.; Conrad, U. *Bio/Technology* **1995,** *13*, 1090.
10. Usha, P.; Rohll, J.B.; Spall, V.; Shanks, M.; Maule, A.J.; Johnson, J.E.; Lomonossoff, G.P. *Virology* **1993,** *197*, 366.
11. Porta, C.; Spall, V.E.; Loveland, J.; Johnson, J.E.; Barker, P.J.; Lomonossoff, G.P. *Virology* **1994,** *202*, 949.
12. McGarvey, P.B.; Hammond, J.; Dienelt, M.M.; Hooper, D.C.; Fu, Z.F.; Dietzschold, B.; Koprowski, H.; Michaels, F.H. *Bio/Technology* **1995,** *13*, 1484.
13. Bullock, W.O. NBIAP News Report, November 1995, p. 6.
14. Guda, C.; Daniell, H. NBIAP News Report, November 1995, p.5.
15. Gibson, J.P. *J. Dairy Sci.* **1991,** *74* , 3258.
16. Pursel, V.G.; Rexroad, C.E. *J. Animal Sci.* **1993,** *71* (S-3), 10.
17. Cruzan, S.M. HHS News Release, P93-40, Recombinant Bovine Somatotropin, U.S. Department of Health and Human Services, Washington, D.C., November 5, 1993.
18. Juskevich, J.C.; Guyer, C.G. *Science* **1990,** *249*, 875.
19. Food and Drug Administration (FDA), Federal Register **1994,** *59*, 6279.
20. Jones, D.D.; Cordle, M.K. *Biotechnology Advances* **1995,** 2, 235.
21. Maga, E.A.; Murray, J.D. *Bio/Technology* **1995,** *13*, 1452.
22. Wilmut, I.; Archibald, A.L.; McClenaghan, M.; Simons, J.P.; Whitelaw, C.B.A.; Clark, A.J. *Experientia* **1991,** *47*, 905.
23. Lee, S.H.; de Boer, H.A. *J. Controlled Release* **1994,** *29*, 213.
24. Wilmut, I.; Archibald, A.L.; Harris, S.; McClenaghan, M.; Simons, J.P.; Whitelaw, C.B.A.; Clark, A.J. *J. Reprod. Fert.*, Suppl. **1990,** *41*, 135.
25. Clark, A.J. *J. Cellular Biochem.* **1992,** *71*, 121.
26. Committee on Opportunities in the Nutrition and Food Sciences, Opportunities in the Nutrition and Food Sciences, Thomas, P.R.; Earl, R., Eds., National Academy Press, Washington, D.C. 1994, p. 40.
27. U.S. Department of Agriculture, Cooperative State Research Service, Report of Workshop on Naturally Occurring Substances in Traditional and Biotechnology-Derived Foods: Their Potential Toxic and Antitoxic Effects, Irvine, CA, USDA, Washington, D.C., July, 1992.
28. Hoy, M.A. *Parasitology Today* **1995,** *11*, 229.
29. Neven, L., in Minutes, Agricultural Biotechnology Research Advisory Committee, Doc. No. 95-06, USDA Office of Agricultural Biotechnology, Washington, D.C. June 26, 1995, p. 11.
30. Plucknett, D.L.; Winkelmann, D.L. *Sci. Amer.*, Sept. 1995, 182.
31. Young, A.L.; Kapuscinski, A.K.; Hallerman, E.; Attaway, D. *Aquaculture and Marine Biotechnology in China*, USDA Foreign Agricultural Service, Washington, D.C., 1995.

32. Organisation for Economic Cooperation and Development (OECD), *Aquatic Biotechnology and Food Safety*, OECD, Paris, 1994.
33. Agricultural Science and Technology Review Board (ASTRB), *Technology Assessments 1993*, U.S. Department of Agriculture, Washington, D.C., 1993, p. 28.
34. Datko, A.; Kaleikau, E.; Heller, S.; Miksche, J.; Smith, G.; Bigwood, D. Probe Newsletter for the USDA Plant Genome Research Program, *4* (3/4), 1, August 1994-January 1995.
35. Stone, B. HHS News Release P94-10 on FLAVR SAVR tomato, U.S. Department of Health and Human Services, Washington, D.C., May 18, 1994.
36. Hoyle, R. *Bio/Technology* **1992,** *10*, 1520.
37. Hamilton, J.O.; Ellis, J.E. *Business Week*, December 14, 1992, p. 98.
38. Rensberger, B. Washington Post, January 12, 1993.
39. Marsa, L. *Eating Well*, *3*, July/August 1993, p. 41.
40. Agricultural Science and Technology Review Board (ASTRB), *Technology Assessments 1993*, U.S. Department of Agriculture, Washington, D.C., 1993, p. 30.
41. Wrubel, R.P.; Krimsky, S.; Wetzler, R.E. *BioScience* **1992,** *42*, 280.
42. Animal and Plant Health Inspection Service, *Federal Register* **1993,** *58*, 17044.
43. Animal and Plant Health Inspection Service, *Federal Register* **1995,** *60*, 43567.
44. Cooperative State Research Service, *Federal Register* **1990,** *55*, 5752.
45. Agricultural Biotechnology Research Advisory Committee (ABRAC), *Performance Standards for Safely Conducting Research with Genetically Modified Fish and Shellfish*, Doc. No. 95-04, USDA Office of Agricultural Biotechnology, Washington, D.C., 1995.
46. Food and Drug Administration (FDA), *Federal Register* **1992,** *57*, 22984.
47. Food Safety and Inspection Service, *Federal Register* **1994,** *59*, 12582.
48. Food Safety and Inspection Service, *Federal Register* **1991,** *56*, 67054.
49. U.S. Congress, Office of Technology Assessment, *A New Technological Era for American Agriculture*, OTA-F-474, U.S. Government Printing Office, Washington, D.C., August, 1992.
50. Adams, M.W.; Perier, F.B.; Kelly, R.M. *Bio/Technology* **1995,** *13*, 662.
51. Adams, M.W.; Kelly, R.M. *Chem. & Engin. News,* Dec. 18, 1995, 32.
52. Wagner, G.; Schmid, R.D. *Food Biotechnol.* **1990,** *4*, 215.
53. Schultz, J.S. *IEEE Engin. Med. Biol.,* March/April 1995, 210.
54. Borman, S. *Chem. & Engin. News*, June 6, 1994, 24.
55. Stix, G. *Sci. Amer.*, Jan. 1994, 149.
56. Young, A.L.; Lewis, C.G. *Pediatric Clinics N. Amer.* **1995,** *42*, 917.
57. van der Wal, P.; Nieuwhof, G.J.; Politiek, R.D., Eds., *Biotechnology for Control of Growth and Product Quality in Swine*, Pudoc Wageningen, Wageningen, Netherlands, 1989.
58. Institute of Medicine (U.S.), Committee on Opportunities in the Nutrition and Food Sciences, *Opportunities in the Nutrition and Food Sciences*, National Academy Press, Washington, D.C., 1994, p. 109.

59 Caragay, A.B. *Food Technol.* **1992,** April 1992, 65.

60. Fitch-Haumann, B. *INFORM* **1993,** *4*, 344.

61. Salt, D.E.; Blaylock, M.; Kumar, P.B.A.; Dushenkov, V.; Ensley, B.D.; Chet, I.; Raskin, I. *Bio/Technology* **1995,** *13*, 468.

62. Hedin, P.A.; Menn, J.J.; Hollingworth, R.M., Eds., *Natural and Engineered Pest Management Agents*, ACS Symposium Series 551, American Chemical Society: Washington, D.C., 1994.

63. Poirer, Y.; Nawrath, C.; Somerville, C. *Bio/Technology* **1995,** *13*, 142.

64. Kelley, H.W. *Always Something New: A Cavalcade of Scientific Discovery*, Miscellaneous Publication No. 1507, USDA Agricultural Research Service, Washington, D.C., 1993.

65. Duvick, D.N. in *Intellectual Property Rights: Protection of Plant Materials*, Pub. No. 21, Crop Science Society of America, Madison, Wisconsin, 1993, p. 21.

66. Cole-Turner, R. *Science* **1995,** *270*, 52.

67. Stone, R. *Science* **1994,** *264*, 495.

68. U.S. Congress, Office of Technology Assessment, *New Developments in Biotechnology: Patenting Life*, OTA-BA-370, U.S. Government Printing Office, Washington, D.C., April 1989.

69. Agricultural Biotechnology for Sustainable Productivity (ABSP), *Intellectual Property Rights: ABSP Workshop Series*, Bedford, B.M., Ed., Agricultural Biotechnology for Sustainable Productivity, Michigan State University, East Lansing, MI, 1994.

70. Caswell, M.F.; Klotz, C.A. Biotechnology and International Organizations, presentation to Pennsylvania Biotechnology Association, Philadelphia, PA, April 18, 1995.

Industrial Applications and Products

Chapter 2

The Potential Impact of Biotechnology in the Food Industry: An Overview

John W. Finley and Saul Scheinbach

Nabisco, Inc., 200 DeForest Avenue, East Hanover, NJ 07936

Biotechnology may be loosely defined as any technology making use of an organism or products derived from an organism. This symposium discusses many of the complex issues of biotechnology as it relates to food with emphasis on applications, control of flavors, and texture. In this chapter we will discuss some examples of the opportunities biotechnology offers a food processor and how some of the barriers that inhibit its exploitation may be overcome. The papers in this volume are excellent examples of how biotechnology is currently impacting the food industry now and in the future.

Ancient man discovered biotechnology when he began to preserve milk by fermenting it into yogurt and cheese. The ancient Sumerians also made the very important discovery that if they fermented grains or grapes with naturally occurring yeast, they could produce delightful beverages. Certainly the bread industry was a by-product of the brewing process!! Gregor Mendel's laws concerning the segregation and independent assortment of "factors" (genes) established the field of genetics which became a springboard for biotechnology. For years natural mutants were utilized to improve all food and agricultural crops through traditional plant and animal breeding. These traditional techniques still have a place in developing our technology. Now with modern biotechnology we can truly engineer microorganisms, plants and animals to become better sources of ingredients, foods or bioreactors to produce specific compounds.

From the perspective of the food industry, biotechnology does offer a multitude of new and exciting opportunities ranging from monoclonal antibody

0097–6156/96/0637–0014$15.00/0

testing for pathogens, to enzymes that improve processing, to new health promoting ingredients, to designer feedstocks with unique nutritional or functional properties. Examples include oils with unique fatty acid compositions, enzymes with greater thermostability, cereals with improved processing or nutritional properties and fruits with improved flavor and storage stability.

As shown in Table I, the pharmaceutical industry is already reaping benefits from the incredible breakthroughs in biotechnology that have allowed the large scale manufacture of products that were previously available in minute amounts or not at all.

The magnitude of pharmaceutical applications of biotechnology will soon be dwarfed by the food and agriculture applications. However, there are concerns about the ultimate impact of biotechnology in food and agriculture. With a myriad of choices, how do we prioritize development and how do we establish the right expertise to facilitate development. Science and technology are not likely to be the largest hurdles, but rather approvals from government agencies, interference from advocacy groups, interpretation in the press and resulting public opinion. We have already seen the positive and negative effects of public reaction, informed or otherwise, concerning recombinant bovine somatotropin (bST) fed to cows.

Plant Biotechnology

We are now starting to see FDA approval of food products from biotechnology. These are listed in Table II. There are a number of plant products currently on the market which have been developed through the application of recombinant DNA technology. Two varieties of tomatoes, mini red peppers and two variations of carrots are currently available in supermarkets. The Flavr-Savr tomato was introduced in early 1994; later the crooked neck squash was approved. Although it is too early to assess the ultimate commercial success of these ventures, initial acceptance of the Flavr-Savr tomato in California was well above expectation.

The polygalacturonase gene that triggers post harvest softening of the tomato is blocked in the Flavr-Savr, thus the product can be ripened on the vine longer and as a result develop more flavor. The product lasts longer in distribution and arrives at the outlets near the height of its flavor. Thus, there is a clear consumer advantage over the typical tasteless winter tomatoes generally found in supermarkets. The benefits with the crooked neck squash are more subtle to the consumer because the modification of the squash was the development of virus resistance. The availability of the squash will increase and the ultimate cost should decrease, so the customer benefit will become more obvious. Perhaps the greatest benefit of the crooked neck squash will be that the squash will require less pesticide. Dr. Weaver (this volume) discusses the application of somaclonal feedback for selection of tomatoes with higher solids. In addition, the technique has the potential to identify tomato variants with high juice consistency, improved or modified flavors and resistance to environmental stresses or disease resistance.

TABLE I

CURRENTLY AVAILABLE MEDICAL APPLICATIONS

FROM BIOTECHNOLOGY

- HUMAN INSULIN
- HUMAN GROWTH HORMONE
- INTERFERONS
- TPA
- CLOTTING FACTOR
- SERUM ALBUMIN
- TUMOR NECROSIS FACTOR
- NERVE GROWTH FACTOR
- RELAXIN

TABLE II

CURRENTLY AVAILABLE FOOD PRODUCTS FROM BIOTECHNOLOGY

- MILK FROM CATTLE RECEIVING bST
- FLAVRSAVR TOMATOES
- CROOKED NECK SQUASH
- IMPROVED CHERRY TOMATOES
- CARROTS
- SWEET MINI-RED PEPPERS
- CHYMOSIN CHEESE
- ASPIRE-NATURAL FUNGICIDE
- NISIN - CHEESE PROTECTION

In the chapter by Hsu and Yang (this volume) the recovery of critical flavor volatiles from cell cultures of coriander and lovage are reported. By using this breakthrough technology it may be possible to isolate large quantities of valuable flavor volatiles from cultured plant cells. Essentially "fermenters" become factories to produce flavors or essences which currently are recovered at very low levels from plant tissue.

There are several products including tomatoes, that will be available in the next three to six years. As pointed out earlier higher solids in tomatoes for sauce, concentrate and catsup production translates to millions of dollars in savings for the processor, potentially better flavor and improved overall product quality. There is growing interest in and opportunity to produce specific oils from new oilseed varieties such as Rapeseed with lower saturated fat or higher levels of medium chain triglycerides (MCT). Economic production of these oils would make MCT oils available for medical applications and higher energy products for sports nutrition. Thus one bioengineered oil could open several new market opportunities.

Limonin and nomilin cause bitter flavor in citrus products. The bitterness is a major problem for the industry. Hasegawa et al. (this volume) have identified three target enzymes involved in limonoid biosynthesis for development of transgenic citrus free of limonin and nomilin bitterness. They are linoleate dehydrogenase, UDP glucose transferase and nomilin deacetylesterase. The isolation of the genes for these enzymes is currently being conducted in order to eventually insert them into cultured citrus cells where they will convert the bitter compounds to non-bitter derivatives. From the cultured cells mature citrus plants will be produced and these plants should produce fruit free of bitter flavor.

Other valuable food and animal feed ingredients can be produced economically through the use of biotechnology. Biotechnology offers us the opportunity to grow microorganisms of commercial value and use them as factories to produce complex biochemical structures. Johnson and Schroeder (this volume) found a strain of the yeast *Phaffia rhodozyma* which produces high levels of astaxanthin. Astaxanthin is a carotenoid which imparts desirable color to the flesh of several species currently being grown in aquaculture. For example, salmon are unable to produce astaxanthin, therefore it must be supplied in the diet. The identification of a mutant of *Phaffia rhodozyma* that produced high levels of astaxanthin became a commercial success and solved a major problem of the aquaculture industry.

Enzymes

One of the major benefits of biotechnology is for the cost effective production of valuable enzymes. In 1987, sales of chymosin alone were a half a billion dollars. Kuraishi (this volume) reports some of the potential applications of transglutaminase, an enzyme recently commercialized by Ajinomoto. The enzyme catalyzes the acyltransfer between the γ-carboxyamide group of glutamine and the ε-aminogroup of lysine crosslinking proteins with a ε-(γglutamyl)lysyl peptide bond. The ramifications of this crosslink are that gel strength and elasticity of protein matrices can be improved. From a commercial standpoint, meat can be bound together without the addition of salt or phosphate. The crosslinking also improves the firmness of yogurt. This work serves as an excellent example of the application of biotechnology to produce large quantities of an enzyme at a reasonable cost that can offer many new and unique opportunities to the food industry.

Enzymes that are more specific with higher levels of activity and greater heat stability offer opportunities for improved process efficiency and longer life in reactors. The enzymes for production of high fructose corn syrup; namely alpha amylase, glucoamylase and glucose isomerase are still some of the most important enzymes in the food industry. The profitability of high fructose corn syrup could be improved if these enzymes were more heat stable. The isolation of bacteria from deep ocean vents offers a new pool of heat stable (up to 130° C) enzymes that could eventually lead to major breakthroughs in enzymes for industrial processing. Lipoxidases offer interesting potential for deliberately oxidizing lipids in wheat. These oxidizing lipids would then act as dough conditioners instead of bleaching agents such as bromate which are currently being used.

Fats and Oils

Designer fats are a rapidly growing aspect of the business. Fats with new or unique functionality are produced by interesterification. The use of lipases to catalyze these changes allows much more specificity thus many new opportunities. It should also be noted that if specific lipases are available more low calorie fats such as caprenin or salatrim can be produced economically.

It is very likely that in the next few years we will see the FDA approval of more and more plants that have been improved using recombinant DNA technology. Table III is not meant to be comprehensive but does show the diversity of improvements being attempted in a variety of plants and fish. From the Table it can be seen that the range of plants being modified is expanding significantly. Current areas of emphasis are flavor improvement in tomatoes and peas, butter fats from rapeseed, higher solids in tomatoes and potatoes. Weaver (this volume) discusses strategies that have been developed to increase the solids content of tomatoes as well as improve the color, flavor, juice consistency, size, shape and shelf life. The modest increase of 1% solids in processing for tomatoes has huge significance for the processor. First, energy savings would be realized because of reduced energy required for concentration. Second, the lower heat impact also results in pastes and concentrates with superior physical properties and better flavor.

TABLE III

PRODUCTS ON THE NEAR HORIZON FROM BIOTECHNOLOGY

- RAPID GROWING SALMON
- IMPROVED TOMATOES
- HIGH SOLIDS TOMATOES, POTATOES
- HIGH STEARIC RAPE OILS - SHORTENING AND FRYING
- MCTs FROM RAPESEED
- LOW SATURATED FATS FROM RAPESEED
- PEST RESISTANT
  - CORN
  - BANANA
  - MELONS
  - WHEAT
  - COTTON

Pesticides

Aside from their usual food value, plants in general are rich sources of antimicrobials and insecticides. Cutler and Hill (this volume) review the natural abundance of these materials and show how they can be altered by biotransformation. Most of these materials are biodegradable and many are not synthesized until their production is induced by the pest invading the plant. The authors also describe a case study of a fungicide of potential commercial value which is currently being field tested.

Many laboratories are working diligently to improve resistance against everything from viruses to insects. It is safe to say that in one lab or another improved disease resistance through biotechnology is being studied for every crop of any agricultural importance. The goal is to substitute the wide application of broad spectrum synthetic pesticides with narrower spectrum biological compounds made by the crop, reducing the environmental impact of agriculture. For example, tolerance of the herbicide glyphosphate has been introduced into soy, corn, rapeseed and sugar beet. Glyphosphate is effective at inhibiting the growth of most plants (weeds) at extremely low concentrations. Plants with introduced resistance are not affected, thus fields can be treated and the glyphosphate will selectively kill the weeds but the resistant crop grows very well. More importantly, the concentrations of glyphosphate are below any animal or human toxicity threshold, and it is degraded rapidly by soil microorganisms. The result is a no-residue pesticide. Also available this year is a corn hybrid genetically altered to produce a toxin that kills the corn borer and not other animals. As the public continues to clamor for lower levels of pesticides or none at all, one of the most significant benefits of biotechnology is the development of plants which have superior resistance to pathogens. A word of caution is that in our zeal to impart pest resistance we must be sure that we do not have a cure that is worse than the disease. For example, many phytochemicals which provide disease resistance can exhibit significant toxicity. We therefore must be cognizant of the amount that is expressed in the plant, and how the risk of the enhanced or new pesticide compares to the conventional spray on pesticide.

When products generated by biotechnology are introduced into the market it is important to identify the benefits and the benefactor. The crooked neck squash is the first of several members of the melon family where improved technology for virus resistance has been successfully applied. We would certainly hope and anticipate that viral resistance can and will be applied to other plant species particularly in the melon family. The ultimate success of this program will help reduce the overall application of pesticides, particularly to fresh produce. The reduced use of pesticides is an example of a benefit for the environment, the grower, the distributors, the processors and ultimately the consumers. The environmental impact of reduced use of chemical pesticides should be enough to have the environmentalists enthusiastically align behind the program. In addition, the grower benefits from reduced cost, eliminating or reducing labor and the cost of the pesticides. One could debate the toxicological risks of synthetic pesticides vs. the equally toxic pesticides produced by transgenic plants. But the latter will be better because they will be produced only when needed and should only affect the target pest. For example, one way to limit unnecessary pesticide production by transgenic plants is to transform them with a *Bacillus thuringiensis* (B.T.) toxin gene linked to a promoter which only turns on when insects begin chewing. The incorporation of antipest activity through genetic manipulation benefits the distribution and/or processing of the product because the need for adding pesticide at these stages is eliminated. The consumer benefits through lower cost of product and the perceived benefit of safety because no pesticides are directly added; however, in this example the consumer is the least obvious benefactor. The consumer initially is not likely to see the cost benefit, because to recoup research and development costs some premiums are going to be associated with the new product. However, if products from biotechnology are going to succeed it is our belief that some benefits must be communicated to the consumer. The FlavrSavr tomato clearly had a benefit the consumer could taste, flavor in the tomatoes even in the winter. In contrast, when bST was used to improve yields of milk the consumer saw little benefit. To make matters worse advocacy groups managed to use scare tactics to discourage the use of milk made from cows receiving bST. Biologically, it is extremely unlikely that the bST could pass through membranes from the cow to the milk. Consider that since it is a protein it would be attacked by digestive enzymes in the gastrointestinal tract. Because the views of the advocacy groups were picked up by the press, and the scientific rebuttals received minimal press, consumer acceptance of a substance that has since resulted in substantial profits for many farmers was delayed. The point is, we need to be sure to communicate the benefits and safety to the consumers. As in the marketing of any new product or technology, we must let the consumer see the advantage for himself or face rejection and increased vulnerability and attack by advocacy groups.

Food Processing Aids

Looking at biotechnology from the perspective of the food processor, the needs and opportunities go well beyond yield, shelf stability and pest resistance. After all the majority of agricultural commodities that are produced for food are utilized by the processed food industry which has been applying biotechnology since ancient times. Consider the historical efforts for food preservation and the seeking out and selecting of improved microorganisms for cheese making, wine production and bread baking. Traditionally, this has been done by selection of naturally-occurring mutants of appropriate microorganisms. One such example is provided by the on-going efforts to produce strains of lactic acid bacteria with improved phage resistance in cheese making. Now we have the opportunity to more specifically identify, select and engineer the desired changes. We can also take advantage of naturally produced substances such as nisin, produced by lactic acid bacteria, which can kill gram positive organisms thereby preserving the cheese.

Until recently the pharmaceutical industry has led the way in the use of micro-organisms to produce desired materials such as antibiotics, insulin, growth factors and hormones. Now, specific enzymes important to the food processing industry are becoming available through recombinant DNA technology.

Cheese

The chymosin story is a good example of some of the difficulties that can be encountered in the commercial application of food products from biotechnology. Chymosin (or rennin as it is known in the dairy industry) is a proteolytic enzyme used to initiate the coagulation of milk in the cheese making process. Traditionally the enzyme was recovered from macerated 4th stomach of a suckling calf, thus, it was available in limited quantities and was relatively expensive. In addition it required a single, but very extensive, commitment by the calf! (Give one to the animal rights group) Because of its wide use it was economically feasible to produce it from recombinant microorganisms. The preparation of the RNA from the calf abomasum and identification of the preprochymosin was fairly straightforward. The pre-sequence is required for secretion of the protein by facilitating movement through the Golgi complex. The pro-sequence, which keeps the enzyme in its zymogen form, is removed by autohydrolysis of the prochymosin in an acid environment such as the stomach.

Since E. coli does not perform these steps, the gene had to be cloned and expressed as mature chymosin. But overproduced enzyme accumulated intracellularly in a denatured form as inclusion bodies. However, chymosin could be recovered by lysing the cells and performing an alkaline renaturation. After the appropriate engineering was completed, it had to be demonstrated that the amino acid sequence of the microbially produced chymosin was identical to the calf chymosin, as part of the FDA approval process. Clearly it was worth the effort. Today, at least 50% of U.S. cheese is produced using chymosin from bioengineered microorganism. An additional benefit is the fact that microbial chymosin has been made kosher so that high quality kosher cheeses can now be produced using the enzyme. Cheesemakers thus were able to significantly increase market share.

Baked Goods

Food processors are exploring the application of many other enzymes which will have equal or greater impact than chymosin. For example, proteases have been used in baking for control of dough viscosity. An example is the production of crackers as shown in Figure 1.

Currently available proteases such as papain have a broad specificity and may not work optimally at a convenient pH in the specific product. In the case of cracker production it would be ideal to have an enzyme which is active at pH 4 during the sponge stage but is inactive during dough proofing at pH 7. This is about the opposite of papain and this example brings up another important issue. In the modern R&D environment it is not likely these problems will be solved by a single hero. In order to best attack the problem of developing a more specific protease, a team should be formed where members would include everyone from the molecular biologists through the workers on the production line to identify the key goals and criteria for success of the enzyme development project. Enzymologists, food processors and molecular biologists need to join forces identifying precise applications, assuring that the needs of the entire cracker making process are addressed by defining the specific activities required, transforming the activities to useful producer organisms producing the enzyme in quantity and purifying the enzyme. This avoids the classic trap of solving one problem and generating two new ones in the process. Optimized proteases may be specific for site of hydrolysis, pH range and temperature for a given product. This would allow for controlled hydrolysis and flavor development at the sponge stage.

Another family of enzymes with significant potential for food processors are the pentosanases. Pentosans bind enormous quantities of water in dough systems and contribute significantly to viscosity in fruit juices. Controlled application of pentosanases allows the processor to more accurately control viscosity in dough systems by regulating the pentosan level.

Reducing pentosans in dough systems, such as cracker doughs, reduces viscosity and improves the ability to bake out the moisture, thus improving

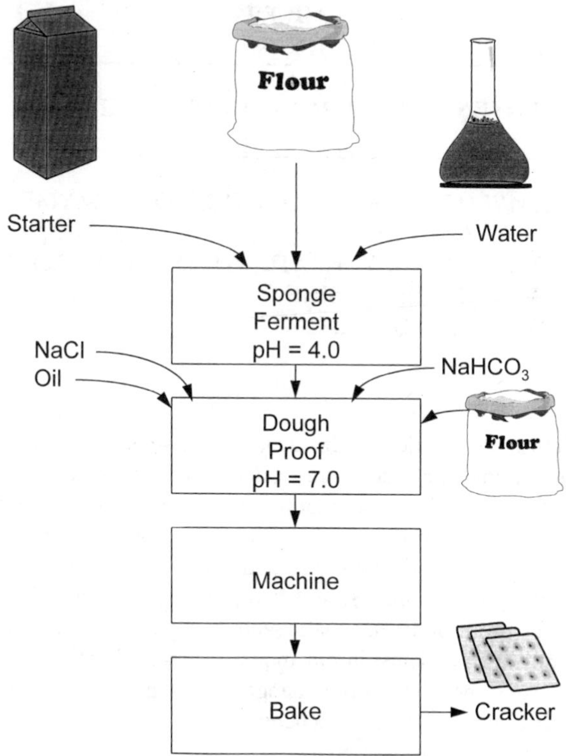

Figure 1. Cracker Production

product quality and consistency. Viscosity control in doughs becomes extremely critical in low-fat products. Normally fat softens cookie or cracker doughs so they can be machined. Without fat, cookie and cracker doughs resemble extremely tough bread dough. These tougher low-fat doughs cannot be machined or form products without the application of appropriate enzyme systems. Consumer demands for high quality low-fat products is growing exponentially. We must develop the appropriate technologies to support the product development of these new foods. Modern biotechnology can supply the enzymes necessary to accomplish this difficult task.

Without a doubt, the greatest opportunity for biotechnology to impact the process food industry is through the development of new or improved raw materials. Some examples for wheat are shown in Table IV.

We would like to again emphasize the importance of cross-functional teamwork in the design, development and assessment of modified raw materials. Because the potential interactions are so complex, the enzymologist, the food technologist and the molecular biologist must work extremely closely. The food

TABLE IV

POTENTIAL IMPROVEMENT OF WHEAT
• HIGHER AMYLOSE FOR PRODUCTION OF ENZYME RESISTANT STARCH
• LOWER LEVELS OF POLYUNSATURATED FATTY ACIDS
• PROTEIN WITH HIGHER LYSINE
• IMPROVED PROTEIN QUALITY

technologist must clearly define the needs or desires for modification. For example, consider the opportunity to produce wheat with low polyunsaturated fatty acids. In the production of low fat products, the wheat lipids are spread over the surface of the wheat flour. With no other fats to protect them, the polyunsaturated lipids are susceptible to hydrolysis and oxidation by lipases and lipoxygenase during processing. In finished product, the lipids are prone to non-enzymatic oxidation. Both enzymatic and non-enzymatic oxidation produce unacceptable off flavors. Thus, in our model for cross functional teamwork, the food technologist defines the desired change as gaining oxidative stability in the flour and dough system. Now the biochemist must work with him to consider the alternatives to achieve the desired goal.

First, one might think about blocking the activity of lipase and/or lipoxygenase. Both enzymes are likely to be required for germination of the wheat, thus, a seed might be developed devoid of the enzymes but the grain might be infertile. An alternative approach is to block the stearate desaturase or oleate desaturase activities in the wheat. Such a change is less likely to be lethal to the embryo. These changes are complicated by the fact that wheat is a hexaploid so several sites on the wheat genome may have to be modified to attain the desired effect. The next step is to identify the site of desaturase genes, sequence them, develop the correct antisense DNA molecule and insert it into the genome so it can transcribe the RNA to block enzyme production. Such changes are the challenge for the molecular biologist on the team. Such team efforts are critical for other equally desirable modifications of wheat. For example, if starch could be modified such that during processing it would resist digestion by normal gastrointestinal tract enzymes, while still being subjected to limited degradation in the colon, significant benefits would be achieved. This "resistant" starch would provide fewer calories, making the wheat flour more useful in production of low calorie ingredients. The fact that the starch can still be partially fermented into fatty acids by colonic bacteria is beneficial for the health of the mucosal wall and potentially increases resistance to colon cancer development. Efforts are currently underway to develop such starch modifications in corn and once that is accomplished the challenge to molecular biologists is to attempt to make similar changes in wheat.

Wheat proteins are unique in their ability to form doughs. It is these doughs which allow the formation of the matrix for bread and crackers. Control of the dough rheology has long been a target of wheat geneticists. Now we are developing a much better understanding of how each of the component proteins in gluten impact the rheology of the dough. The next step is to produce plants that will yield controlled quantities of the desired proteins, resulting in bread doughs with superior functionality.

The production of higher lysine varieties of wheat and corn has been a goal of plant breeders for generations and successful corn varieties have been produced by conventional methods. With the application of modern molecular biology it should be possible to incorporate and force expression of genes that will produce lysine-rich proteins in wheat. Although not critical in most North American diets, such varieties of wheat could be particularly useful in the export market or, if combined with drought resistance, the varieties would offer potential in Africa.

There have already been a number of successes in expression of modified vegetable oils from canola. Extending the ability to produce designer fats in soy could have enormous economic and potential health ramifications. Consider the potential of incorporating desirable fatty acids such as conjugated linoleic acid into the Sn-2 position of soy oil. Simply controlling the fatty acid distribution in soy oil offers interesting potential. For example, if we could reduce the amount of polyunsaturate while increasing stearic acid, the resultant fats would act like partially hydrogenated fats in margarines and shortenings without hydrogenation. Avoiding hydrogenation of the oil circumvents concerns about the production of trans fatty acids.

We can all see the apparently unlimited opportunities for biotechnology. With the appropriate teamwork, the technical hurdles can and will be overcome. What are the barriers? The barriers that we see are 1) who bears the cost for development of commodity products; 2) public perception of benefit vs. risk; and 3) the regulatory environment, particularly labeling.

Economic Issues

Firstly, concerning the issue of research and development costs of specifically modified commodity crops, we can use the example of wheat with a modified lipid profile devoid of polyunsaturated fatty acids. When such a crop is developed, who owns the technology? Clearly if Company A pays to develop the technology it should be able to recover the R&D costs and hopefully make a profit. If the modified wheat is developed by a seed producer, he is likely to have an advantage the first year but once the wheat is in the distribution system, how is the seed producer's germ plasm going to be protected? Secondly, there is the complex issue of keeping the new grain separate from "unengineered grains" and clearly identified. If the development is supported by a specific food processor, how can the new grain be kept for the sponsor's use? One answer traditionally has been contract farming. There are some notable successes with contract farming but generally the problems outweigh the gains. From the vantage point of a large user of a commodity like wheat, a separate distribution system that will keep modified wheat distinct from unmodified wheat must be developed and implemented. If a user is going to make the necessary investment to develop a genetically modified wheat or a similar commodity product, he must have some protection or assurance that he can recapture the investment. Some assurance must be put in place to encourage the processor to make the investment in the research. We would warn that without the input from processors, the products with greatest potential could be overlooked. Even worse, undesirable characteristics could find their way into the system. For example, disease resistance for Russian rye was incorporated into bread wheat. The new gene improved yield and provided excellent disease resistance. The proteins however were difficult and the doughs became so sticky that they were unusable. As a result of the outcry by end users (bakers) the hard wheat line with the modifications have been discontinued. Again the case is made for cooperation from geneticists, to grower, distributor and end user in the food industry. We are all on the team and to take full advantage of the new technology we must work as a team.

We are all in business for profit. The bottom line is that we must be able to sell our wonderful new products from biotechnology for a reasonable cost. If the public is to purchase these products there must be a benefit. The public good can include economics, environmental health or other areas. But it must be perceived by the consumer as a benefit. Let's look at a few examples of some current products from biotechnology, the advantages and the so called public good.

First, looking at Antisense technology in fruits and vegetables, for example the Flavr-Savr tomato, there is a clear public advantage, shelf life and flavor are improved. There appears to be enough benefit that such products can command a premium price. There is also at least the hope that it might increase fruit and vegetable consumption.

In the case of chymosin, the cheese producer has the advantage of lower cost and a more consistent source of enzyme. In addition to cost savings in cheese production, the use of recombinant chymosin opened a new market with superior quality kosher cheese.

A subtle but important benefit of biotechnology is the use of genetic probes and tests based on the polymerase chain reaction to screen for and identify pathogens in foods. The economics could result in $40 billion/year cost savings as a result of reduced illness. Clearly, these are public benefits from many of the biotechnology-derived products that are currently available and will become available in the next century. As more sophisticated products become available, the benefits should continue to increase. Biotechnology very likely offers the best answer to respond to society's demands for plentiful new food varieties. These foods must be appetizing, nutritious, safe and healthy with minimal environmental impact for consumers. The success or failure of biotechnology in the food industry for the next millennium will be based on the consumer's confidence in the safety of these products and the ability to answer questions about safety with scientific objectivity to the satisfaction of the regulatory agencies.

Food Safety

In the United States, the FDA is responsible for regulating the introduction of foods, drugs, pharmaceuticals, and medical devices into the marketplace. The safety of both crop foods and food ingredients that include flavors and additives must be thoroughly assessed before they can be licensed for human consumption. The FDA has had a well established, although not foolproof, system for the approval of new foods and food products for some time. Critics of the FDA, however, have argued that it tends to favor the interests of industry and is too lenient in enforcing its own regulations. Both the FDA and the food industry, which is represented by the International Food Biotechnology Council (IFBC), have argued somewhat convincingly that new regulations are not required for foods and foodstuffs that are developed by recombinant DNA technology because the FDA already requires that any unlicensed food or food ingredient, regardless of how it is produced, must be assessed for safety with respect to toxicity, allergenicity, and impurity testing. If genetic modifications by either selective breeding or recombinant DNA methodologies change the composition of an accepted food or foodstuff, then the company, after the safety of the product is proved, must inform the consumer through labeling that the new product differs from the traditional one. As an example, let's return to chymosin.

To ensure a reliable, convenient, and, possibly, cheaper industrial supply of chymosin, one of the chymosin genes was cloned and expressed, and the product was harvested from E. *coli* K-12. When a petition requesting permission to use recombinant chymosin for the commercial production of cheese was presented to the FDA, it was necessary to decide what criteria should be required for the approval process. Because there has been a long history of using rennet containing chymosin in the cheese-making industry, the FDA reasoned, that if the recombinant chymosin was identical to the naturally occurring chymosin, then excessive testing was not necessary. In essence, the petitioner had to show that the recombinant chymosin was identical to the chymosin of rennet. To substantiate this, restriction mapping, DNA hybridization, and DNA sequencing were used to establish that the DNA sequences of the cloned and native chymosin gene were identical. Moreover, the recombinant chymosin had the same molecular weight as purified calf chymosin, and the biological activities of the two forms of the enzyme were the same.

Next, it was essential to establish that the recombinant chymosin preparation was safe. The company showed that, as part of its purification process, recombinant chymosin is extracted from inclusion bodies cell debris, and other impurities, that include nucleic acids. Although the presence of minute amounts of E. *coli* K-12 cells in the final preparation of chymosin is undesirable, numerous studies have established that this strain is nontoxigenic and nonpathogenic to humans. To ensure that the recombinant chymosin preparations did not contain an unexpected toxin, the results of animal testing showed no adverse effects. After compiling all the information, the FDA concluded that the recombinant chymosin could be licensed for commercial use without labeling.

The FDA stated its view of labeling biotechnology-derived foods in 1992. It said that labels must appear if a food derived from a new plant variety differs from its traditional counterpart such that the common or usual name no longer applies or if a safety or usage issue exists.

For example, new allergenicity without a label could be misleading. However, the genetic method used to develop the new variety is immaterial because no evidence exists that new bioengineered foods are different with respect to safety and therefore should not be labeled. Furthermore, if the safety of the a transgenic plant, for example, ever became problematic, it would simply not be marketed. This was recently the case for a transgenic soybean that became allergenic by receiving a high methionine protein gene from brazil nuts.

Recently the National Food Processors Association (NFPA) stated that the industry feels that application of biotechnology to food products does not imply any need for special labeling (and, therefore, regulations). The association has said that it will use existing rules to govern food label information for products from biotechnology. No additional labeling information is required unless the products nutritional value is changed, the new food contains a known allergen or the common or usual name of the food no longer applies. The NFPA further opposes actions at any government level that result in pre-market approval of food derived from biotechnology if it delays or restricts market entry. It also urges the FDA to take the lead in setting a uniform and consistent nutritional policy for food biotechnology, also with the hope that an international accord can be reached so that new policies do not generate trade barriers.

In summary, biotechnology affords us with unique and exciting opportunities. It is our hope that we can embrace and apply the technology in the competitive world market place. The applications in food and agriculture are essentially unlimited. The opportunity for biotechnology to yield financial impact is dependent on consumer acceptance and regulation without strangulation. Manufacturers and scientists must work together to develop appropriate changes, while keeping regulators in the loop. Early discussions with regulatory bodies and open sharing of data will help the producer as well as the regulator expedite commercialization of the new technology. If we maintain responsible development, promote maximum safety, avoid overregulation and overreaction by adversary groups, biotechnology will substantially improve our diet, health and well-being.

If we can promote an environment of teamwork, clear communication between the scientific community and the public, fair regulation and responsible education, food biotechnology will result in a completely new food industry in the future.

Chapter 3

The Usefulness of Transglutaminase for Food Processing

Chiya Kuraishi, Jiro Sakamoto[1], and Takahiko Soeda

Food Research and Development Laboratories, Ajinomoto Co., Inc., 1–1 Suzuki-cho, Kawasaki-ku, Kawasaki-shi, 210 Japan

Transglutaminase (TG) is an enzyme that catalyzes the crosslinking reaction between glutamine residues and lysine residues in protein molecules. This crosslinking results in many unique effects on protein properties through the ϵ-(γ-glutamyl)lysyl peptide bonds. Ajinomoto Co., Inc., and Amano Pharmaceutical Co., Ltd., have found a new transglutaminase in a microorganism (MTG) and are the first in mass production and commercialization of this enzyme. This transglutaminase has various effects on the physical properties of food proteins, gelation capability, thermal stability, water-holding capacity, etc.

Transglutaminase (E.C. 2.3.2.13) (TG) is an enzyme that catalyzes an acyl-transfer reaction between the γ-carboxyamide group of peptide- or protein bound glutaminyl residues and primary amines (*1*) (Fig. 1-a). When transglutaminase acts on protein molecules, they are cross-linked and polymerized through ϵ-(γ-glutamyl)lysyl peptide bonds [ϵ-(γ-Glu)Lys bond] (Fig. 1-b). In the absence of suitable primary amines or in the case that the ϵ-amine of lysine is blocked by certain chemical reagents, e.g., citraconic anhydride, it is possible to make water act as the acceptor, and the glutamyl residue changes to a glutaminyl residue by deamidation through transglutaminase reaction (*2*) (Fig. 1-c). There is no appropriate blocking reagent that is safe and edible, and this deamidation reaction is not utilized in food industry. Transglutaminase is widely distributed in the nature and has been found in various animal tissues, fish, plants, and micro-organisms (*3-6*). For example, transglutaminase in human blood is known as factor XIII. It takes part in a human blood clotting system that forms crosslinks

[1]Current address: Ajinomoto Europe Sales GMBH, Stubbenhuk 3, 20459 Hamburg, Germany

0097–6156/96/0637–0029$15.00/0

of fibrin molecules and stabilizes fibrin polymer. The transglutaminase-mediated crosslinking of protein units cause drastic physical changes not only in biological systems but also in protein-rich foods. Many studies are being carried out to use the enzyme for food industry as a protein modifier having unique effects. But most of these studies have been done by using mammal transglutaminase which was too expensive to utilize on an industrial scale. Recently, for the first time by anyone in the world, we have succeeded in mass production and commercialization of transglutaminase by a fermentation method in association with Amano Pharmaceutical. This new transglutaminase from microorganism (MTG) is generally very useful in the food industry.

Properties of Microorganism Transglutaminase

Measuring Transglutaminase Activity. Enzymic activity of transglutaminase is measured by the hydroxamate procedure with N-carbobenzoxy-L-glutaminylglycine (Z-Gln-Gly) (Fig. 2). The enzymic activity unit is defined as the amount causing the formation of 1 μmole of hydroxamic acid in one minute at 37°C (*7*).

Enzymological Properties of MTG. MTG is active over a wide range of temperature and the optimal temperature is 50°C. It is stable between pH 5-9, which is the pH range for most food processing (*4*). As contrasted with mammal transglutaminase (guinea pig liver transglutaminase), MTG is characterized by its calcium independent activity (*4*) (Table I). MTG is able to react without adding calcium ion so that it is easier to handle and more practical to use in food processing.

The ϵ-(γ-Glutamyl)lysyl peptide bonds

The Effect on Physical Properties. The most effective reaction of MTG is the crosslinking through ϵ-(γ-Glu)Lys bonds (Fig. 1-b). When ϵ-amino groups of lysine residues act as acyl acceptors, ϵ-(γ-Glu)Lys crosslinks are formed. The crosslinking reaction may be both intermolecular and intramolecular and causes significant physical property changes in protein-rich foods. The ϵ-(γ-Glu)Lys bonds are covalent bonds which are stable unlike ionic bonds and hydrophobic bonds. Therefore, even the few ϵ-(γ-Glu)Lys bonds in foods have a profound effect on the physical properties.

Nutritional Values. There are some chemicals, e.g., glutaraldehyde, known to be able to act as crosslinking reagents, but these are not allowed in foods because of food safety considerations. Polymerization by enzymes is much more mild because biological transformations are considered safer than chemical reactions. Moreover, with respect to nutritional values, the bioavailability of ϵ-(γ-Glu)Lys crosslinks has been reported by several researchers (*8-10*). The nutritional value of MTG-treated casein has been examined, and it has been confirmed that the crosslinked proteins have no adverse effect and can be absorbed in the body (*11*). The analytical method for estimating ϵ-(γ-Glu)Lys bonds in foods was investigated and it has been confirmed that ϵ-(γ-Glu)Lys bonds exist in a number of general

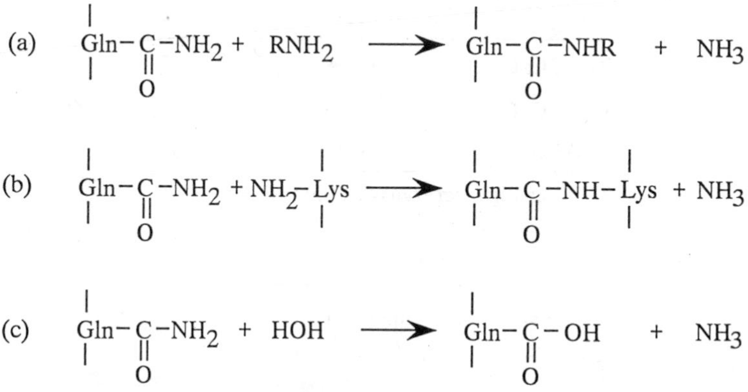

(a) $\text{Gln}-\underset{\|}{\overset{}{C}}(=O)-NH_2 + RNH_2 \longrightarrow \text{Gln}-C(=O)-NHR + NH_3$

(b) $\text{Gln}-C(=O)-NH_2 + NH_2-\text{Lys} \longrightarrow \text{Gln}-C(=O)-NH-\text{Lys} + NH_3$

(c) $\text{Gln}-C(=O)-NH_2 + HOH \longrightarrow \text{Gln}-C(=O)-OH + NH_3$

Figure 1. General reactions catalyzed by transglutaminases.
(a) acyl-transfer reaction
(b) crosslinking reaction
(c) deamidation

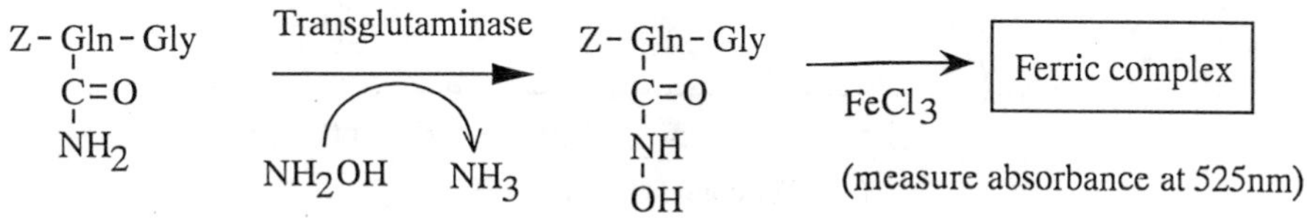

$\text{Z-Gln-Gly}\,[C{=}O,\ NH_2] \xrightarrow[NH_2OH \to NH_3]{\text{Transglutaminase}} \text{Z-Gln-Gly}\,[C{=}O,\ NH,\ OH] \xrightarrow{FeCl_3}$ Ferric complex

(measure absorbance at 525nm)

Figure 2. Hydroxamate procedure.

Table I. Calcium Independent Activity of MTG

	MTG	GTG[a]
0 mM $CaCl_2$	100%	0%
1 mM $CaCl_2$	100%	39%
5 mM $CaCl_2$	99%	100%

[a] Guinea pig liver transglutaminase

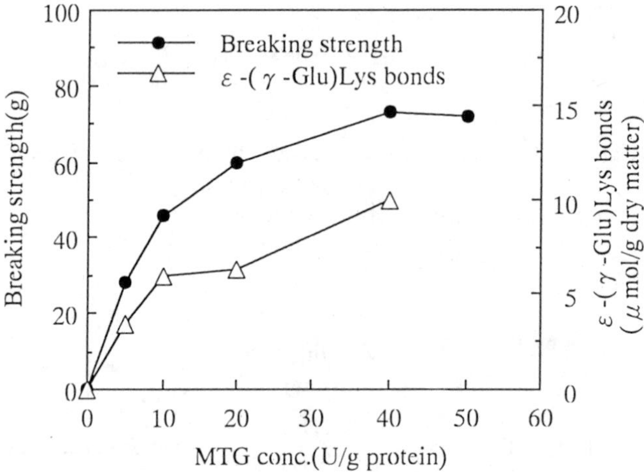

Figure 3. Effect of MTG treatment on gel strength of soy protein and amount of ε-(γ-Glu)Lys bonds of soy protein incubated at pH 7.0 at 37°C for 1 hr. Protein concentration: 10 (w/w%).

foods (*12*). Thus, foods containing ε-(γ-Glu)Lys bonds are already being consumed.

Changes through ε-(γ-Glu)Lys bond formed by Transglutaminase. ε-(γ-Glu)-Lys bonds cause unique effects on proteins. Those effects are very useful for many kinds of foods containing proteins and help to increase their commercial value.

Gelation Capability. Even a protein solution that can not form a gel by itself usually will turn into a gel if transglutaminase is used to form crosslinks in such proteins. In the case of protein that has gelation capabilities, the protein gel becomes firmer through the transglutaminase treatment. A caseinate solution will not form a gel by itself. When MTG is added to a caseinate solution at 5 units/g protein, a gel is formed from the liquid phase after 1 hr incubation at pH 7 at 37°C. It is known that soy protein and myosin will gel with heat treatment. If transglutaminase is added, such protein solutions will form gels without heat treatment (*13*). These changes in gelation capabilities are due to ε-(γ-Glu)Lys bond formation. The firmness of the gel from soy protein isolate (SPI), prepared by the MTG treatment at pH 7 at 37°C, is related to the number of ε-(γ-Glu)Lys bonds (Fig. 3). The breaking strength of the gel is increased as the number of ε-(γ-Glu)Lys bonds increase. It is interesting to note that if too much MTG is added, the firmness of the gel is decreased, an unexpected result. A gel formed with a ten percent caseinate solution treated with MTG 20 units/g protein at 37°C at pH 6.5 is weaker than a gel formed with MTG 15 units/g protein (Fig. 4). There may be a limit to the ability of ε-(γ-Glu)Lys bonds to improve gel strength, and perhaps an excess of ε-(γ-Glu)Lys bonds may weaken the gel structure. Such a gel formed with an excess of ε-(γ-Glu)Lys bonds by use of transglutaminase becomes weak and has less water-holding capacity. Sometimes syneresis, a separation of liquid from a gel, is observed. We hypothesized that excess ε-(γ-Glu)Lys bonds would inhibit uniform develpment of the protein network for entraining sufficient water.

Viscosity. When a protein is polymerized and increases in molecular weight, a solution of it usually increases in viscosity. Therefore, a protein solution treated with MTG increases in viscosity. Caseinate solutions treated with various concentrations of MTG at pH 6.5 at 55°C for 30 minutes were heated to deactivate the enzyme and to stop the crosslinking reaction. After freeze-drying, the MTG-treated caseinate powder was rehydrated, and the viscosity of caseinate solution was determined. The caseinate gel gets more viscous at higher enzyme concentrations (Fig. 5). Formation of cross-linked caseinate was confirmed by SDS-PAGE (data not shown). The monomer fractions of caseinate diminished or disappeared, and polymer fractions (dimers and trimers) increased. Polymer fractions which did not enter the running gels were formed by MTG treatment at above 3 units/g protein. With more reaction (longer time, more enzyme added, higher temperature, etc.) a caseinate solution can be transformed to a gel. Sols of various viscosities, not gels, can be prepared by controlling the enzyme reaction conditions, i.e. reaction time, temperature, concentration of enzyme added, etc.

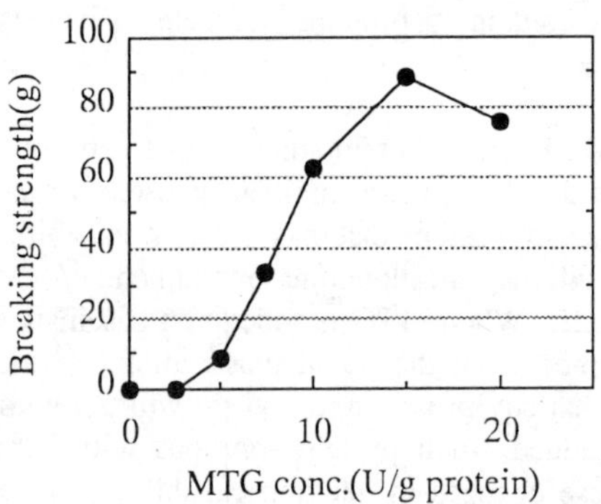

Figure 4. Effect of MTG on gel strength of caseinate solution incubated at pH 6.5 at 37°C for 1 hr. Protein concentration: 10 (w/w%).

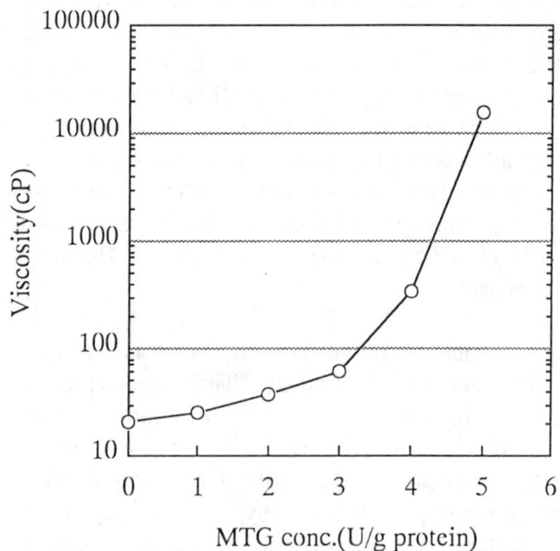

Figure 5. Viscosity of MTG-treated sodium caseinate.

Thermal Stability. As described above, ϵ-(γ-Glu)Lys bonds formed by transglutaminase reaction are covalent bonds and stable even with temperature changes. The thermal stability of protein gel structure transglutaminase-treated is improved by ϵ-(γ-Glu)Lys bonds formed inter/intra protein molecules. Gelation of gelatin is a phenomenon dependent on thermal changes, and the gels are thermoreversible. The gelatin gel is stabilized by hydrophobic bonds, so if it is heated to certain temperatures, it turns to a sol or solution state. If a gelatin gel is treated with MTG, a few ϵ-(γ-Glu)Lys bonds are introduced into its structure, and its thermostability is significantly improved. Jelly products stable to high temperatures can be produced. Table II shows the thermal stability data by MTG treatment. A gelatin gel difficult to melt even at 120°C can be prepared depending on the gelatin concentration and enzyme reaction conditions.

Water-holding Capacity. A gel formed with ϵ-(γ-Glu)Lys bonds with MTG has improved water-holding capacities. Gelatin gel holds water molecules in its protein matrix structure, and it is possible to form stable aqueous gel even at 2% protein concentration. Gelatin gel is a good representative of a water-holding protein gel. Food gels, such as sausages and yogurts, often have problems of syneresis, separation of a solution from the gel, so that gelatin is usually added to such food gels to improve water-holding capacities. Food gels treated with transglutaminase, which forms more stable covalent bonds, ϵ-(γ-Glu)Lys bonds, are able to hold more water in spite of temperature changes or physical shocks. For example, yogurt, which is an acid milk gel formed by gradual acidification with a lactic starter, has some problems of serum separation with a change of temperature or physical impact. As shown in Fig. 6, yogurt which is made from MTG-treated milk has a larger capacity for holding water, and the whey syneresis is prevented. With use of transglutaminase, food gel products with good water-holding capacities can be produced without adding gelatin.

Applications in the Food Industry

Binding Food Pieces (Restructured Meat). The function of transglutaminase, which is capable of crosslinking protein molecules, is to bind pieces of food together. In order to bind pieces of meat, it has been necessary to restructure meat. These restructured products have been developed to utilize low-value, small meat pieces and to enhance their market value. The conventional way is to bind meat pieces to each other with salt-extracted (cured) muscle protein, with heat treatment if necessary. Recently, a new useful binding system was developed by utilizing MTG. Meat or fish pieces can be bound together using only the enzyme. Binding can be improved by using transglutaminase and caseinate (Fig.7). Caseinate treated with transglutaminase becomes more viscous, and viscous caseinate acts as a stable glue which can be used to hold together different foodstuffs. It is possible to bind meat cubes or thin fish fillets and to make larger beef steaks or larger fish fillets. Heating and/or freezing is not needed to get good binding with transglutaminase. Reformed meat/fish products can be prepared for distribution in the chilled state. Uniform products from various sized or shaped meats and fish portion by using MTG leads to effective utilization and added-value enhancement of food resources.

Table II. Thermostability of MTG-treated Gelatin Gel

MTG reaction time	Gel appearance after incubation for 20 minute at 100 ℃
20 minute	Fluid
30 minute	Weak gel
60 minute	Gel

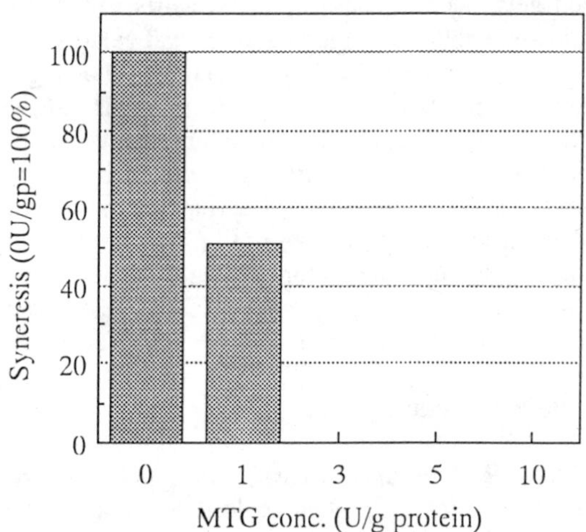

Figure 6. Effect of MTG treatment on water-holding capacity of the acid milk gel (set-type yogurt). MTG was added to reconstituted milk solution and incubated at 25°C for 2 hr. After enzyme reaction, the lactic starter was added, and the acid milk gel, set-type yogurt, was prepared after 4-6 hr fermentation.

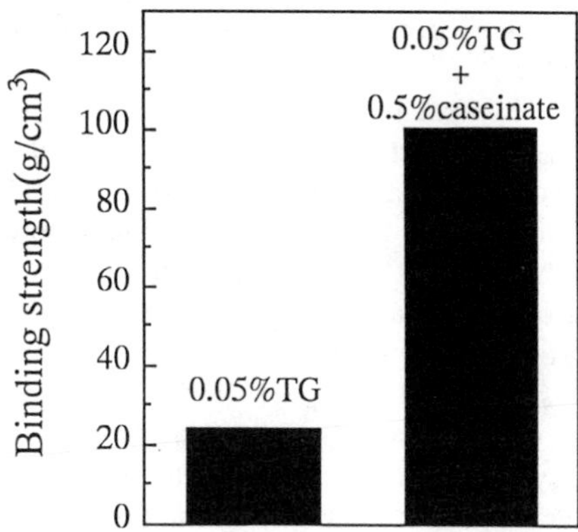

Figure 7. Effect of MTG and caseinate on binding strength of restructured meat. Raw pork meat cubes that had MTG and caseinate added were put into a mold and incubated at 5°C for 2 hr. Reformed meat was sliced and the binding strength was measured.

Health Demands. In processing of prepared foods, especially meat products, salt and/or some phosphates is/are usually added to increase water-holding capacity, binding, consistency, and to improve texture. Recently, as health demands increase, prepared foods with reduced salts or phosphates are distributed widely and have glutted the food industry. However, these healthy foods with reduced salts or phosphates have undesirable texture and physical properties. With the use of transglutaminase for preparing these "healthy" meat products, binding capacity, water-holding capacity, and viscosity are improved through ϵ-(γ-Glu)Lys crosslinks. For example, the breaking strength of the low-salt sausage, in which the salt content was reduced from 1.7% to 0.4%, decreased by 20% as compared with the control sausage containing 1.7% salt. But the addition of MTG 2 units/g protein recovered the decreased quality of the texture of the low-salt sausage and enhanced the breaking strength equal to that of the control sausage. Healthy prepared foods can be enjoyed which have higher qualities than those previously prepared.

Improvement of Yogurt Quality. Milk proteins are also good substrates for transglutaminase. As previously stated, the problems of serum separation of yogurt can be solved by adding transglutaminase, which improved the water-holding capacity of the gel (Fig. 6). MTG treatment improves qualitites other than syneresis. Gel strength of set-type yogurt and viscosity of stirred yogurt are much more acceptable sensorially. Because transglutaminase treatment improves gel strength and increases the viscosity of protein solutions (sols), MTG gives good physical qualities and sensory properties to set-and stirred yogurt. When the yogurt was treated with MTG 3 units/g protein at 25°C for 2 hr, the breaking strength of the set-type yogurt was increased from 15 g/cm^3 to 32 g/cm^3. Excess enzyme treatment, however, does not produce the same effect. The set-type yogurt treated with MTG 10 units/g protein was weaker than the gel treated with 5 units/gram protein. This result suggests that the excessive formation of ϵ-(γ-Glu)Lys bonds in intra/intermolecular proteins may inhibit the network formation of the yogurt gel. The optimal reaction conditions are at 25°C for 2 hr, 1-5 unit/g protein of MTG. Stabilizers are often used during the manufacture of yogurt to enhance and maintain the desirable characteristics in yogurt. Their mode of action in yogurt includes two basic functions: the binding of water and increase in viscosity. The effect of MTG, preventing syneresis and improving viscosity, is that of a stabilizer.

Many unique characteristics in protein properties are obtained by introducing ϵ-(γ-Glu)Lys bonds into proteins by the use of transglutaminase. In the research of protein gel structure and conformation, the formation of ϵ-(γ-Glu)Lys bonds provides many interesting topics to be discussed. In Japan, the use of our microbial transglutaminase, MTG, is dramatically increasing as an innovative ingredient for food processing because of the recognition of its unique characteristics. We are expanding the applications of transglutaminase and hope that the use of MTG will contribute to the global food industry.

Literature Cited

1. Folk, J.E. *Adv. Enzymol.* **1983**, *54*, 1-56.
2. Motoki, M.; Seguro,K.; Nio, N.; Takanami, K. *Agric. Biol. Chem.* **1986**, *50*, 3025-3030.
3. Folk, J.E. *Ann. Rev. Biochem.* **1980**, *49*, 517-531.
4. Ando, H.; Adachi, M.; Umeda, K.; Matsuura, A.; Nonaka, M.; Uchio, R.; Tanaka, H.; Motoki, M. *Agric. Biol. Chem.* **1989**, *53*, 2613-2617.
5. Yasueda, H.; Kumazawa, Y.; Motoki, M. *Biosci. Biotech. Biochem.* **1994**, *58*, 2041-2045.
6. Icekson, I.; Apelbaum, A. *Plant Physiology,* **1987**, *84*, 972-974.
7. Folk, J.E.; Cole, P.W. *J. Biol. Chem.* **1966**, *241*, 5518-5525.
8. Raczynski, G.; Snochowski, M.; Buraczewski, S. *Br. J. Nutr.* **1975**, *34*, 291-296.
9. Finot, P.-A.; Mottu, F.; Bujard, E.; Mauron, J.; In *Nutritional Improvement of Food and Feeds Proteins*; Friedman, M., Ed.; Plenum: London, 1978; pp 549-570.
10. Friedman, M.; Finot, P.-A. *J. Agric. Food. Chem.* **1990**, *38*, 2011-2020.
11. Seguro, K. et. al., *J. of Nutr.* submitted for publication.
12. Sakamoto, H.; Kumazawa, Y; Kawajiri, H.; Motoki, M. *J. Food. Sci.* **1965**, *60*, 416-419.
13. Nonaka, M.; Tanaka, H.; Okiyama, A.; Motoki, M.; Ando, H.; Umeda, K.; Matsuura, A. *Agric. Biol. Chem.* **1989**, *53*, 2619-2623.

Chapter 4

Biotechnology of Astaxanthin Production in *Phaffia rhodozyma*

Eric A. Johnson and William A. Schroeder

Departments of Food Microbiology; Toxicology; and Bacteriology, Food Research Institute, University of Wisconsin, 1925 Willow Drive, Madison, WI 53706–1187

Carotenoids have recently received considerable interest because of their potential in delaying or preventing degenerative diseases such as heart disease, cancer, and aging. Astaxanthin is the principal carotenoid pigment of several animals of economic importance in aquaculture including salmon. Salmon cannot synthesize astaxanthin and it must be supplied in their diet. This requirement has created a market for astaxanthin of about 100 million US dollars. The yeast *Phaffia rhodozyma* synthesizes astaxanthin as its primary carotenoid and has attracted interest as a biological source of the pigment. Astaxanthin production has been improved by the isolation of mutants that produce higher levels of astaxanthin (≥3,000 μg g^{-1}), and by an understanding of the biosynthesis and function of carotenoids in the yeast. Astaxanthin protects the yeast against singlet oxygen and peroxyl radicals. The carotenoids appear to be associated with lipid globules within the yeast that translocate to the cytoplasmic membrane as the yeast ages, possibly providing protection against age-related destructive reactions caused by activated oxygen species. In this chapter, the physiology of astaxanthin production is described in *P. rhodozyma* and strategies are described for improving industrial synthesis of this important carotenoid.

Among the agricultural industries, aquaculture is the fastest growing sector and is expected to grow into a $40 billion/year industry by the year 2000 (*1*). In the 1980's there was tremendous growth in the salmon and shrimp aquaculture industries. Salmon aquaculture increased from about 10,000 metric tons in 1980 to 300,000 metric tons in 1991 and may reach 500,000 metric tons in 1995. Much of the growth in aquaculture will occur in developing nations, and should contribute to the health of their economies.

The growth of the aquaculture industry has created a relatively large market for the carotenoid pigment astaxanthin (3,3'-dihydroxy-β,β-carotene-4,4'-dione ($C_{40}H_{52}O_4$); Figure 1), which must be supplied in the diets of farmed salmon since animals lack the ability to synthesize carotenoids. Astaxanthin is provided in salmon feeds to impart the vivid reddish-orange coloration characteristic of salmon flesh. In additon to its ornamental benefits, astaxanthin may also have essential biological functions since it and related carotenoids are effective scavengers of singlet oxygen and

0097–6156/96/0637–0039$15.00/0

peroxyl radicals (*2,3*). Although the commercial market for astaxanthin is privately held, it is currently estimated to be $80-100 million U.S. dollars per year, and will increase as the aquaculture industry expands. The high value of astaxanthin reflects its relatively complex chemical structure, and the presence of two chiral centers at the 3 and 3' positions of the molecule.

Considerable effort has been expended by the biotechnology industry in developing natural sources of astaxanthin. Interest in biological sources of astaxanthin reflects a trend by consumers to prefer natural products instead of those made by chemical synthesis. The freshwater alga *Haematococcus pluvialis* (*4*) and the yeast *Phaffia rhodozyma* (*1*) have attracted the most interest as potential biological sources of astaxanthin. Each source has its benefits and limitations. *H. pluvialis* produces high quantities of astaxanthin (up to 50,000 $\mu g\ g^{-1}$), but commercial manufacture is difficult due to lengthy photoautotrophic freshwater cultivation that is subject to bacterial and protozoan contamination, and extensive downstream processing of the harvested cells to liberate the astaxanthin which is present in cysts (*5*). *P. rhodozyma* has attracted considerable commercial interest because it grows rapidly to high cell densities in industrial fermentors, and because yeasts have a rich tradition as safe and effective feed ingredients. Several companies have developed strains and fermentation systems that give *Phaffia* products that contain $\geq$3000 μg astaxanthin - g^{-1} yeast. These hyperproducer strains, however, grow more slowly than their wild-type parents and also accumulate high levels of carotenoid intermediates. The U.S. Food and Drug Administration has not permitted the use of *Phaffia* products containing more than 4% carotenoid intermediates (*6*). Additionally, practical difficulties remain concerning commercial fermentations. These include the low optimal growth temperature of *Phaffia* (22° C), and the decreased yield of astaxanthin in defined media (*1*). Consequently, other microbial sources of astaxanthin are also being evaluated including marine halobacteria (*7*) and eubacteria (*8-10*) but these are in the early developmental stages.

The research efforts in our laboratory have been directed towards understanding the function of astaxanthin in *P. rhodozyma* in order to provide a rational strategy for improving yeast strains by positive genetic selections. During the past five years, we have found evidence that astaxanthin biosynthesis is regulated by singlet oxygen ($^1O_2$) and peroxyl radicals, and that an important function of the astaxanthin pathway is to protect the yeast against toxic oxygen metabolites generated in its natural environment and by intracellular oxidative metabolism. These findings have provided a rational approach to improved strain development and fermentation conditions for astaxanthin production through biotechnology.

Role of Carotenoids in Protecting *Phaffia rhodozyma* Against Singlet Oxygen and Peroxyl Radicals

P. rhodozyma was initially isolated by Herman J. Phaff and Japanese collaborators in the 1960's from sap fluxes of the wounds of various species of *Betula* (Birch) and related trees. The yeast has generally been isolated in mountainous regions of northerly latitudes at high elevation, and generally in cold temperatures (*11*). Its unusual physiological features, in particular its ability to synthesize astaxanthin as its primary carotenoid, were recognized during the 1970's (*12-14*). Many strains of this yeast have also been isolated in recent years from various regions in Russia (*15*). *P. rhodozyma* belongs to the taxonomic group of heterobasidiomycetous yeasts, a diverse group of fungi of basidiomycetous origin that live at least part of their life cycle in a single cell yeast like state (*16,17*).

P. rhodozyma's limited habitat of wounded trees at high altitudes indicates that the yeast is highly adapted to its ecological niche, and environmental and chemical properties of the habitat select for its colonization. At high altitudes, incident ultraviolet radiation and ozone (O_3) concentrations would be high, which could result in high concentrations of $^1O_2$ and other reactive oxygen metabolites. Singlet oxygen is quite toxic to microorganisms (*18*). Singlet oxygen is generated in forest air by the reaction of ozone with plant leaves (*19*). Secondary metabolites in certain species of plants react with ultraviolet light or ozone to generate $^1O_2$, which may prevent parasitic infections (*20*). Other reactive oxygen metabolites would also be increased in the high altitude ecology. Hydrogen peroxide is produced in forest air by reaction of O_3 with terpenes (*21*). Lipid peroxyl radicals may also be formed in this environment, and these are thought to play a role in forest decline (*21*). Carotenoids are effective quenchers of $^1O_2$, catalytically detoxifying $^1O_2$ without destruction of the carotenoid (*22*). Since carotenoids are known to specifically quench $^1O_2$, and also to react with H_2O_2 and peroxyl radicals, we investigated whether astaxanthin protected *P. rhodozyma* against excited oxygen species.

Reactive oxygen species are formed in all aerobic cells and can destroy various cell components including nucleic acids, proteins, and membranes (*23,24*). These destructive reactions are believed to contribute to degenerative diseases such as arteriosclerosis, cataracts, cancer and aging in higher eukaryotes including humans (*24,25*). The biological chemistry of excited oxygen species is very complex and involves many different chemical species generated by nonenzymatic and enzymatic reactions. Two of the most strongly reactive oxygen species, $^1O_2$ and possibly hydroxyl radical (OH·), are generated from superoxide radical (O_2^-) and hydrogen peroxide (H_2O_2) (*26,27*). O_2^- and H_2O_2 formation in the cell is usually associated with spillover from cytochrome cellular respiration or by cyanide-insensitive respiration (alternative oxidase pathways) mediated by P-450 cytochromes, flavin oxygenases, and peroxisomal metabolism (*24*). Many groups of fungi and plants possess alternative oxidase pathways (*28-30*).

Alternative oxidase activity in plants and fungi is dependent on environmental, developmental, nutritional, and other signals (*30*). We found that while the total level of oxygen uptake (mmol O_2 min^{-1} mg $cells^{-1}$) was relatively constant during the growth cycle of *P. rhodozyma*, cyanide-insensitive O_2 uptake became dominant as the cell population entered stationary phase (Fig. 2). This shift in respiration and probable increased formation of excited oxygen species correlated with an increase in pigmentation of the cells (*31*).

A relationship between toxic oxygen species and carotenogenesis was supported by a unique selection for carotenoid hyperproducing mutants. Growth for 1 to 2 months of *P. rhodozyma* on agar plates containing antimycin A, an inhibitor of cytochrome respiration, gave rise to highly pigmented vertical papillae from the yeast colonies (*32*). Isolation of these papillae yielded yeast strains that produced considerably more astaxanthin than the wild-type parent strains. Surprisingly, these mutants were more sensitive to respiratory inhibitors including thenoyltrifluoroacetone and antimycin A. It was proposed that blockage of the respiratory chain resulted in the increased formation of excited oxygen species, possibly O_2^-, H_2O_2, and $^1O_2$ which in turn triggered astaxanthin biosynthesis (*32*).

A relation between oxidative stress and carotenogenesis was further supported by study of *P. rhodozyma*'s defenses against O_2^- and H_2O_2. Eukaryotic cells are known to possess various detoxification systems for O_2^-, H_2O_2, and peroxyl radicals including superoxide dismutase (SOD), catalase, and glutathione peroxidase. SOD

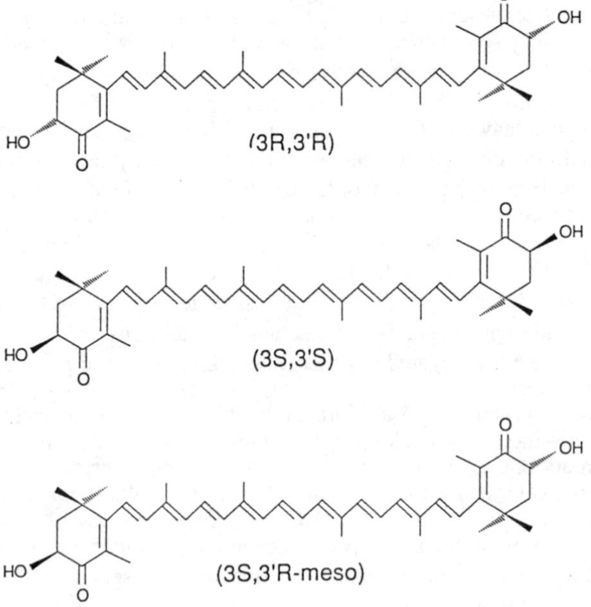

(3R,3'R)

(3S,3'S)

(3S,3'R-meso)

Figure 1. Structure of astaxanthin and its R and S configurational isomers.

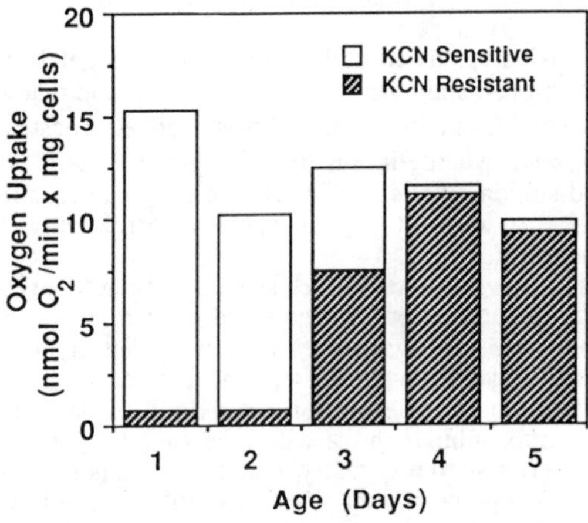

Figure 2. Oxygen uptake by *P. rhodozyma* strain UCD-67-385 and sensitivity to 1.3 mM KCN as a function of culture age. Reproduced with permission from reference *37*. Copyright 1995 American Society for Biochemistry and Molecular Biology.

catalyzes the reduction of O_2^- to H_2O_2 and O_2, while catalase and peroxidases form H_2O and O_2 from H_2O_2 and peroxides. Three different classes of SOD have been identified in eukaryotic cells, each class containing a different transition metal cofactor (Mn, Fe, or Cu). The three classes of SODs also differ in cellular location: Mn-SOD has been demonstrated in mitochondria, whereas Fe-SOD and Cu/Zn-SOD are cytosolic (*33,34*). Interestingly, staining of native electrophoresis gels of *P. rhodozyma* cell extracts indicated that only the mitochondrial Mn-SOD is present, and that the yeast apparently lacks the cytosolic Fe- and Cu-forms of SOD (*31*). Catalase activity was also present in low quantities in *P. rhodozyma* compared to *Saccharomyces cerevisiae*. These observation suggested that carotenoids may partly compensate for the lack of SOD and low activity of catalase in the cytosol.

Hydrogen peroxide and O_2^- can react in the cytosol to form more toxic oxygen species, including OH·,$^1O_2$, and lipid peroxides. Since peroxyl radicals react with carotenoids through free radical mechanisms (*35*), we evaluated whether exposing cells to H_2O_2 and O_2^- would select for pigmented yeasts. Treating yeast cultures with 1 mM H_2O_2 and 0.1% paraquat (an O_2^- generator) was more effective in selecting for carotenoid hyperproducers than selections with either compound alone (*31*). The resistance to H_2O_2 and O_2^- also showed an age dependency of the yeast population. Susceptibility to H_2O_2 correlated with the cellular content of carotenoid in young exponentially growing populations, and carotenoid hyproducing strains in the growth phase were more resistant to H_2O_2 than the wild-type. However, when growth slowed and the yeast population entered into stationary phase, the carotenoid hyperproducers but not the wild-type cells showed a sharp decrease in H_2O_2 resistance. This difference in tolerance could not be attributed to differences in levels of catalase activity, and the biochemical explanation for the dramatic drop in H_2O_2 resistance in the hyperproducer remains to be explained. Since H_2O_2 resistance in growing cultures depended on their carotenoid content, selections were attempted. Repeated exposures of young cultures to H_2O_2 gave a slight positive selection for increased carotenoid content (*31*), supporting the role of carotenoids in protecting cells against peroxyl radicals.

During oxidative metabolism, mitochondria can release damaging quantities of O_2^- (*33*), and we therefore evaluated whether carotenoids provided protection against O_2^-. Superoxide anion was generated by exposing yeast cultures to duroquinone (DQ), which catalyzes the one electron reduction of O_2 to O_2^-. In young cells where the carotenoid content is low, both the wild-type and a carotenoid hyperproducing strain were highly sensitive to DQ (*31*). As the cultures aged and entered into stationary phase, resistance to DQ increased in both cell populations, although the resistance was more pronounced in the hyperproducer, possibly due to the higher content of carotenoids in the mutant. Qualitative anaysis of SOD stains on electrophoresis gels did not show an increase in SOD during stationary phase.

Since increased O_2^- generation appeared to correlate with increased carotenoid biosynthesis, we examined the effect of DQ exposure on carotenoid levels and composition. Exposure of cultures to DQ increased the levels of carotenoid produced by about 40%, and also increased the relative proportion of xanthophylls and diminished the levels of carotene precursors (*31*). In addition to producing more carotenoids, the increased proportion of xanthophylls would provide greater resistance to oxidative stress since xanthophylls are generally more effective antioxidants than carotenes (*2*). This pattern of increased levels of carotenoids and a higher proportion of xanthophylls also takes place as cultures age. The astaxanthin pathway may function in part to prevent aging of yeast and possibly to supply antioxidant capacity to their progeny. Microscopic examination of autofluorescence supported that carotenoids are translocated to the young buds during reproduction (An, G.-H.; Johnson, E. A.; unpublished data).

The observation that exposure to O_2^- enhanced carotenogenesis suggested that gene expression of carotenogenic enzymes may respond to oxidative stress. We hypothesized that O_2^- may not be the actual signal since it is not known to react directly with carotenoids, and the combination of H_2O_2 and O_2^- was more effective than either compound alone in stimulating carotenogenesis. Since the reaction of H_2O_2 and O_2^- has been shown to generate $^1O_2$, a very reactive oxygen species, we investigated whether $^1O_2$ exposure could enhance carotenogenesis and select for hyperproducers (*36,37*). The effect of $^1O_2$ on carotenogenesis was evaluated using $^1O_2$ generators including the food dye rose bengal (RB) (*38*). Exposure of *P. rhodozyma* wild type or hyperproducer to RB and light for 24 and 48 h resulted in induction of carotenogenesis. Increasingly higher exposure levels during 48 hour incubations led to a selection for the more highly pigmented strain in cultures inoculated with a 1:1 mixture of wild-type and hyperproducing strains (Figure 3). Selection by RB was enhanced in the presence of the carotenoid biosynthetic inhibitor thymol (*36*), apparently because the hyperproducer was more resistant to carotenoid biosynthesis blockade. Selection for carotenogenesis was decreased by the $^1O_2$ quencher DABCO (*36*).

Singlet oxygen is also generated by the plant secondary metabolite α-terthienyl (α-T). When activated by 366 nm light, α-T produces $^1O_2$ in high yield which can be nematocidal (*39*). Exposure of a mixture of *P. rhodozyma* strains that differed in carotenoid content to α-T and 366 nm light for ≤5 min resulted in selection of the most highly pigmented strain (*36*). In contrast to RB, induction of carotenogenesis by sublethal levels of $^1O_2$ generated by α-T required only a brief (≤5 min) exposure, suggesting that $^1O_2$ triggered gene expression in the astaxanthin pathway (*37*).

Although effective selection of *Phaffia* astaxanthin-hyperproducers and maintenance of strain stability was achieved by light-generated $^1O_2$, photochemical generation of $^1O_2$ is probably not practical in industrial fermentations. We investigated practical means by which $^1O_2$ could be generated using dark chemical reactions (*40*). Acidified HOCl (pH 5.0) is known to spontaneously decay and generate $^1O_2$ (*41*). However, exposure of *P. rhodozyma* to acidified HOCl did not give consistent selection of hyperproducers (*40*). These inconsistent results with HOCl may have been caused by killing of yeasts by species other than $^1O_2$ (*42*). To improve the selection, HOCl was reacted with H_2O_2 to enhance the yield of $^1O_2$ (*41*). Exposure of *Phaffia* strains to equimolar amounts of H_2O_2 and HOCl for periods of only 5 min greatly favored survival of carotenoid hyperproducing strains. Additionally, this selection was largely relieved by the $^1O_2$ quencher DABCO supporting the importance of carotenoids in protecting against $^1O_2$.

Although selection using H_2O_2 and HOCl was effective, production of an animal feed using these compounds may lead to toxic products. An alternative approach is to generate $^1O_2$ from O_3. Ozone and ozonated water have been investigated in the poultry and soft drink industries as sanitizing agents (*43*). Although the microbiocidal mechanisms have not been elucidated, it is probable that ozone spontaneously decomposes to generate $^1O_2$ (*19*). When O_3 gas was sparged through a culture containing a mixture of *P. rhodozyma* strains, the most highly pigmented strains predominated. This selection was extremely effective when cells were suspended in a buffer without other organic compounds, which increased the effective concentration of

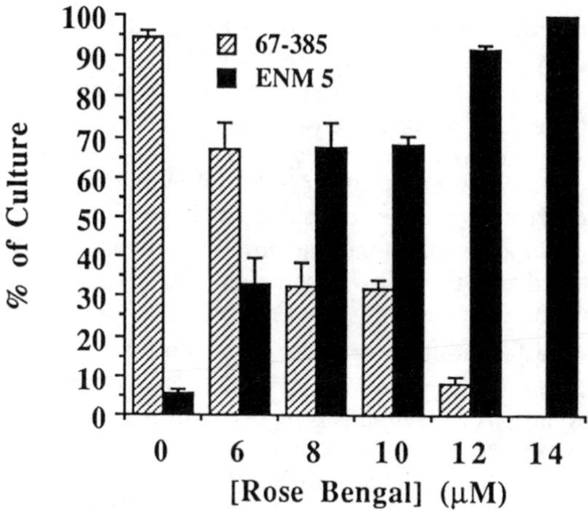

Figure 3. Selection for carotenoid hyperproducing strain (ENM-5; ~2700 μg carotenoid g^{-1}) over the wild-type (UCD-67-385) with rose bengal and visible light. Reproduced with permission from reference *36*. Copyright 1995 Society for Industrial Microbiology.

O_3. When cells were suspended in a defined media, the efficacy of the selection was slightly decreased, presumably due to the presence of certain amino acids known to detoxify ozone. DABCO relieved the O_3 selection in a dose dependent manner. Ozone exposure decreased the level of astaxanthin and increased the level of β-carotene. This alteration in carotenoid content could be reversed by maturation of the cells in media free of O_3 (*40*).

Regulation of the Astaxanthin Pathway in *P. rhodozyma* by $^1O_2$ and Peroxyl Radicals

Our results support the hypothesis that $^1O_2$ may trigger astaxanthin biosynthesis, and that carotenoids serve an antioxidant role in the yeast by reacting with peroxyl radicals. Generation of $^1O_2$ may explain the increase in carotenoid formation mediated by O_2 and blue light in *P. rhodozyma* (44,45). Short light pulses of 15 minutes resulted in transient increases in astaxanthin levels in *P. rhodozyma*, but the highest carotenoid yields were obtained by continuous illumination.

The results presented in this chapter strongly support the hypothesis that $^1O_2$ is a signal for induction of carotenogenesis in *P. rhodozyma*. The mechanisms by which $^1O_2$ triggers carotenoid synthesis is not known. An intriguing mechanism of carotenoid regulation by blue light and $^1O_2$ has been proposed (46): (a) blue light interacts with protoporphyrin IX in the membrane and generates $^1O_2$; (b) $^1O_2$ then reacts with a membrane bound regulatory protein(s) that initiates a cascade resulting in increased synthesis of carotenogenic enzymes; (c) when carotenoid levels rise they quench $^1O_2$ and as a result, expression is consequently repressed. This interesting proposal involves a regulatory protein whose structure is altered by $^1O_2$.

Besides induction by $^1O_2$, our laboratory has also found evidence that the astaxanthin pathway may be regulated by feedback inhibition by astaxanthin. When *P. rhodozyma* was exposed to H_2O_2 for 3 hr, the total carotenoid level did not change significantly but the level of astaxanthin (the end product) decreased and the levels of carotene biosynthetic intermediates increased (*37*). However, the magnitude of the change in carotenoid concentrations using H_2O_2 as an oxidant was nominal, possibly due to the presence of catalase in the cell and the hydrophilic affinity of H_2O_2. A more substantial decrease in astaxanthin and increase in carotenes was observed when *P. rhodozyma* was treated with t-butylhydroperoxide (tBOOH), which is lipophilic and is also a suicide substrate for catalase (*47*). Treatment of *P. rhodozyma* for 6 h with 8 mM tBOOH caused astaxanthin to decrease from 186 μg g^{-1} to 134 μg g^{-1}, while β-carotene increased from 17 μg g^{-1} to 77 μg g^{-1} (*37*). These data support the hypothesis that carotenoid biosynthesis in *Phaffia* is is regulated by end product inhibition.

Reactive oxygen species have recently been found to be important signaling molecules in many cellular responses (*48*). Besides their known negative roles of damaging cell components, and their positive attribute of defense against microorganisms, they have also been implicated as intracellular messengers (*48*). It was emphasized by Khan and Wilson (*48*) that compared to other molecules in electronically excited states, $^1O_2$ has an exceptionally long lifetime, and that $^1O_2$ reacts fast and selectively with many unsaturated organic molecules including carotenoids. Singlet oxygen may be a common unifying trigger for carotenogenesis in many organisms, but further work is needed to elucidate its role. *P. rhodozyma* may serve as an excellent microbial model to study the regulation of carotenogenesis and the possible functions of carotenoids in cell aging. A sexual cycle has recently been described for *Phaffia* (*15*), which should enable biochemical studies to be complemented by genetic analysis of carotenoid synthesis and function.

Cellular Location of Carotenoids

Since astaxanthin is not excreted, its location and compartmentalization within the cell is important in understanding its function and in developing hyperproducing mutants. Our laboratory has used Laser Confocal Fluorescence Microscopy (LCFM) to investigate the distribution and concentration of carotenoids in *P. rhodozyma* (*49, 50*). Experimental conditions were devised in which the autofluorescence of cells was correlated with their carotenoid content. Carotenoids appeared to be located in lipid globules in the cytoplasm. As the cells aged, the lipid globules containing carotenoids fragmented and were translocated to cytoplasmic membraneous regions and also migrated into developing buds (progeny). These results suggested that the production and translocation of carotenoids in lipid globules is probably a mechanism for protecting the yeast and its progeny against reactive oxygen species generated during metabolism and aging.

A maximum cellular content of carotenoids could be estimated since there was a quantitative relationship between total carotenoid content and average fluorescence intensity of a culture. A maximum yield of between 14,600 and 19,000 μg g^{-1} was obtained. However, there was considerable heterogenity in autofluorescence among individual cells which suggests that the yield could be improved. Maximum synthesis could also be improved by directed transport, esterification of astaxanthin, or excretion as described below.

Investigations in other laboratories have also supported the hypothesis that carotenoid biosynthesis may be compartmentalized and transport associated. In preliminary investigations, lipid globules and cytoplasmic membranes were isolated by differential centrifugation (Peschek; R, personal communication, 1992). The purity of the fractions was supported by characteristic marker enzymes. Interestingly, about 80% of the astaxanthin was found in the membrane fractions and only 20% in the lipid fraction. In contrast, 70% of β-carotene was found in the lipid fraction. These data show that carotenoids are compartmentalized in *P. rhodozyma*, and that maturation of cells is associated with translocation of carotenoids to the cytoplasmic membrane, where they may be oxidized to astaxanthin. This compartmentalization also suggests that the desaturase and cyclase activities are associated with lipid globules, while the C-3 and C-4 oxidase functions may be carried out by cytoplasmically associated proteins. Compartmentalization of enzymes and translocation and storage of carotenoids is a difficult yet potentially rich research area. Understanding carotenoid translocation could be valuable for strain development in *Phaffia*.

The autofluorescent properties of carotenoids enabled the development of a method to enrich for carotenoid hyperproducers by fluorescence activated cell sorting (FACS). When excited by an argon laser ($\lambda = 488$ nm) astaxanthin autofluoresces ($\lambda = 545$ nm). By using the proper wavelength filters to screen out autofluorescence from other cellular components including other carotenoids, it is possible to identify individual cells with high astaxanthin levels. These astaxanthin hyperproducing mutants are then sorted from the general population and subsequently plated for individual colonies. By this method, it has been possible to select one hyperproducer in a total population of 10,000 cells (*49*). Since cell populations can be examined at rates of ~5,000 cells per second, this technology could be very powerful for the isolation of microorganisms containing a variety of autofluorescent intracellular secondary metabolites or surface bound molecules that could be tagged with fluorescent markers.

A consideration in the development of astaxanthin hyperproducing strains of *P. rhodozyma* is the solubility of carotenoids in lipid. Compared to many other yeasts, *P. rhodozyma* is relatively lipid-rich and contains a total fat content of 20 to 30% (w/w) depending on the strain and culture conditions. The solubility of carotenoids in lipid is generally very low: at 30° C the solubilities of β-carotene in corn oil and olive oil are only 0.08% and 0.1% respectively (*51*). Assuming a lipid concentration in the cell of 25% and a carotenoid solubility of 0.1%, the maximum concentration of β-carotene in cell biomass would be 250 $\mu g\ g^{-1}$ cells. Numerous methods exist to increase carotenoid concentration in commercial lipid preparations (*51*) and the presence of carotenoid hyperproducing strains indicate that a solubility limit of 0.1% can be surpassed in *Phaffia*. However, limited carotenoid solubility in the cell and phase separation in the cytoplasm (*52*) may be a major limitation in strain improvement. One mechanism to increase astaxanthin solubility may be through esterification. Interestingly, in the alga *Haematococcus*, which produces 50,000 $\mu g\ g^{-1}$ astaxanthin, only 1% of the astaxanthin (500 $\mu g\ g^{-1}$) is in the free form, and 99% is esterified with fatty acids (*53*). These results suggest that developing strains of *P. rhodozyma* with the capacity to esterify astaxanthin could lead to carotenoid hyperproducers. Alternatively, redirecting transport (*54*), using permeabilized cells (*55*), or developing biotransformation systems (*56*) could also increase the yield and improve the economics of astaxanthin production.

Modeling of Astaxanthin Formation in *P. rhodozyma*

A theoretical mathematical model of the astaxanthin biosynthetic pathway was developed by Kelly (*57*). Assuming an atomic formula for cell biomass of $C_8H_{18}O_4N$, a theoretical equation for astaxanthin biosynthesis was proposed:

$$3.83 \text{ mol glucose} + 0.954 \text{ mol ammonia} + 4 \text{ mol } O_2 = 0.213 \text{ mol astaxanthin} + 0.954 \text{ mol biomass } (C_8H_{18}O_4N) + 6.82 \text{ mol } CO_2 + 12.67 \text{ mol } H_2O$$

The model indicates that oxygen is a key limiting nutrient and that 21 mol O_2 are required for each mol astaxanthin produced. Also, as O_2 becomes limiting, biomass production is favored at the expense of astaxanthin biosynthesis. During astaxanthin biosynthesis, 1.75 mol O_2 are consumed per mol of glucose consumed while only 1.53 mol O_2 are required per mol of glucose for biomass production. This model suggests that the isolation of mutants with both high astaxanthin yields and high growth rates will prove very difficult.

Conclusion and Perspective

Our studies have supported the conclusion that *P. rhodozyma* and probably other heterobasidiomycetous yeasts have evolved the capacity to produce carotenoids to protect against reactive oxygen species. *P. rhodozyma* appears to be deficient in detoxification enzymes for reactive oxygen species compared to ascomycetous yeasts such as *Saccharomyces*, and they compensate for this deficiency by producing carotenoids and possibly other protective systems. Singlet oxygen appears to be a key signalling compound that triggers carotenoid biosynthesis. Due to the rudimentary nature of *Phaffia* as a biological system for studying carotenogenesis, some of the conclusions drawn in this article are based on cause and effect. More research is needed to elucidate the molecular mechanisms of astaxanthin biosynthesis and the role of carotenoids and other secondary metabolites in the evolution of this unique group of eukaryotes.

Literature Cited

1. Johnson, E.A.; Schroeder, W.A. *Adv. Biochem. Eng. Biotechnol.* **1995**, *53*, 119-178.
2. Hirayama, O.; Nakamura, K.; Hamada, S.; Kobayashi, Y. *Lipids.* **1994**, *29*, 149-150.
3. Oliveros, E.; Murasecco-Suardi, P.; Braun, A. M.; Hansen, H.-J. In: *Carotenoids. Part A. Chemistry, Separations, Quantitation, and Antioxidation; Methods in Enzymology, Vol. 213*; Academic Press, San Diego, **1992**; pp. 420-429.
4. Grung, M.; D'Souza, F. M. L.; Borowitzka, M.; Liaaen-Jensen, S. *J. Appl. Phycol.* **1992**, 165-171.
5. Bubrick, P. *Bioresource Technol.* **1991**, *38*, 237-239.
6. Anonymous. *Food Chem. News.* **Nov. 6, 1995**, 35-36.
7. Calo, P.; Demiguel, T.; Sieiro, C.; Velasquez; J. B.; Villa, T. G. *J. Appl. Bacteriol.* **1995**, *79*, 282-285.
8. Misawa, N.; Satomi, Y.; Kondo, K.; Yokoyama, A.; Kajiwara, S.; Saito, T.; Ohtani, T.; Miki, W. *J. Bacteriol.* **1995**, *177*, 6575-6584.
9. Kajiwara, S.; Kakizono, T.; Saito, T.; Kondo, K.; Ohtani, T.; Nishio, N.; Nagai, S.; Misawa, N. Plant Molec. Biol. **1995**, *29*, 343-352.
10. Yokoyama, A.; Miki, W. *FEMS Microbiol. Lett.* **1995**, *128*, 139-144.
11. Phaff, H. J.; Miller, M. W.; Yoneyama, M.; Soneda, M. In: *Proc. 4th IFS: Ferment. Technol. Today*; Terui, G., Ed.; Kyoto Society of Fermentation Technology, Osaka, Japan, **1972**; pp 759-774.
12. Miller, M. W.; Yoneyama, M.; Soneda, M. *Int. J. Syst. Bacteriol.* **1976**, *26*, 286-291.

13. Andrewes, A. G.; Phaff, H. J.; Starr, M. P. *Phytochem.* **1976**, *15*, 1003-1007.
14. Andrewes, A. G.; Starr, M. P.; *Phytochem.* **1976**, *15*, 1009-1011.
15. Golubev, W. I. *Yeast.* **1995**, *11*, 101-110.
16. Oberwinkler, F. *Stud. Mycol.* **1987**, *30*, 61-74.
17. *Heterobasidiomycetes: Systematics and applied aspects*; Boekhout, T.; Fell, J. W., Eds.; Studies in Mycology, Centraalbureau voor Schimmelcultures, Delft, The Netherlands, **1995**; vol. 38.
18. Restaino, L.; Frampton, E. W.; Hemphill, J. B.; Palnikar, P. *Appl. Environ. Microbiol.* **1995**, *61*, 3471-3475.
19. Kanofsky, J. R.; Sima, P. *J. Biol. Chem.* **1991**, *266,* 9039-9042.
20. Kyo, M.; Miyauchi, Y.; Fujimoto, T.; Mayama, S. *Plant Cell Rep.* **1990**, *9*, 393-397.
21. Becker, K. H.; Brockamann, K. J.; Bechara, J. *Nature* **1990**, *346*, 256-258.
22. Bradley, D. G.; Min, D. B. *Crit. Rev. Food. Sci. Nutr.* **1992**, *31*, 211-236.
23. Stadtman, E. *Science.* **1992**, *257*, 1220-1224.
24. Ames, B. N.; Shigenaga, M. K.; Hagen, T. M. *Proc. Natl. Acad. Sci. USA.* **1993**, *90*, 7915-7922.
25. Krinsky, N. I. *J. Nutr.* **1989**, *119*, 123-126.
26. Krinsky, N. I. *TIBS.* **1977**, *2*, 35-38.
27. Halliwell, B.; Gutteridge, J. M. C. *Meth. Enzymol.* **1990**, *186*, 1-85.
28. Shiraishi, A.; Fujii, H. *Agric. Biol. Chem.* **1986**,*50*, 447-452.
29. McIntosh, L. *Plant Physiol.* **1994**, *105*, 781-786.
30. Vanlebergh, G. C.; Vanleberge, A. E.; McIntosh, L. *Plant Physiol.* **1994**, *106*, 1503-1510.
31. Schroeder, W. A.; Johnson, E. A. *J. Gen. Microbiol.* **1993**, *139*, 907-912.
32. An, G.-H.; Schuman, D. B.; Johnson, E. A. *Appl. Environ. Microbiol.* **1989**, *55*, 116-124.
33. Fridovich, I. *Annu. Rev. Biochem.* **1983**, *23*, 239-257.
34. Moore, M. M.; Breedveld, M. W.; Autor, A. P. *Arch. Biochem. Biophys.* **1989**, *270*, 419-431.
35. Burton, G. W.; Ingold, K. U.; *Science.* **1984**, *224*, 569-573.
36. Schroeder, W. A.; Johnson, E. A. *J. Industr. Microbiol.* **1995**, *14*, 502-507.
37. Schroeder, W. A.; Johnson, E. A. *J. Biol. Chem.* **1995,** *270,* 18374-18379.
38. Lamberts, J. J. M.; Neckers, D. C. *Tetrahedron.* **1985**, *41,* 2183-2190.
39. Kyo, M.; Miyauchi, Y.; Fujimoto, T.; Mayama, S. *Plant Cell Rep.* **1990**, **9**, 393-397.
40. Schroeder, W. A.; Calo, P.; DeClerq, M. L.; Johnson, E. A. **1996**, submitted for publication.
41. Khan, A. U.; Kasha, M. *Proc. Natl. Acad. Sci. USA.* **1994**, *91*, 12362-12364.
42. McKenna, S. M.; Davies, K. J. A. *Biochem. J.* **1988**, *254*, 685-692.
43. Greene, A.K.; Few, B.K.; Serafini, J.C. *J. Dairy Sci.* **1993**, *76*, 3617-3620.
44. Meyer, P.S.; DuPreez, J.C. *System. Appl. Microbiol.* **1994**, *17*, 24-31.
45. An, G.-H.; Johnson, E. A. *Ant. van Leeuwenhoek.* **1990**, *57*, 191-203.
46. Hodgson, D. A.; Murillo, F. J. In *Myxobacteria II*; Dworkin, M.; Kaiser, D.; Eds.; American Society of Microbiology, Washington, D. C., **1993**; pp 157-181.
47. Pichorner, H.; Jessner, G.; Ebermann, R. *Arch. Biochem. Biophys.* **1993**, *300(1)*, 258-264.
48. Khan, A. U.; Wilson, T. *Chemistry and Biol.* **1995**, *2*, 437-445.
49. An, G.-H.; Bielich, J.; Auerbach, R.; Johnson, E. A. *Bio/Technol.* **1991**, *9*, 70-73.

50. An, G.-H. Thesis, University of Wisconsin, Madison, **1991**.
51. Nishinomiya, H. M.; Itami, M. T.; Kato, M. U. S. Patent # 3,227,561. **1966**.
52. Walter, H.; Brooks, D. E. *FEBS Lett.* **1995**, *361*, 135-139.
53. Renstrom, B.; Borch, G.; Skulberg, O. M.; Liaaen-Jensen, S. *Phytochemistry.* **1981**, *20,* 2561-2564.
54. Brodelius, P.; Pedersen, H. *Trends Biotechnol.* **1995**, 30-36.
55. Felix, H.; Brodelius, P.; Mosbach, K. *Anal. Biochem.* **1981**, *116*, 462-470.
56. Salter, G. J.; Kell, D. B. *Crit. Rev. Biotechnol.* **1995**, *15*, 139-177.
57. Kelly, S. E. Thesis, University of Wisconsin, Madison, **1990**.

Chapter 5

Antimicrobial, Insecticidal, and Medicinal Properties of Natural Product Flavors and Fragrances

Horace G. Cutler[1], Robert A. Hill[2], Brian G. Ward[2], B. Hemantha Rohitha[2], and Alison Stewart[3]

[1]Agricultural Research Service, U.S. Department of Agriculture, Richard B. Russell Center, 950 College Station Road, Athens, GA 30613
[2]HortResearch, Ruakura Research Centre, East Street, Hamilton, New Zealand
[3]Department of Plant Science, Lincoln University, P.O. Box 84, Canterbury, New Zealand

The primary use of natural product flavors and fragrances has been in food, perfume and, historically, the preservation of flesh, for example, Egyptian mummies. A large body of evidence has been building over a period of decades that indicates thier possible use as antimicrobials, insecticides or antifeedants, and medicinals. With respect to the latter, the overuse of medicinal antibiotics has led to widespread resistance of pathogens. This, in turn, has elicited a reevaluation of botanicals to control non-life threatening infections. A number of examples of insecticides and antifeedants are examined in addition to some fungicides. Special emphasis is placed on the natural product fungicide 6-pentyl-α-pyrone, which is also a flavoring agent.

Because of the images that flavors and fragrances conjure in our minds, the thought that they may have very practical use as antimicrobials, insecticides or insect repellants, and medicinals is treated with a good deal of scepticism. Whenever the subject is broached, one finds that the listener automatically stops paying attention because, as if by some hypnotic suggestion, the mind has wandered. While this is generally the case, the reason for this apparent aberration can be analyzed fairly easily, for nearly all human beings, and many other animals, possess olfactory memory. We remember, from early childhood, the perfumes and essences worn by those that we have loved and those that we have disliked. In the latter case, the odor of an enemy may, years later, make one nauseous when introduced in a non-threatening environment. Another reason for the lack of attention is because historical mental trapdoors are sprung and visions of Marco Polo traveling with his father to China along the Silk Road to bring back spices, and other treasures, flood the mind. The search for the Northwest Passage, which led to the discovery of America, was solely motivated by the lust for precious metals, silk, and aromatic spices. Last, but not least, the use of aromatics, especially botanical pitch, in the preservation of mortal remains and animals to aid in the arduous journey to the next world was common practice in ancient Egypt. It was precisely these botanicals that made each person a *sol invictus* (an unconquerable sun), personified in Egyptian religion as Ra, the Sungod, and later, the Roman god *Sol Invictus,* whose feastday the

Published 1996 American Chemical Society

Christians adopted, for obvious reasons, as Christmas. To emphasize the point, so common were mummies in ancient Egypt that in the 19th century they were broken up and used as fuel to fire the boilers of railroad locomotives. The trade in spices and botanicals must have been immense, just for funereal purposes.

Another reason for the apparent scepticism concerning the practical use of flavors and fragrances as medicinals and agrochemicals is because of another dichotomy that arises when these substances are considered. The first line of reasoning is that the natural product aromatics are already used in the food and beverage market and exclusively in the perfume industry. Both use legions of materials, many of which are high value added products and the market is, in a sense, self contained. That is, there is a thriving market well into the forseeable future and, therefore, no real purpose is served by expanding these borders. The second approach makes the claim that while there may be compounds, or mixtures of compounds, that are suitable candidates for agrochemical and medical use, are any of these practical and efficacious? Furthermore, the literature is replete with examples of potentially useful substances, but how many have withstood rigorous testing?

There is, however, a new underlying element that makes a case for the use of flavors and fragrances especially as agrochemicals. That is the inexorable move, on the part of the public, away from persistent chemicals to those that are considered to be highly biodegradable. In addition, one has to consider that this class of natural products has been used for several hundred years by millions of people without any apparent deleterious effects. This further suggests that they need to be thoroughly examined not only for further exploitation, but also academically to determine their mode of action.

In this presentation, it is our aim to examine certain areas for which some realistic applications have taken place, at least insofar as insecticidals and antimicrobials are concerned, and to demonstrate their efficacious effects.

Medicinals

A number of natural product flavors and fragrances that fall into this category have recently been described in a short review (1). There are, however, a few medicinals from this group that are active as insecticides and fungicides against phytopathogens. Their medicinal properties are described here and their other properties will be discussed in the relevant sections of this chapter.

The essential oils of several species of the culinary herb basil have been examined for their biological activity against animal pathogens including *Epidermophyton floccosum, Microsporum canis,* and *Trichophyton mentagrophytes.* Of the four species tested, *Ocimum americanum, O. basilicum, O. gratissimum* and *O. sanctum,* the most active was the oil from *O. gratissimum* which completely inhibited the growth of the test organisms at 1000 ppm (2). However, the other species did induce inhibitory effects but, as we shall see later, *O. sanctum* had moderate, specific insecticidal activity.

A question that should be addressed at this point is the origin of the essential oils and the processing methods. First, the method of extraction, with a few exceptions, is steam distillation, but just how different the proportionalities of the compounds are from batch to batch is not elaborated. Neither are the possible changes from season to season taken into account. No consideration has been made as to the possible synergism between the components, if any. And last, it is common knowledge that flavor profiles obtained on extracts made by supercritical fluid extraction show many more products than are obtained by steam distillation. That is, certain compounds are not captured by steam distillation. While these finer points may not be of interest to the flavor market because the public has been conditioned to accept those materials presently available, the implications that these non-captured volatiles may have other uses should not be ignored.

Thymol (Figure 1[**1**]) is another natural product that has found mutiple uses. A plant that grows in the Himalayas, in the Tarai region, is *Trachyspermum ammi* from which fruit was harvested, washed with 70% ethanol and rinsed with water. Steam distillation yielded an essential oil of which thymol was the major component. The oil was inhibitory to the growth of *E. floccosum, M. canis,* and *T. mentagrophytes* at 900 ppm, on topical application, while thymol was equally as active at 1000 ppm. The fruit is used as a folk medicine against intestinal ailments including diarrhoea, flatulence and indigestion, but most importantly as a treatment against cholera (3) on the subcontinent of India.

ß-ionone (Figure 1[**2**]) is a fairly widespread compound occuring in many plants, the classical example being English violets, *Viola odorata,* especially those that grow on the west coast of England, in Devonshire. The scent from this species is so pervasive that a small bouquet placed in a room will pleasantly permeate the air. But the compound is one of the components of green tea which is used in voluminous quantities in Japan and which is accepted as having curative properties. Analyses have shown that ß-ionone falls in eighth place on a list of the ten major flavor components (4). A later study of mate tea, made from the Holly family member, *Ilex paraguayensis,* showed that the second most abundant component was ß-ionone. When it was tested against thirteen medically important pathogens *in vitro* it was especially active against *Propionibacterium acnes* (5), an organism that causes dermatomycosis. Violets are also part of medicinal folklore and the whole plants, especially the roots, are made into a tea that is considered to be highly useful in soothing irritated membranes and respiratory problems. The American violet does not have the perfume of the English species, but then no analyses have been made to determine the amounts of ß-ionone in the American type, even though it is also used for medical purposes (6).

Limonene (Figure 1[**3**]) also occurs with regularity throughout the plant kingdom in several species and it also exhibits diverse biological activity. One source of the compound is the spice cardamom, *Elattaria cardomomum,* and it has been shown to be very effective in controlling *P. acnes* (7). Other sources of limonene include members of the genus *Citrus* and the essential oil of grapefruit, *Citrus paradisi* has been shown to be active against the phytopathogen *Botrytis cinerea* , though lemon, *C. aurentifolia,* is inactive. Again, it should be emphasized that many natural products are highly target specific and, in addition, it is fair to speculate that synergism may occur between individual components in essential oils.

Tea tree oil, obtained from *Melaleuca alternifolia,* has been shown to be highly active against *Candida albicans,* which causes 'thrush' in infants and other diseases, and, in addition, excellent control of vaginitis caused by the same organism. Bacterial impetigo, caused by *Staphylococcus aureus* and *Streptococcus sp.* has also been treated successfully. The constitution of the oil has been analyzed and it contains 1,8-cineole 9.1%, *p*-cymene 16.4%, terpinen-4-ol 31.2% and α-terpineol 3.5% (Figure 1 [**4,5,6,7**]).

Insecticides and Antifeedants

Essential oils may play a significant role in the suppression and eradication of insects, especially in stored products in closed containers. If the atmosphere can be treated with these natural products and a certain equilibrium maintained in storage until distribution of the commodity in the marketplace, success can be attained. Furthermore, upon opening the container the natural product is dissipated and the stored product may not have an off-flavor.

One very practical example is the use of tumeric rhizomes, *Curcuma longa,* which are used in India and Pakistan to repel insects in rice (8). Basmati rice, considered to be among the finest in the world for its gastronomic properties, is

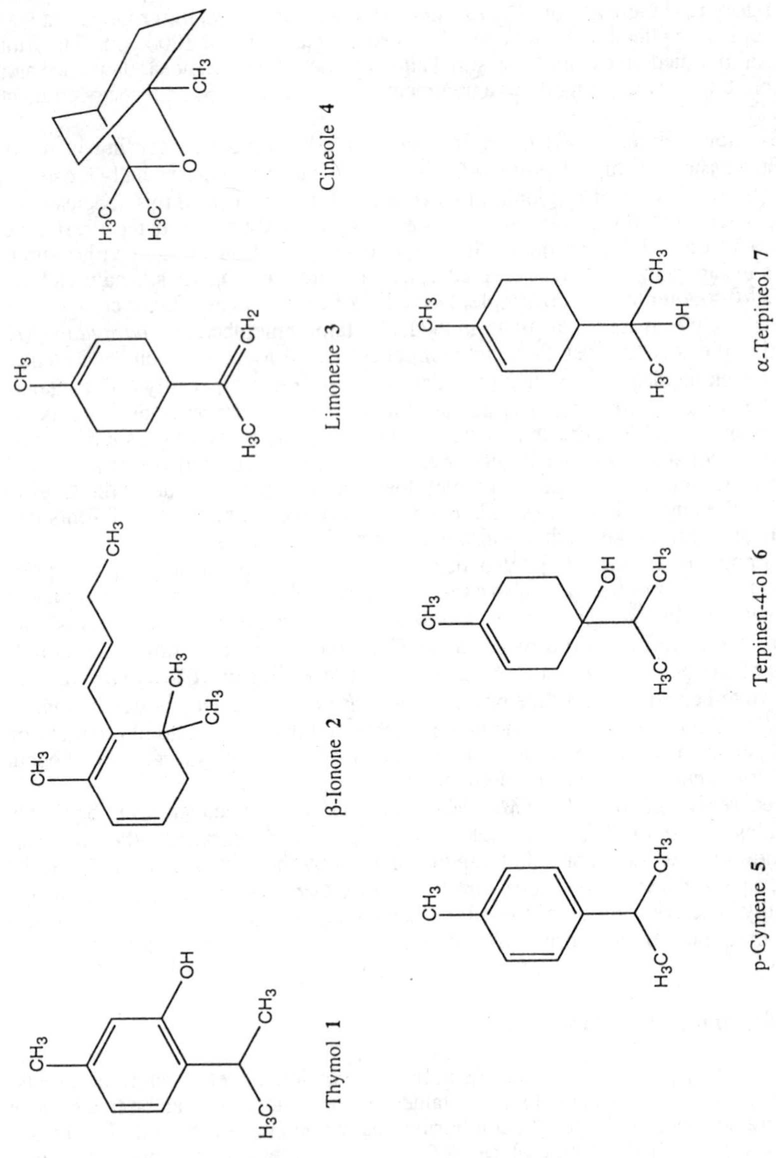

Thymol 1

β-Ionone 2

Limonene 3

Cineole 4

p-Cymene 5

Terpinen-4-ol 6

α-Terpineol 7

Figure 1. Some compounds referred to in this work.

commonly treated with tumeric powder to deter insects and the method appears to be highly successful (9). Analytical examination shows that tumeric contains tumerone and ar-tumerone (Figure 2[**8,9**]) which are insect repelling agents (10). Two advantages are gained from this teatment, apart from ensuring that insect infestation does not take place: first, the tumeric is removed if the rice is washed before cooking and second, if the rice is not washed then a slightly spicey flavor is imparted to the cooked product.

A major insect of apples is the Light Brown Apple Moth (LBAM), *Epiphyas postvittana.* The insect, which was introduced to New Zealand from Australia, occurs not only in orchards but also appears in stored apples. To complicate the life cycle there are twenty different weed species that are alternate hosts for the insect, thus ensuring survival of the species, and it is a quarantine pest in New Zealand. Trials were conducted on LBAM using eleven different essential oils to evaluate the insecticidal properties. Ten larvae, at the second instar stage, were introduced into 500 mL jars and 7 cm filter papers impregnated with 150 μL of essential oil were individually suspended in each jar and the tops sealed. Jars were then placed under artificial light with a 16 hour daylength, at 20 ^{0}C, and the experiment was triplicated. Percentage mortality was calculated 24 hours after exposure to the volatiles and the percentage mortality data were Arcsin transformed before evaluation by analysis of variance. Mean mortality was compared using Dunnett's tables (11). The following materials were examined: eucalyptus *(Eucalyptus globulus)*, cypress *(Cupressus sempervivens)*, camphor *(Cinnamomum camphora)*, spearmint *(Mentha cardiaca)*, white thyme (*Thymus vulgaris*), pheasant grass (*Litsea cubeba),* sassafras (*Sassafras albidum)*, anethol, red thyme (*T. vulgaris),* aniseed *(Pimpinella anisum),* and boronia (*Boronia megastima).*

All the materials tested were effective with the exception of boronia (Figure 4). Eucalyptus induced a 93% mortality, while cypress gave 90%, camphor 83% (Figure 2[**10**]), spearmint 78%, white thyme 75%, pheasant grass 50%, sassafras 50%, anethole (Figure 2[**11**]) 35%, red thyme 34%, and aniseed 28% ($P < 0.05$). It is interesting to note the disparity between white thyme and red thyme which indicates that either there is a significant increase in one bioactive compound in one cultivar versus the other, or that a specific ratio of compounds in the mixture is more effective than another.

Another storage pest of considerable importance is the Flour Beetle, *Tribolium confusum,* which attacks not only wheat flour, but all types of meal. Again, essential oils were examined for insecticidal properties in trials. Ten adult beetles were sealed in 500 mL jars in which were suspended 7 cm filter papers impregnated with 150 μL of individual essences. The jars were placed under artificial light, with a 16 hour daylength, at 20 ^{0}C for 24 hours and then the percentage mortality was recorded and the data statistically analyzed (11). The experiment was triplicated. The essential oils examined were: eucalyptus, camphor, white thyme, red thyme, basil *(Ocimum sanctum)*, peppermint *(Mentha piperita),* pheasant grass, chamomile *(Matricaria chamomilla),* pepper *(Piper nigrum),* cedarwood *(Juniperus virginiana),* kukui *(Aleuriges moluccana)*-cold pressed oil, walnut *(Juglans niger),* and flaxseed *(Linum usitalissimum).* Of these, eucalyptus induced 100% mortality, camphor 100%, white thyme 95%, red thyme 30%, basil 28%, peppermint 28%, pheasant grass 10%, and chamomile 7% ($P < 0.05$) (Figure 5).

In comparing the results of the two sets of experiments, it will be seen that red thyme was 41% less effective against LBAM than white thyme and in controlling Flour Beetle there was, in comparison, a 65% difference between the two botanical cultivars. Furthermore, the effectiveness of the essential oils drops off rapidly so that only eucalyptus, camphor, and white thyme may be considered as useful candidates. Oddly, black pepper was inactive in these assays, while in earlier work certain components of this commodity have been shown to have strong insecticidal properties. The earliest report being in 1943 (12), which was followed not only by later reports of insecticidal

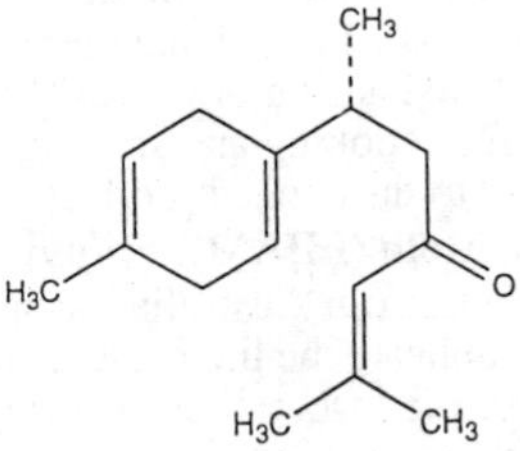

Tumerone **8**

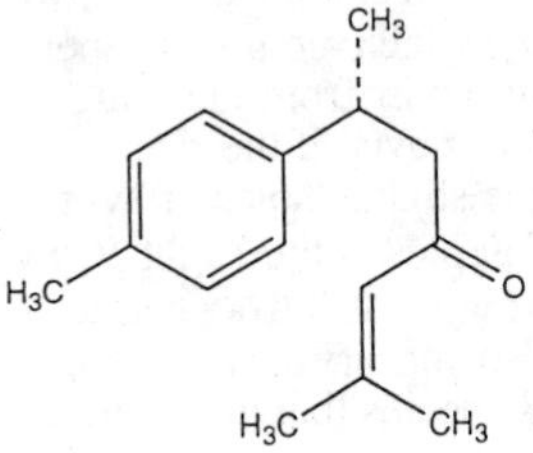

ar-Tumerone **9**

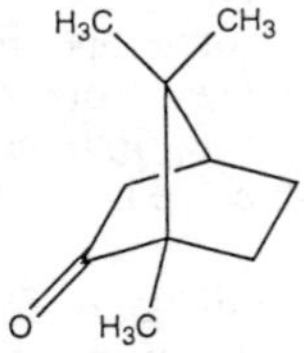

Camphor **10**

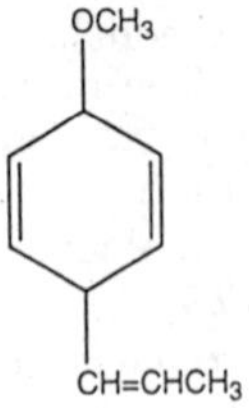

Anethole **11**

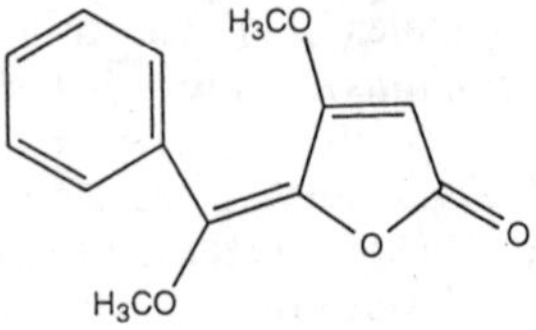

Z-Fadyenolide **12**

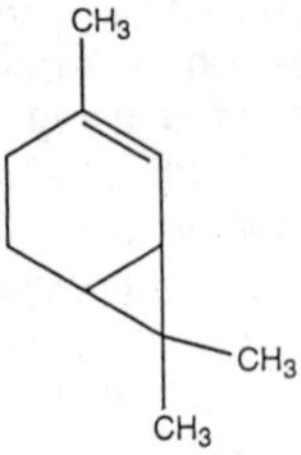

2-Carene **13**

Figure 2. Some compounds referred to in this work.

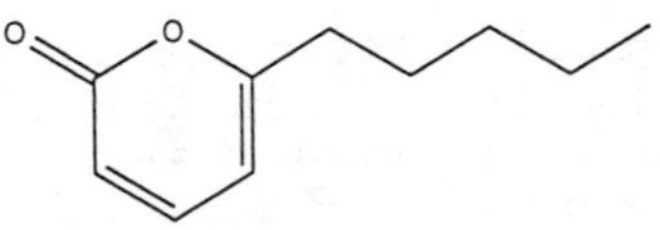

6-Pentyl-α-pyrone **14**

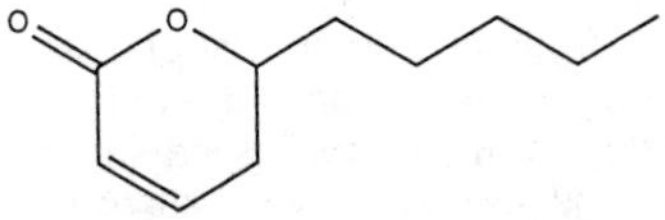

Massoilactone **15**

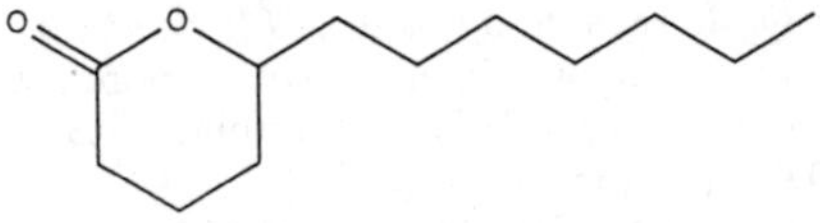

δ–Dodecalactone **16**

Figure 3. Some compounds referred to in this work.

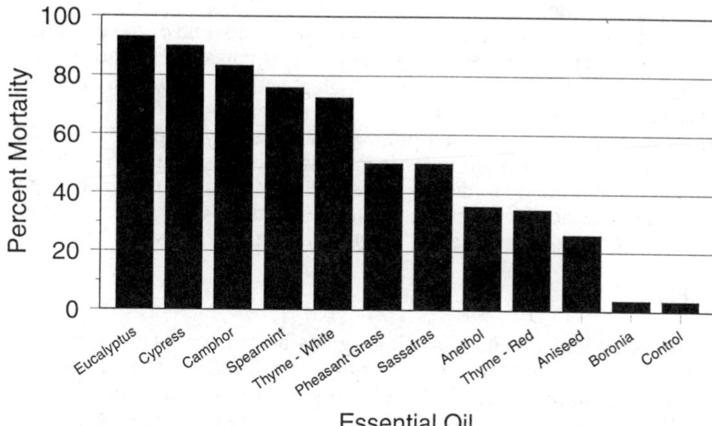

Figure 4. Effects of essential oils on light brown apple moth, *Epiphyas postvittana*.

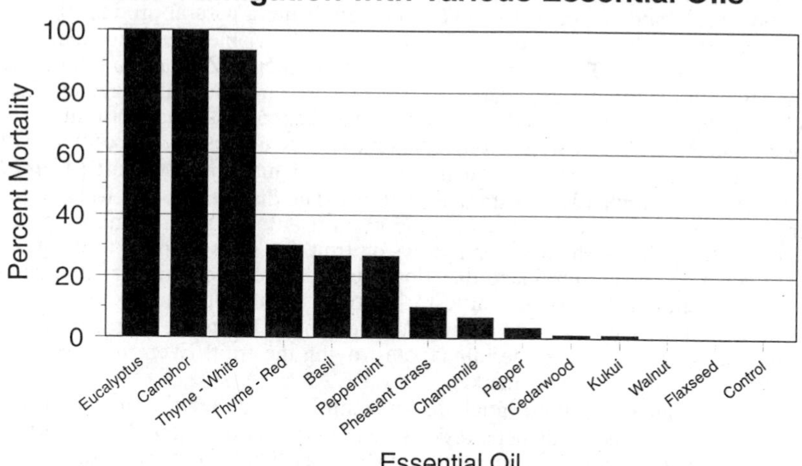

Figure 5. Effects of essential oils on flour beetle, *Tribolium confusum*.

activity (13) but also by chemical derivatization of *Z* and *E* -fadyenolide (Figure 2[**12**]), isolated from *P. fadyenii,* to produce synthetic compounds that were active against *T. confusum.* But it appears that the components of black pepper were not tested against this particular insect. The fact that black pepper was not active against *T. confusum* but was active against the housefly *Musca domestica* again serves to illustrate the target specificity of natural products.

In examining the two figures it will be noted that some of the essential oils are not common to both sets of experiments and that is because the negative data, for the most part, have not been presented. The biological activity of *O. sanctum* has already been mentioned in the section on medicinals, as has thymol.

A question that arises at this point is the necessity of using essential oils as opposed to using plants, or plant parts from which the essences are obtained. If the example of the use of tumeric to deter insects from infesting Basmati rice points to one lesson, it is that tumeric rhizomes are far less expensive than pure tumerone or ar-tumerone, and that the present system is practical and cost effective.

Antimicrobials

A number of essential oils and flavoring agents have been used experimentally to control certain phytopathogens. For example, volatiles of crushed tomato leaves, *Lycopersicon esculentum,* were analyzed and shown to contain limonene, 2-carene (Figure 2[**13**]), 2-hexenal, hexanal, 2-nonenal and nonanal, and they were individually tested against the hyphae of certain microorganisms. Two of these compounds, 2-hexenal and hexanal, were active against *Alternaria alternata* and *Botrytis cinerea* so that 0.23-4.6 µmole/L air and 0.32-12.8 µmole/L air was all that was needed, respectively, of each of these compounds to control *B. cinerea* (14).

While several other examples of antimicrobial activity are available it is necessary, in the interest of space, to concentrate on the control of *B. cinerea* , a serious phytopathogen both in the field and in post harvest storage facilities. The intention is to show that two different laboratories working independently, separated by the distance of half-a-world, came together to solve a problem using a natural product flavoring agent to control *B. cinerea* and to produce a commercially viable, safe fungicide.

In the early 1980's, a program was instituted in New Zealand to replace hard pesticides with readily biodegradable natural products and biocontrol agents. These agents were to be developed to protect domestic and export markets, both timber and horticultural. A number of serious phytopathogens were addressed, one of which was *B. cinerea* , a fungus that was causing particular damage in kiwifruit *(Actinidia chinensis)* both in orchards and storage facilities. The disease appears to be present during most growing seasons, but in some years it is more severe than others. To compound the problem, the organism grows on fruit that have dropped to the ground and abundant spores are produced that float freely in the air providing an abundant source of inoculum. At time of harvest, the fruit is torn from the peduncle (stem) creating a landing site and port-of-entry for the spores. Fruits are stored at 4 ^{0}C in packing sheds and the spores germinate, destroying the fruit. Even mild infections make the product unsuitable for market.

Initial evaluations of essential oils for control of *B. cinerea* were conducted *in vitro.* Eighteen millimeter diameter disks were individually treated with 30 µL of each extract, then placed on malt agar in petri dishes that had been heavily seeded with spores of *B. cinerea* . These were incubated at 20 ^{0}C, readings were taken daily, for five consecutive days, of the clear zones surrounding each filter paper disk. The following essential oils were examined: melissa *(Melissa officinalis)*, lemongrass *(Cymbogon citratus),* lavender cv. Mt. Blanc *(Lavendula augustifolia)*, peppermint, pennyroyal *(Mentha pulegium)* , rose *(Rosa* sp.*),* thyme, spearmint, bois de rose *(Rosa* sp.*),* tea tree *(Melaleuca alternifolia),* petigrain *(Citrus sinensis),* jasmine *(Jasminum grandiflorum),* neroli *(Citrus aurantium* sub. *aurantium),* wintergreen *(Gaultheria procumbens),* hyssop *(Hyssopus officinalis),* sassafras, grapefruit *(Citrus paradisi),* sage *(Salvia officinalis),* majoram *(Oreganum syriacum),* myrrh *(Commiphora guidotti),* rosemary *(Rosemarinus officinalis),* nutmeg *(Virola surinamensis),* patchouli *(Pogostemon cablin),* sandalwood *(Santalum album),* ginger *(Zingiber officinale),* pimento *(Pimenta dioica),* frankincense (*Boswellia sacra),* vertivert *(Andropogon muricagus),* fennel *(Foeniculum vulgare),* and lemon *(Citrus aurantifolia).*

The results (Figure 6) show that a large number of the extracts were effective in controlling the germination of *B. cinerea* spores, the least effective ones on the list being sage, majoram, myrrh, rosemary, nutmeg, patchouli, sandalwood, ginger, pimento, frankincense, vertivert, fennel, and lemon. Tea tree oil, which was briefly addressed in the Medicinal section, was inhibitory although lemon oil, which contains limonene, was inactive.

Concurrent with these bioassays were experiments using live strains of *Trichoderma harzianum* to control phytopathogens. Among the treatments were formulations for use as biocontrol agents consisting of pastes, mixtures on organic substrates and other proprietary preparations. These were applied as paints to infected lesions, and incorporated into the rhizosphere to successfully control a number of diseases such as *Armillaria novae zealandia* , which destroys Monterey pine *(Pinus radiata),* and kiwifruit. Other treatments included injections into tree trunks, with excellent results. Treatment of stored kiwifruit purposely inoculated with *B. cinerea* was also successful when the biocontrol agent, in the live state, was used. However, the chemical agent that acted as the fungicide was not known.

In the late 1970's, work was in progress isolating biologically active natural products from fungi for potential agrochemical use. One of the accessions was a strain of *Trichoderma harzianum* found growing luxuriantly on a pile of Slash Pine logs *(Pinus elliotii)* and in a bioassay directed fractionation, using the etiolated wheat coleoptile *(Triticum aestivum),* 6-pentyl-α-pyrone (6-PAP) (Figure 3[**14**]) was isolated. Not only did the metabolite inhibit the growth of wheat coleoptiles, but it showed antifungal properties against *Chaetomium cochlioides* 195, *C. cochlioides* 189, *C. cochlioides 193,* and *Aspergillus flavus.* Furthermore, when it was diluted to ratios as low as 1:40, in an inert carrier, it effectively controlled *A. flavus,* which produces the aflatoxins in stored products. While this was an important discovery, the information was not verbally disclosed until the summer of 1983 at the 186th meeting of the American Chemical Society (15). The complete work was published in 1986 (16). Significantly, it was first synthesized for use as a flavoring agent (17). Later, it was found in natural peach essence (18), and in extracts of *Trichoderma viride* (19), but the biological properties had not been described. The fact that it had a Flavors and Extracts Manufacturers Association (FEMA) number made it an attractive candidate for development as a fungicide, especially for use in stored products.

Eventually, the biocontrol work and the chemical investigations on *T. harzianum* between New Zealand and the United States meshed, so that in 1993 large scale trials were instituted with 6-PAP on stored kiwifruit that had been artificially inoculated with *B. cinerea* at harvest. In both sets of trials, the fruits were hand harvested, inoculated with spore suspensions of the pathogen in solution at the point of detachment from the stem. Then, 6-PAP was added at 1 day following harvest (Trial 1), at 2 days (Trial 2) and at 3 days (Trial 3) (Figure 7). Application rates for 6-PAP were 25 μL (shown as 25% in Figure 7) and 100 μL (shown as 100% in Figure 7) using acetone as the carrier in the diluted treatment. The reason for staggering the days of application was because there is always some proliferation of the newly exposed

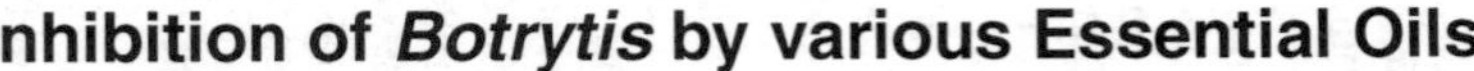

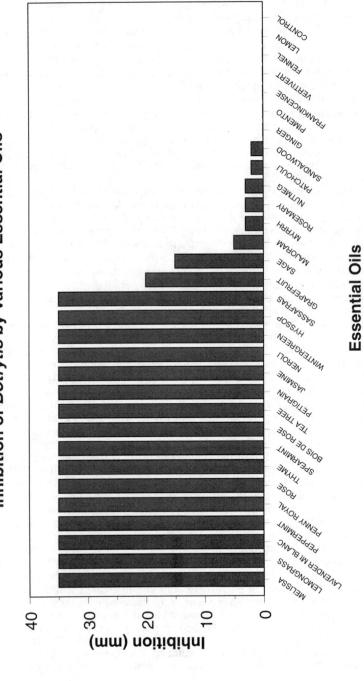

Figure 6. Inhibition of *Botrytis cinerea* by essential oils.

tissue when the fruit is torn from the vines and the corking that takes place offers some natural protection to the fruit. Ronilan, a commercial fungicide, was included as an internal standard although fungicides presently available on the market are not allowed to be used at this point for market bound fruit. Examination of the results from only two of the experiments that were conducted in 1993 showed that after 14 weeks in storage at 4 ^{0}C the 6-PAP treated fruit was fungus free, with the exception of 25% 6-PAP which showed slight infection (< 2%). By comparison, 55% of the controls were infected in Trial 1 and those in Trials 2 and 3 were infected 15 and 8%, respectively. Ronilan only moderately controlled fungal infection (Figure 7). A second experiment (Figure 8) showed similar results with 6-PAP although it will be noted that the fungal damage in the controls was more intense in Trial 2 . Again this variable was due to the corking over effect that is not controllable. In both cases the fruit was in storage for over three months.

In 1994, a second series of trials were conducted with 6-PAP, but the concentration ranges were extended to 10, 25, 50, and 100 μL per kiwifruit using acetone as the carrier (Figure 9 and 10). Additionally, massoilactone (Figure 3[**15**]) and δ-dodecalactone (Figure 3[**16**]), which are congeners of 6-PAP, were included in the first set of experiments (Figure 9). ß-ionone, a compound already patented for antifungal properties (20), was included in the second set of experiments. Both sets of kiwifruit were inoculated with *B. cinerea* and then treated with the flavors and essences 4 hours, or 51 hours later. In the first experiment, fruit treated with 6-PAP were fungal free after 30 weeks in storage at 4 ^{0}C, and massoilactone was also quite active in suppressing the phytopathogen (Figure 9). In the second experiment, 6-PAP was again highly effective in controlling *Botrytis* rot after 32 weeks in cold storage, but ß-ionone was ineffective (Figure 10). It is necessary to emphasize that in these experiments, kiwifruit treated with 6-PAP were still as firm, plump and juicy as the day that they were harvested, one year following treatment and kept in cold storage. Furthermore, taste panels agreed that the flavor, texture and appearance of the treated fruit was excellent. In the 1993-1994 trials a total of 3,000 kiwifruit were treated. None of the congeners of 6-PAP were as long lasting in these experiments, even though they had shown promise in early *in vitro* bioassays.

Recent large scale trials, conducted in 1995, have confirmed the efficacy of 6-PAP treatments against *Botrytis* and confrim that the material has practical application as a safe fungicide. These results will be discussed in a further publication.

Extensive *in vitro* experiments have been conducted on the effects of 6-PAP on *B. cinerea* at all stages of development. A sampling of these experiments is now considered. Among the earliest was observations of the effect of the metabolite on hyphal growth at various concentrations. A 5 mm mycelial plug from the edge of a 7-day old colony of *B. cinerea* was placed side upwards on a potato-dextrose agar plate and immediately, a 5 μL droplet containing specific concentrations of 6-PAP was added to the plug with dilutions being made in acetone. Five concentrations of 6-PAP were chosen. Plates were then sealed, inverted, and incubated at room temperature in normal light. All experiments were done in duplicate. Results were noted every day for ten days and growth was expressed as a percentage of the controls.Complete suppression of hyphal growth was observed with 6-PAP at concentrations ranging from 10-15%, even after 10 days, but 5% gave virtually no control after 7 days (Figure 11).

Another stage of development of *B cinerea* was challenged with 6-PAP and this was the sclerotial stage. Sclerotia from the microorganism were produced on potato-dextrose agar plates. The inoculated plates were left for five to six weeks in the dark and the sclerotia were harvested by progressive sieving through mesh sizes of 2000, 250, and 50 μm. The sclerotia were put into 15% sodium hypochlorite for 1 minute to surface sterilize them, and then in sterile distilled water for 5 minutes prior to use. Of the 120 sclerotia employed in each set of experiments, 24 were dipped in sterile

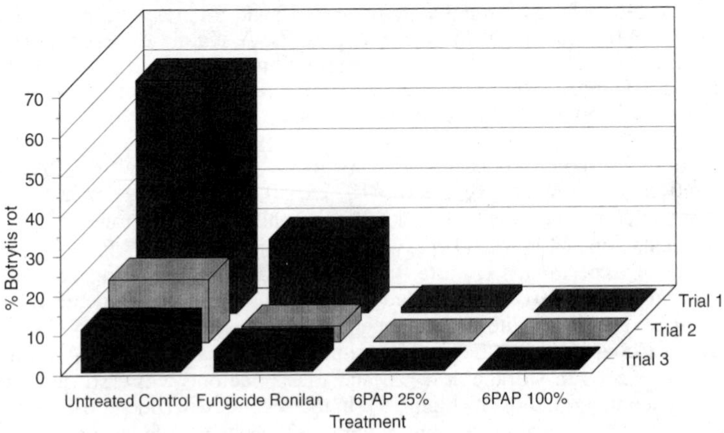

Figure 7. Effects of 6-pentyl-α-pyrone on *Botrytis* rot, 1993 trial, part 1.

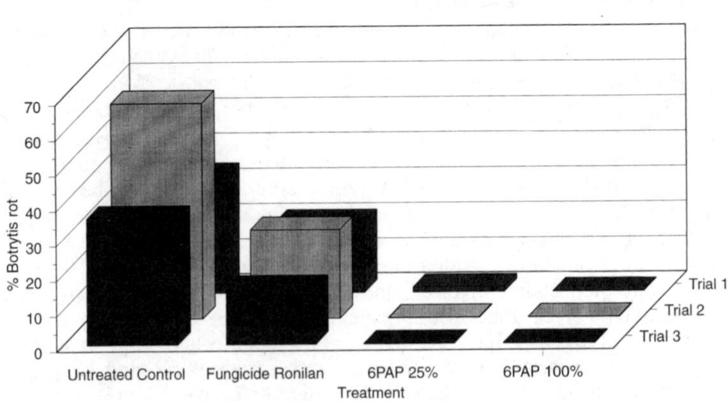

Figure 8. Effects of 6-pentyl-α-pyrone on *Botrytis* rot, 1993 trial, part 2.

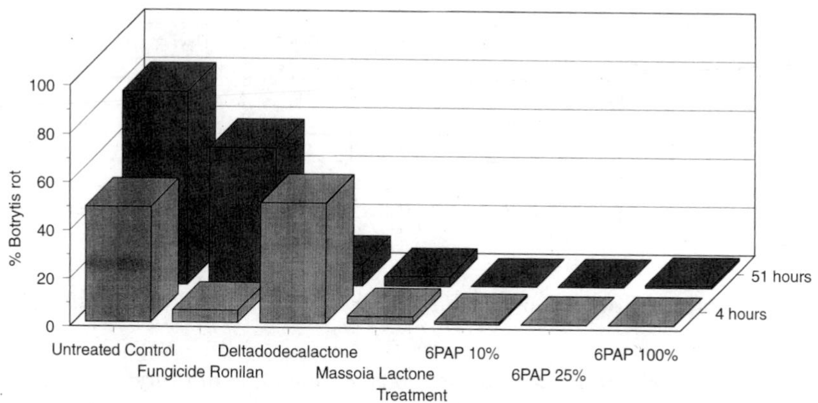

Figure 9. Effects of 6-pentyl-α-pyrone on *Botrytis* rot, 1994 trial, part 1.

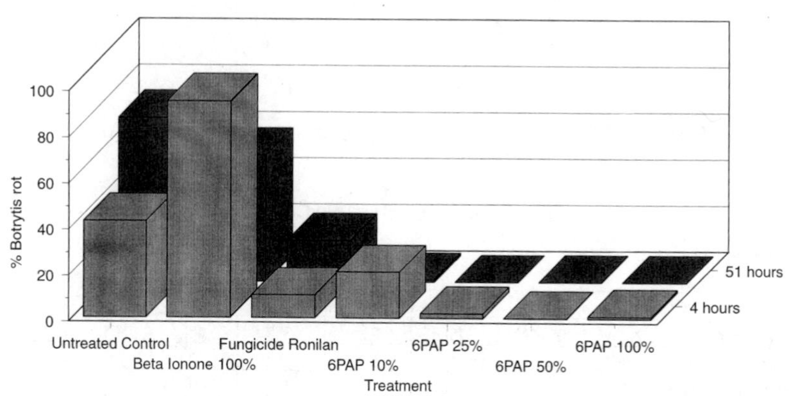

Figure 10. Effects of 6-pentyl-α-pyrone on *Botrytis* rot, 1994 trial, part 2.

The effect of higher 6-PAP concentration on hyphal growth of *Botrytis*.

Experiment 1: Growth as a percentage of control.

6-PAP conc. (%)	Day 1	Day 2	Day 3	Day 4	Day7	Day 10
5	0	0	0	0	50	97
10	0	0	0	0	0	0
15	0	0	0	0	0	0
20	0	0	0	0	0	0
25	0	0	0	0	0	0

Figure 11. Effects of 6-pentyl-α-pyrone on hyphal growth of *Botrytis*.

The effect of 6-PAP on sclerotial germination of *Botrytis*.

Experiment 2: Number of germinates (out of 24).

Day	6-PAP concentration				
	0% (SDW)	0%A	1%	2%	5%
2	0	1	0	0	0
3	1	3	0	0	0
4	1	3	1	0	0
8	3	3	0	0	0

Figure 12. Effects of 6-pentyl-α-pyrone on sclerotia of *Botrytis*.

Germination of *Botrytis* Conidia.

Experiment 3c.

%	6-PAP concentration							
germination	0% (W)	0% (A)	1%	2%	3%	4%	5%	10%
after one day	>95	>95	<10	20	0	0	0	0
after five days	>95	>95	>90	90	90	0	0	0

Figure 13. Effects of 6-pentyl-α-pyrone on conidia of *Botrytis*.

distilled water (SDW) only, 24 in 100% acetone (A), 24 in 1% 6-PAP and so on up to 5% 6-PAP. The results indicate that 2% and 5% 6-PAP were sufficient to control sclerotial germination (Figure 12).

Conidia of *B. cinerea* were also treated with 6-PAP. A liquid spore suspension was prepared by pipetting approximately 5 mL of sterile distilled water onto sporulating colonies followed by gentle scraping with a sterile inoculating loop to release spores into the water which were pipetted off the plate and passed through a muslin lined funnel. The spores were diluted in Czapek-Dox liquid medium to a concentration of 1 x 10^5 spores/mL. A 250 μL droplet of potato-dextrose agar was spread onto a microscope slide and immediately, before the agar set, a 7.5 μL droplet of 6-PAP solution was pipetted onto the agar and the carrier acetone evaporated, leaving the metabolite suspended in the agar. When the agar had set, a 7.5 μL droplet of the spore suspension was added to the agar surface, the microscope slide placed in a humid environment, and incubation took place at room temperature. Sterile distilled water (W) and acetone (A) were used as controls. The 6-PAP concentrations, arrived at as the results of several previous experiments, were 1, 2, 3, 4, 5, and 10%. The results (Figure 13) show that 4% 6-PAP successfully controlled germination of conidia after five days. Thus, all of the stages of the life cycle were controlled by applications of the metabolite.

The activities of massoilactone and δ-dodecalactone will be the subject of a future report.

Conclusion

There are a number of flavors and essences that may be developed for use as agrochemicals and medicinals. While some are very potent against specific targets, they may not have a broad spectrum of activity and this means that some serious changes in thinking and application have to take place before they are universally accepted. That trade-off may mean the introduction of more benign control agents which will have less harmful impact on the environment. Much work remains to be done, however, not only in the identification of the biologically active constituents, but also in the area of synergism.

Literature Cited

1. Cutler, H.G. *Agrofood Industry Hi Tech* **1995**, *6*, 19-23.
2. Singh, S.P.; Singh, S.K.; Tripathi, S.C. *Indian Perfumer* **1983**, *27*, 171-173.
3. Singh, S.P.; Dubey, P.; Tripathi, S. C. Mykosen **1986**, *29*, 37-40.
4. Kubo, I.; Muroi, H.; Himejima, M. *J. Agric. Food Chem.* **1992**, *40*, 245-248.
5. Kubo, I.; Muroi, H.; Himejima, M. *J. Agric. Food. Chem.* **1993**, *41*, 107-111.
6. Lust, J. *The Herb Book*; Bantam Books: New York, **1976**, pp 203-204.
7. Kubo, I.; Himejima, M.; Muroi, H. *J. Agric. Food Chem.* **1991**, *39*, 1984-1986.
8. Sreenivasamurthy, V.; Krishnamurthy, K. *Food Sci.t (Mysore).* **1959**, *8*, 284-288.
9. Olojede, F.; Engelehardt, G.; Wallnafer, P.R.; Adegoke, G.O. *World J. of Microbiol. and Biotech.* **1993**, *9,* 605-606.
10. Su, H.C.F.; Robert, H.; Jilani, G. *J. Agric. Food Chem.* **1982**, *30*, 290-292.
11. Dunnett, C.W. *J. American Statistical Assoc.* **1955**, *50*, 1096-1121.
12. Harvill, E.K.; Hartzell, A.; Arthur, J.M. *Contrib. Boyce Thompson Inst.* **1943**, *13*, 18-92.

13. Nair, M.G.; Mansingh, A.P.; Burke, B.A. *Agric. Biol. Chem.* **1986**, *50*, 3053- 3058.
14. Hamilton-Kemp, T.R.; McCracken, Jr., C.T.; Loughrin, J.H.; Andersen, R.A.; Hildebrand, D.F. *J. Chem. Ecol.* **1992**, *18,* 1083-1091.
15. Cutler, H.G. In *Bioregulators, Chemistry and Uses*; Ory, R.L.; Rittig, F.R., Eds.; ACS Symposium Series 257, American Chemical Society: Washington DC, **1984**; pp 167.
16. Cutler, H.G.; Cox, R.H.; Crumley, F.G.; Cole, P.D. *Agric. Biol Chem.* **1986**, *50*, 2943-2945.
17. Nobuhara, A. *Agric. Biol. Chem.* **1969**, *33*, 1264-1269.
18. Sevenants, M.R.; Jennings, W.G. *J. Food Sci.* **1971**, *36*, 536.
19. Collins, R.P.; Halim, A.F. *J. Agric. Food Chem.* **1972**, *20*, 437-438.
20. US Patent #4,231,789 to Okii et al., issued 1980.
    US Patent #4,474,816 to Wilson, D.M. Jr., and R.C. Gueldner, issued 1984.

Chapter 6

Characteristic Odorants of Wasabi (*Wasabia japonica matum*), Japanese Horseradish, in Comparison with Those of Horseradish (*Armoracia rusticana*)

Hideki Masuda, Yasuhiro Harada, Kunio Tanaka, Masahiro Nakajima, and Hideki Tabeta

Okayama Laboratory, Ogawa & Company, Limited, 1–2, Taiheidai, Shoo-cho, Katsuta-gun, Okayama-ken, 709–43, Japan

The volatile components of both wasabi and horseradish formed by the hydrolysis of thioglucosides with myrosinase are known to possess a strong pungency. The difference between wasabi and horseradish is a green odor. The ratio of the concentration to the odor threshold, C/T, is called the aroma value. Sensory response is logarithmically proportional to the amount of stimulus. In this study, the log of the aroma values, log(C/T), for 14 isothiocyanates have been used instead of their conventional aroma values. ω-Alkenyl isothiocyanates, which possessed higher log(C/T) values than those of horseradish, contributed to the green odor of wasabi. Furthermore, the yields and the values of log(C/T) of the isothiocyanates catalyzed by myrosinase were affected by pH and temperature.

Myrosinase (thioglucoside glucohydrolase) is present in plants belonging to the *Cruciferae* family (*1,2*). The thioglucosides involved in wasabi, Japanese horseradish, and horseradish are hydrolyzed by myrosinase to form the volatile compounds, including isothiocyanates, thiocyanates, nitriles, etc. (*3*). Among these compounds, the isothiocyanates have been recognized as the characteristic flavor compounds because of their pungency. Many investigators have reported the amounts of volatile isothiocyanates in wasabi and horseradish (*4-6*). Allyl isothiocyanate, the main component of wasabi and horseradish, exhibits the most pungent odor (*7,8*). As for horseradish, the major component after allyl isothiocyanate, phenethyl isothiocyanate, is known to be sufficiently characteristic (*9, 10*). The remarkable difference between wasabi and horseradish is the green odor. The volatile compounds of wasabi have a more greenish note than those of horseradish. Recently, it has been suggested that ω-methylthioalkyl isothiocyanates contribute to the green odor of wasabi (*11,12*). Furthermore, the odor threshold

0097–6156/96/0637–0067$15.00/0

value and odor description of allyl isothiocyanate and three kinds of ω-alkenyl isothiocyanates, the major components next to allyl isothiocyanate in wasabi, have been reported (*13*). However, a detailed study on the contribution of odor of each isothiocyanate in wasabi compared to that in horseradish has not yet been made.

The properties of myrosinase purified from mustard seed (*14, 15*) and wasabi (*16*) were studied. The pH and temperature activity, and the pH and temperature stability of wasabi myrosinase were examined in detail. However, the effects of pH and temperature on the yields of the isothiocyanates have not yet been reported.

This study focuses on the determination of the characteristic odorants of wasabi in comparison with those of horseradish by calculation of the logarithmic ratio between the concentration of isothiocyanate and its odor threshold value. The relationship between the yields of the isothiocyanates formed by the action of wasabi myrosinase and the conditions of hydrolysis are presented.

Experimental

Stems of wasabi (1 kg) and horseradish (1 kg) collected in Japan and New Zealand, respectively, in January 1995. They were crushed, allowed to stand for 1 h at 25°C and extracted with 5 x 5 L of dichloromethane. Each solution was concentrated to 4.9 g and 6.7 g, respectively, using a rotary evaporator (35°C/300 mmHg).

In the study of the influences of pH and temperature, stems of wasabi (1.26 kg) collected in Japan in June 1995, were freeze-dried at 0.1 mmHg for 72 h. The freeze-dried stems (206 g) were crushed to form wasabi powder. The powder (10 g) was hydrolyzed in the phosphate buffer under different pHs (4, 7 and 9) and at different temperatures (3°C and 25°C). Each suspension was extracted with 3 x 50 ml of dichloromethane and concentrated using a rotary evaporator (35°C/300 mmHg).

The concentrations of 14 isothiocyanates were determined by GC. A Hitachi G5000 fitted with an FID was used. A DB-1 (30 m x 0.25 mm i. d.) fused-silica capillary column was employed. Operating conditions were as follows: initial oven temperature, 60°C, then to 250°C at 3°C/min and held for 30 min; injector temperature, 250°C; carrier gas, 0.5 ml/min N_2. Peak areas were obtained with a Hitachi D-2500 Chromato-Integrator. To estimate the concentrations of the components, an internal standard, phenyl isothiocyanate, was used.

The ω-alkenyl isothiocyanates, except for allyl isothiocyanate, were prepared by the isomerization of the corresponding ω-alkenyl thiocyanates (*17*). The ω-alkenyl isothiocyanates were converted to the corresponding ω-methylthioalkyl isothiocyanates (*18*). The other isothiocyanates were purchased from commercial sources and purified by vacuum distillation.

The odor threshold values in a water solution of the odorants were estimated by the 2/5 test (*19*). The ten panelists picked both of the flasks containing the odorous solution. Each dilution step was $1\sqrt{10}$ the concentration of the previous one.

The flavor dilution (FD) factors, 2^n, were obtained using the aroma extract dilution method developed by Grosch and co-workers (*20, 21*). The extract was diluted stepwise with dichloromethane in the volume ratio of 1:1 until odorous compounds were no longer detected by GC sniffmg.

Results and Discussion

The remarkable difference between wasabi and horseradish was the concentrations of the ω-alkenyl isothiocyanates (**5-8**) (Table I), the aryl isothiocyanates (**10**) and (**11**) (Table I), and the ω-methylthioalkyl isothiocyanates (**12-14**) (Table I). The odor threshold values of the ω-alkenyl isothiocyanates (**5-7**) were lower than the other isothiocyanates. In addition, these ω-alkenyl isothiocyanates had about half the value of that of allyl isothiocyanate (**2**) (Table I). The alkyl isothiocyanates (**1**, **3** and **4**) (Table I) possessed a chemical odor but no pungent note. Hence, it seems that those compounds make relatively small contributions to the characteristic odor of wasabi and horseradish. The other isothiocyanates except for(**1**, **3** and **4**), however, had pungent and/or radish-like odors.

The concentration divided by the odor threshold value, C/T is called the aroma value (*22*). Odorants with high aroma values are important contributors to the characteristic flavors. In general, the estimated feel of the stimulus is logarithmically proportional to the objectively measured strength of the stimulus (*23*). Accordingly, in this research we have evaluated the log of the aroma value, log(C/T), instead of the conventional aroma value. The values of log (C/T) of the isothiocyanates versus their retention indices (RI) are shown in Figure 1, top. Allyl isothiocyanate (**2**) had the highest value of log(C/T) in both volatiles (Figure 1, top). As for the ω-alkenyl isothiocyanates (**5-8**), the values of log(C/T) in wasabi were higher than those in horseradish. Above all, (**5**) and (**6**) possessed the highest value after (**2**). On the other hand, the aryl isothiocyanates (**10** and **11**) in horseradish had higher values than those in wasabi. The isothiocyanate (**11**) had highest value except for (**2**). Hence, it seems that the remarkable difference in odor between wasabi and horseradish is due to (**5-8**), and (**10** and **11**). In general, the ω-methylthioalkyl isothiocyanates (**9** and **12-14**) had lower log(C/T) values than those for the other isothiocyanates. The value of log(C/T) for 6-methylthiohexyl isothiocyanate (**13**) in wasabi was the highest of all the ω-methylthioalkyl isothiocyanates. The last note given out by wasabi was similar to the odor of (**13**). Consequently, it is suggested that the character impact compound in the last note of wasabi is attributed to be the presence of (**13**).

The flavor dilution (FD) factor, 2^n, is the maximum dilution value at which the odor is detected by GC sniffing (*24*). The odorant with high FD factor is considered to be potent (*25*). In general, the FD factor is proportional to the aroma value of the compound. Figure 1 (bottom) shows the exponents (n) of the FD factor 2^n, versus their retention indices (RI) of the isothiocyanates. The behavior of each compound in Figure 1 (bottom) was similar to that in Figure 1 (top). Therefore, the values of log(C/T) for each isothiocyanate observed in Figure 1 (top) were found to be proportional to the corresponding exponents (n) of the FD factor 2^n shown in Figure 1 (bottom).

Usually wasabi is prepared by grinding the raw stem at room temperature. The pungency of wasabi decreases rapidly as time passes. The yields of the alkyl isothiocyanates in the different pHs at 3°C are shown in Figure 2 (top). The distinguishing increase in the yields of the alkyl isothiocyanates (**1**, **3** and **4**) took place within 1 minute as a result of the rapid hydrolysis of thioglucoside. The yields of (**1**, **3** and **4**) increased in the order: pH 4 <pH 9 <pH 7. These results can be explained by the behavior of wasabi myrosinase at different pHs (*16*). The

Table I. Concentrations, Odor Threshold Values and Odor Descriptions of Isothiocyantes.

No.	Substituent	Concentration[a] W[c]	Concentration[a] H[d]	T[b]	Odor Description
1	isopropyl	7.6	Tr[e]	0.63	chemical
2	allyl	1880	1570	0.046	strongly pungent, mustard-like
3	*sec*-butyl	13	27	0.09	chemical
4	isobutyl	3.9	0.6	0.13	sweet, chemical
5	3-butenyl	25	5.5	0.017	green, pungent
6	4-pentenyl	31	1.7	0.22	green, pungent
7	5-hexenyl	8.0	0	0.021	green, pungent, fatty
8	6-heptenyl	0.6	0	0.075	green, pungent, fatty
9	3-methyl-thiopropyl	Tr	1.5	0.65	strongly radish-like, pungent
10	benzyl	0	1.6	0.035	chemical, pungent
11	2-phenethyl	Tr	133	0.24	strongly radish-like, pungent
12	5-methyl-thiopentyl	1.5	Tr	0.80	radish-like, pickle-like
13	6-methyl-thiohexyl	4.8	Tr	0.49	radish-like, sweet, fatty
14	7-methyl-thioheptyl	0.9	Tr	0.28	sweet, fatty, radish-like, pickle-like

[a] mg/kg
[b] Odor threshold value, ppm
[c] Wasabi collected in January
[d] Horseradish collected in January
[e] Less than 0.5 mg/kg

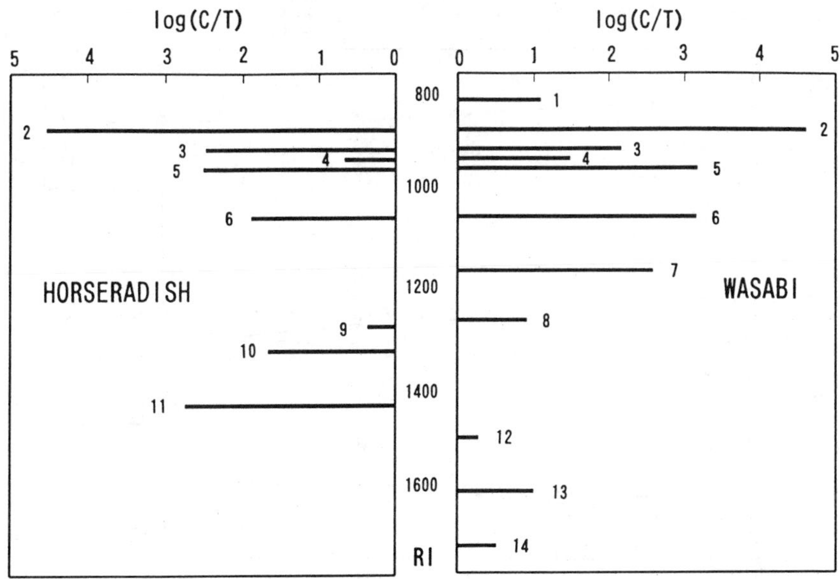

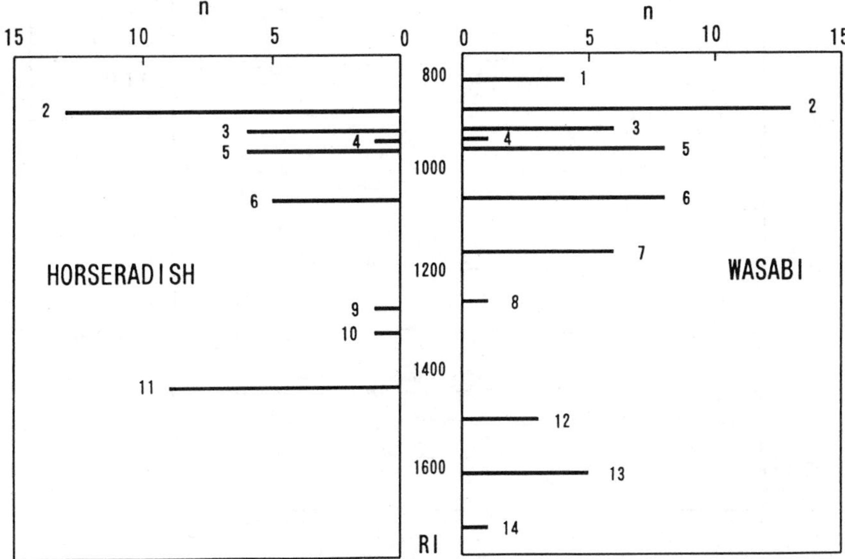

Figure 1. Logarithms of the ratios of the concentrations to the odor threshold values (log(C/T)) (top) and exponents (n) of the FD factor 2^n, (bottom) for the isothiocyanates versus their retention indices (RI).

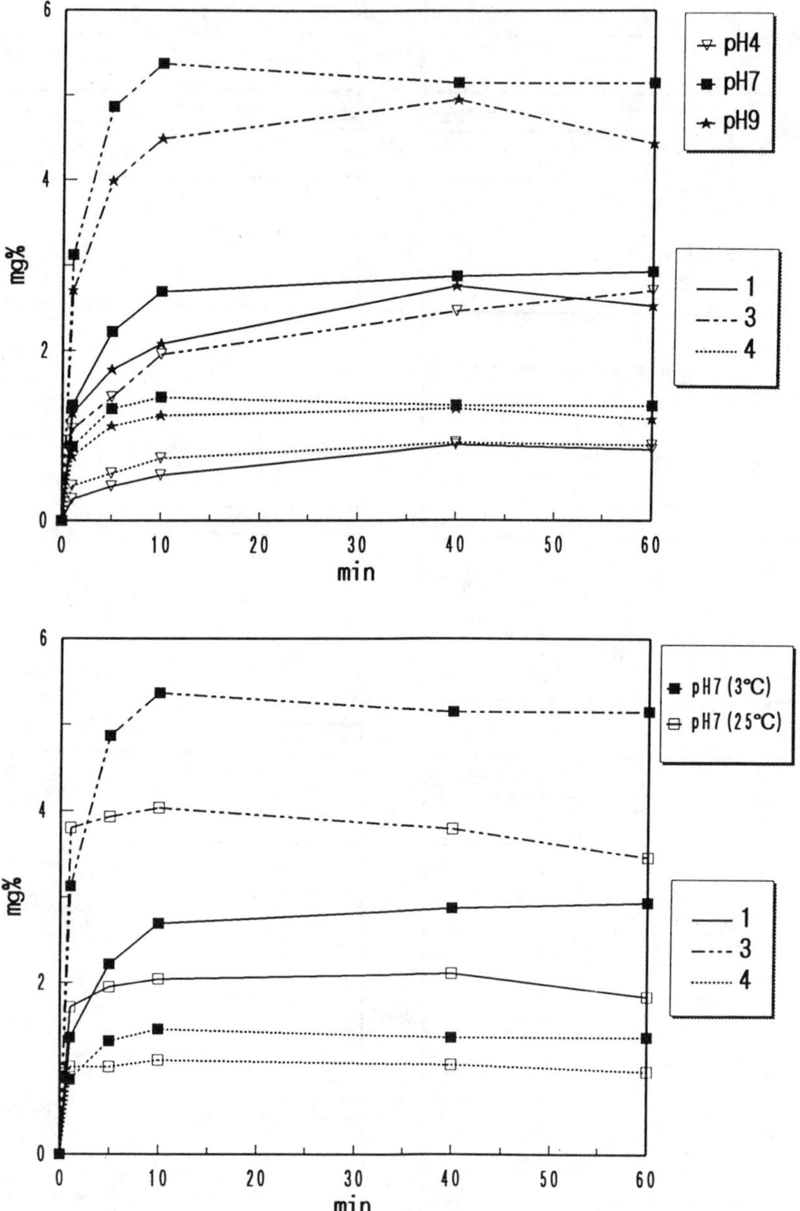

Figure 2. Effects of pH at 3°C (top) and temperature at pH 7 (bottom) on the formation of the alkyl isothiocyanates (**1**, **3** and **4**).

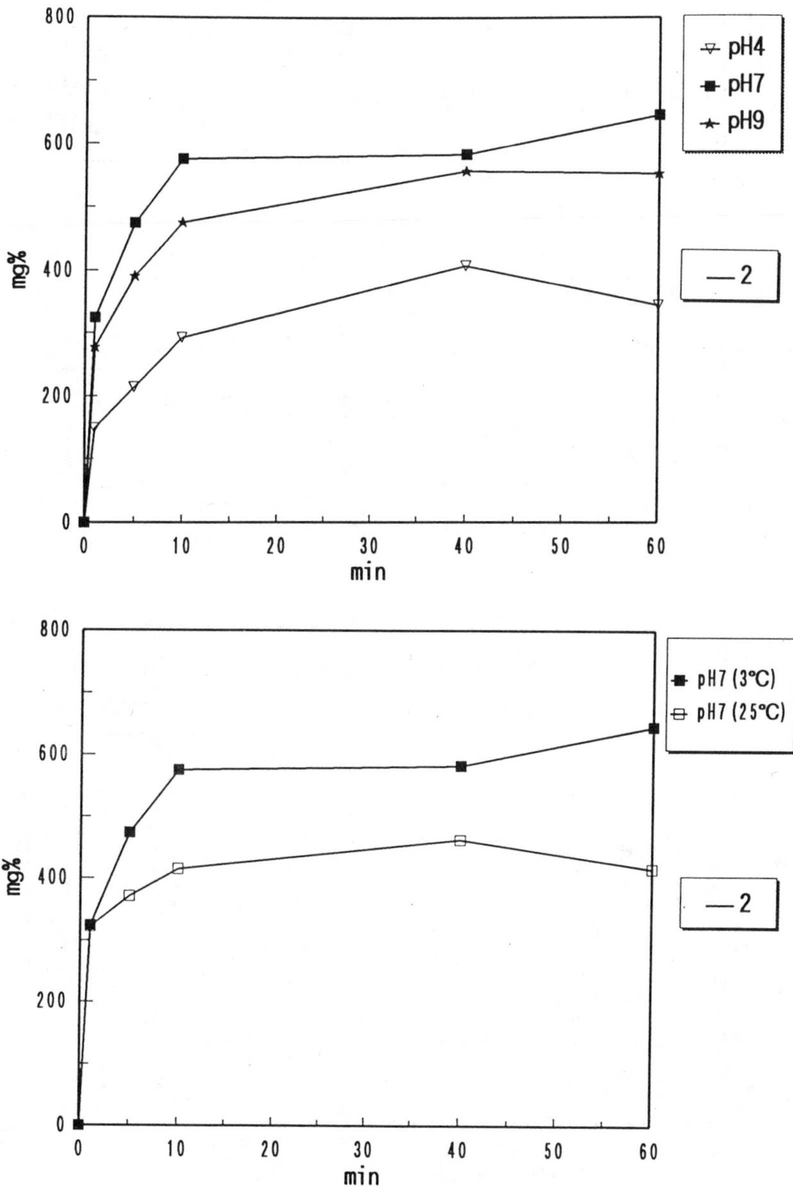

Figure 3. Effects of pH at 3°C (top) and temperature at pH 7 (bottom) on the formation of the allyl isothiocyanate (**2**).

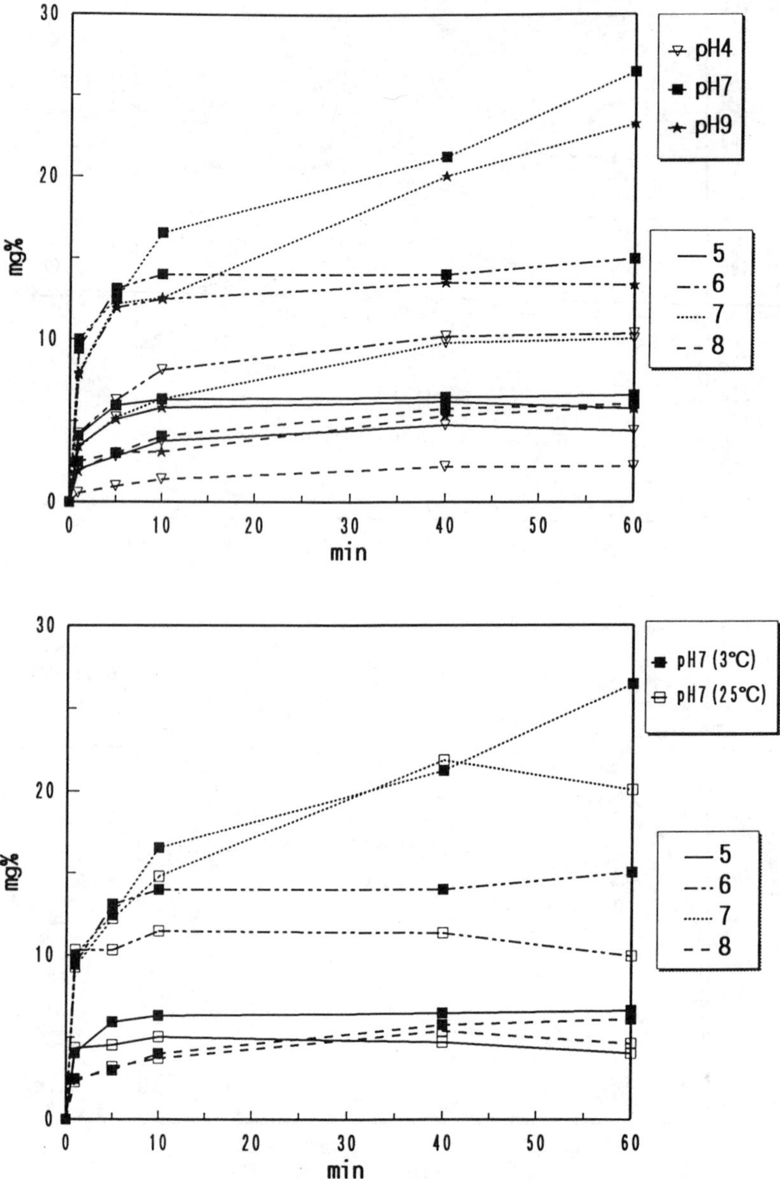

Figure 4. Effects of pH at 3°C (top) and temperature at pH 7 (bottom) on the formation of the ω-alkenyl isothiocyanates (**5-8**).

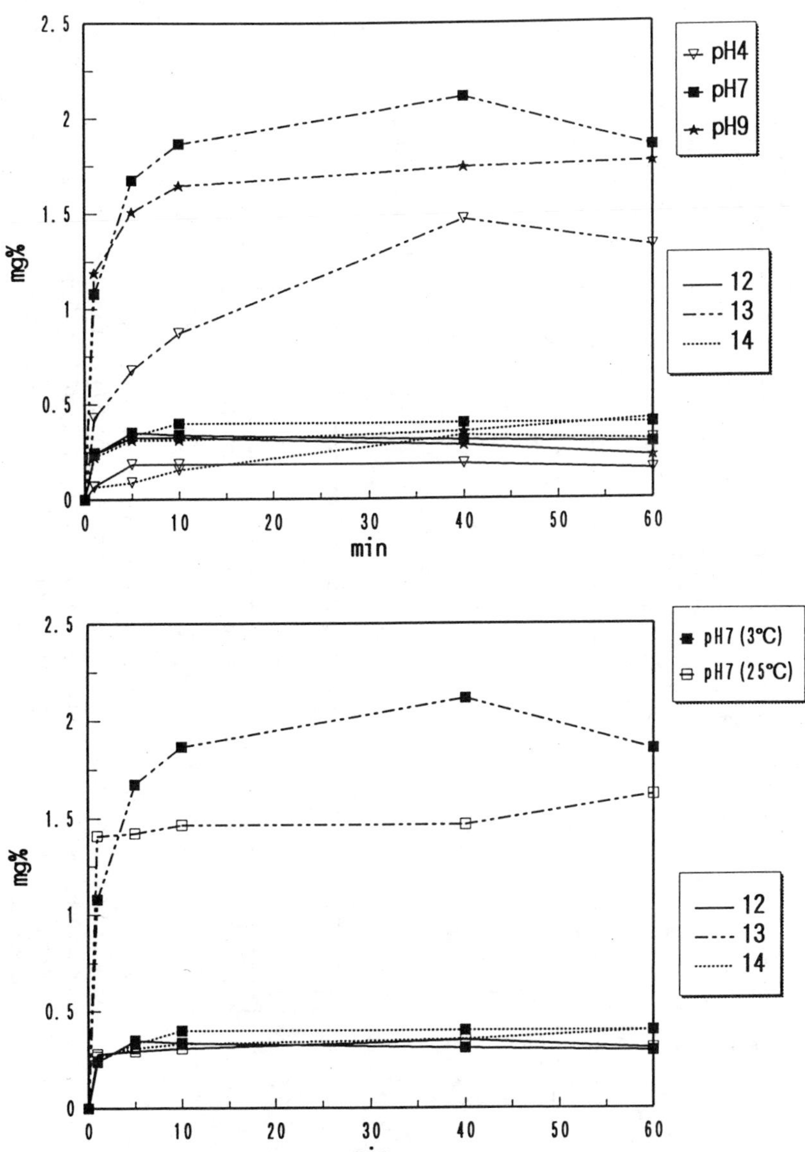

Figure 5. Effects of pH at 3°C (top) and temperature at pH 7 (bottom) on the formation of the ω-methylthioalkyl isothiocyanates (**12-14**).

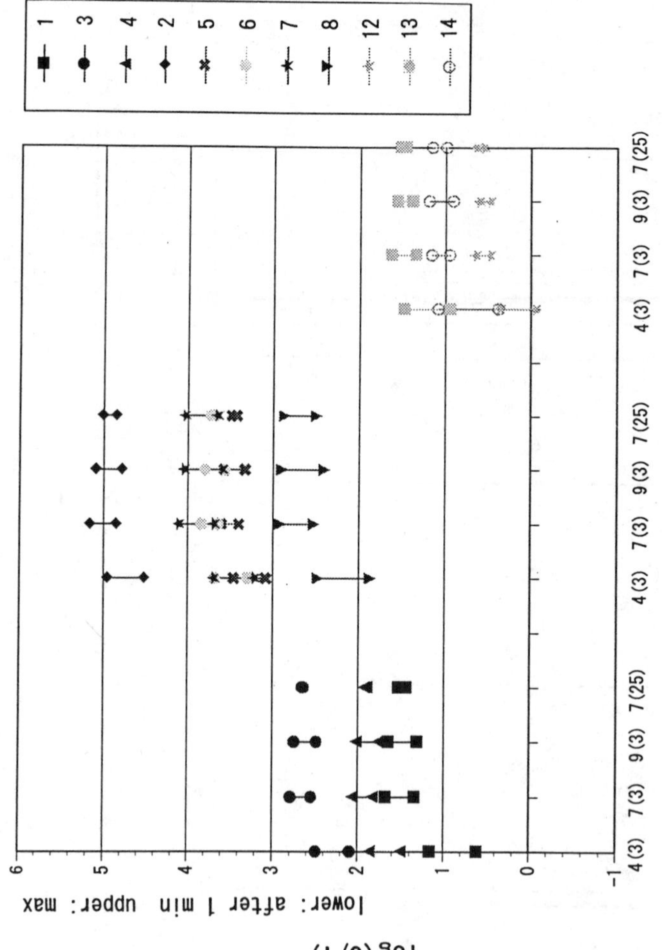

Figure 6. Effects of pH and temperature on the log of the ratios of the concentrations to the odor threshold values, log(C/T), for the isothiocyanates; upper plots: the max values of log(C/T) during hydrolysis, lower plots: the values of log(C/T) after 1 minute of hydrolysis.

changes in the amounts of (**2**), the ω-alkenyl isothiocyanates (**5-8**) and the ω-methylthioalkyl isothiocyanates (**12-14**) at three kinds of pHs were similar to that of alkyl isothiocyanates (top of Figure 3, 4 and 5, respectively). Further, the amounts of (**1**, **3** and **4**) at 3°C were greater than those at 25°C (Figure 2, bottom). The behavior of isothiocyanate (**2**, **5-8** and **12-14**) at two different temperatures were roughly the same as that of the alkyl isothiocyanates (bottom of Figure 3, 4 and 5, respectively).

The relationship between the values of log(C/T) and the hydrolysis conditions of wasabi is represented in Figure 6. The values of log(C/T) obtained after 1 minute of hydrolysis increased in the order of pH 4 at 3°C < pH 9 at 3°C < pH 7 at 3°C < pH 7 at 25°C. In general, the order of values of the max log(C/T) was as follows: pH 4 at 3°C < pH 7 at 25°C < pH 9 at 3 °C < pH 7 at 3°C. The values of log(C/T) under pH 7 at 25°C were the highest at the beginning of the enzymatic hydrolysis. The max log(C/T) obtained under pH 7 at 3°C, however, was found to be highest. In addition, the difference between the max log(C/T) and the log(C/T) after 1 min of hydrolysis increased in the order of pH 7 at 25°C < pH 9 or 7 at 3 °C < pH 4 at 3 °C. The hydrolysis under pH 7 at 25°C, which is the usual condition when eating, gave the smallest difference. Interestingly, the alkyl isothiocyanates (left) (**1**, **3** and **4**), the ω-alkenyl isothiocyanates (middle) (**2** and **5-8**), and the ω-methylthioalkyl isothiocyanates (right) (**12-14**) had similar tendencies in the evaluated log(C/T).

Conclusions

ω-Alkenyl isothiocyanates (**5-8**) have been found to contribute significantly to the top green odor note of wasabi. In addition, the characteristic last note of wasabi is attributed to 6-methylthiohexyl isothiocyanate (**13**). In order to evaluate the contribution toward intensity of odor, the log of the aroma value has been proposed.

The yields and the log of the aroma values of the isothiocyanates hydrolyzed by wasabi myrosinase are significantly affected by pH and temperature.

Acknowledgements

We are grateful to Dr. Hitomi Kumagai for her helpful advice.

Literature Cited

1. Kjaer, A.; Conti, J.; Larsen, I. *Acta Chem. Scand.* **1953**, *7*, 1276-1283.
2. Nagashima, Z.; Uchiyama, M. *J. Bull. Agr. Chem. Soc. Japan* **1959**, *23*, 556-557.
3. *Phytochemical Dictionary* ; Harborne, J. B.; Baxter, H., Editors; Taylor & Francis: Washington, D.C., **1993**; pp 92-99.
4. Kojima, M.; Uchida, M.; Akahori, Y. *Yakugaku zasshi* **1973**, *93*, 453-459.
5. Grob, K, Jr.; Matile, P. *Phytochemistry* **1980**, *19*, 1789-1793.
6. Mazza, G. *Can. Inst. Food Technol. J.* **1984**, *17*, 18-23.
7. Kojima, M.; Ichikawa. I. *J. Ferment. Technol.* **1963**, *47*, 263-267.

8. Ina, K.; Sano, A.; Nobukuni, M.; Kishima, I. *Nippon Shokuhin Kogyo Gakkaishi* **1981**, *28*, 365-370.
9. Ina, K.; Takasawa, R., Yagi, A.; Ina, H.; Kishima, I. *Nippon Shokuhin Kogyo Gakkaishi* **1990**, *37*, 256-260.
10. Gilbert, J.; Nursten, H. E. *J. Sci. Fd Agric.* **1972**, *23*, 527-539.
11. Tokarska, B.; Karwowska, K. *Nahrung* **1983**, *27*, 443-447.
12. Ina, K.; Ina, H.; Ueda, M.; Yagi, A.; Kishima, I. *Agric. Biol. Chem.* **1989**, *53*, 537-538.
13. Masuda, H.; T.; Tateba, H.; Tsuda, T.; Mihara, S.; Kameda, W. *Nippon Nogeikagaku Kaishi* **1989**, *63*, 235.
14. Tsuruo, I.; Yoshida, M.; Hata, T. *Agric. Biol. Chem.* **1967**, *31*, 18-26.
15. Bjoerkman, R.; Loennerdal, B. *Biochimica et Biophysica Acta* **1973**, *327*, 121-131.
16. Ohtsuru, M.; Kawatani, H. *Agric. Biol. Chem.* **1979**, *43*, 2249-2255.
17. Masuda, H.; Tsuda, T.; Tateba, H.; Mihara, S. Japan Patent 90,221,255, **1990**.
18. Harada, Y.; Masuda, H.; Kameda, W. Japan Patent 95,215,931, **1995**.
19. Amoore, J. E. In *Molecular Basis of Odor*; Kugelmass, I. N., Editor; Charles C. Thomas: Springfield, Illinois, 1970; pp 16-22.
20. Blank, I.; Fischer, K. H.; Grosch, W. *Z. Lebensm. Unters. Forsch.* **1989**, *189*, 426-433.
21. Grosch, W. *Trends Food Sci. Technol.* **1993**, *4*, 68-73.
22. Rothe, M.; Thomas, B. *Z. Lebensm. Unters. Forsch.* **1962**, *119*, 302-311.
23. Wright, R. H. In *The Sense of Smell*; CRC Press, Inc.: Boca Raton, Florida, 1982, pp 77-84.
24. Grosch, W. *Flavor Fragrance J.* **1994**, *9*, 147-158.
25. Abbott, N.; Etievant, P.; Issanchou, S.; Langlois, D. *J. Agric. Food. Chem.* **1993**, *41*, 1698-1703.

Chapter 7

Creation of Transgenic Citrus Free from Limonin Bitterness

S. Hasegawa[1], C. Suhayda[1], M. Omura[2], and M. Berhow[3]

[1]Western Regional Research Center, Agricultural Research Service, U.S. Department of Agriculture, 800 Buchanan Street, Albany, CA 94710
[2]Fruit Tree Research Station, Okitsu Branch, Ministry of Agriculture, Forestry & Fisheries, Okitsu, Shimizu, Shizuoka 424–02, Japan
[3]National Center for Agricultural Utilization Research, Agricultural Research Service, U.S. Department of Agriculture, 1815 North University Street, Peoria, IL 61604

Bitterness due to limonin in citrus juices is a major problem for the citrus industry worldwide. Modern gene transfer methodology provides a promising approach for solving this problem. Basic research on the biosynthesis of limonoids in citrus fruit has shown the presence of several promising enzymes, if their activities could be genetically enhanced in or transferred to commercial species of citrus. Accomplishing this could lower the levels of the bitter limonoids in developing fruit. Two of these limonoid metabolic enzymes have been isolated and their amino acid sequences determined. Preparation of gene constructs for genetic transfer to navel orange is in progress.

There are two bitter principles in citrus juices, the limonoids and the flavanone neohesperidosides. Bitterness in intact fruit tissues and freshly prepared juice is caused by flavanone neohesperidosides in citrus cultivars related to pummelo, such as naringin in grapefruit and neohesperidin in the Seville orange (1). The gradual development of bitterness in citrus juices such as navel orange after juicing, referred to as "delayed bitterness" is caused by limonoids (2). This delayed bitterness differentiates limonoid bitterness from flavanone neohesperidoside bitterness. The limonoid bitterness is usually due to limonin (**1**, Fig. 1), which is present in all citrus species. Some species of *Citrus* do not contain a sufficient quantity of limonin in their juices to have a bitterness problem.

Fruit tissues do not normally contain bitter limonin, but instead contain the nonbitter precursor of limonin, limonoate A-ring lactone (**2**). Limonoate A-ring lactone is the predominant limonoid aglycone present in fruit tissues of most citrus species, and it is gradually converted to limonin after the juice is extracted. This conversion proceeds under acidic conditions below pH 6.5 and is accelerated by the

0097–6156/96/0637–0079$15.00/0

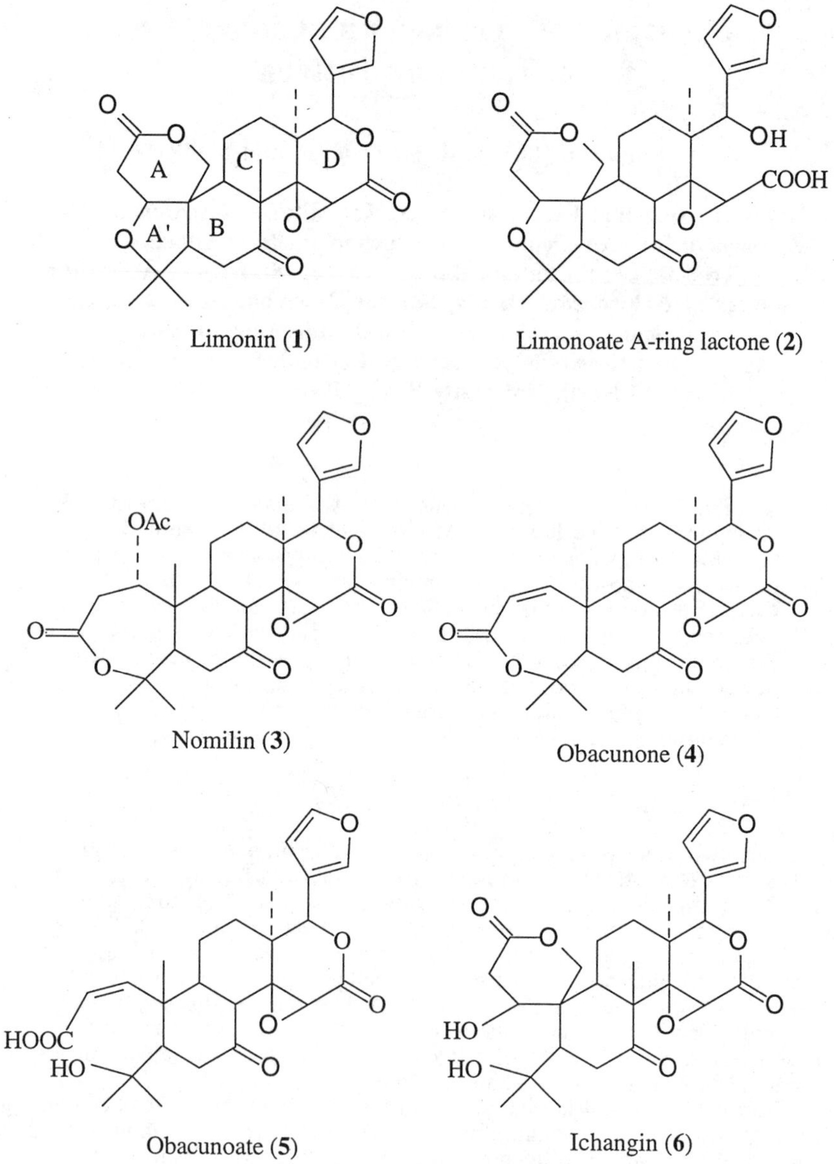

Limonin (**1**)

Limonoate A-ring lactone (**2**)

Nomilin (**3**)

Obacunone (**4**)

Obacunoate (**5**)

Ichangin (**6**)

Figure 1. Structure of limonoids

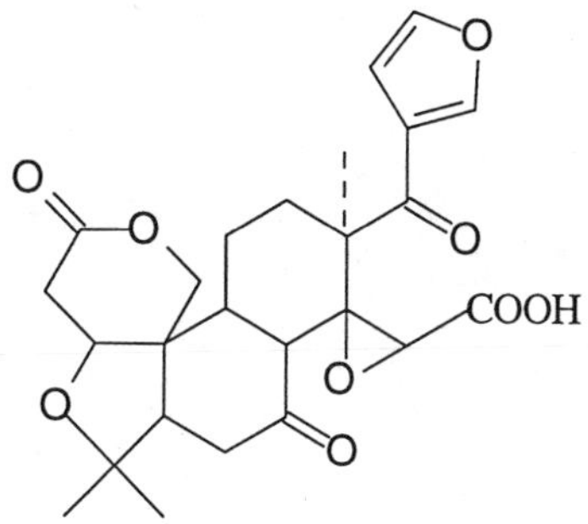

17-Dehydrolimonoate A-ring lactone (**7**)

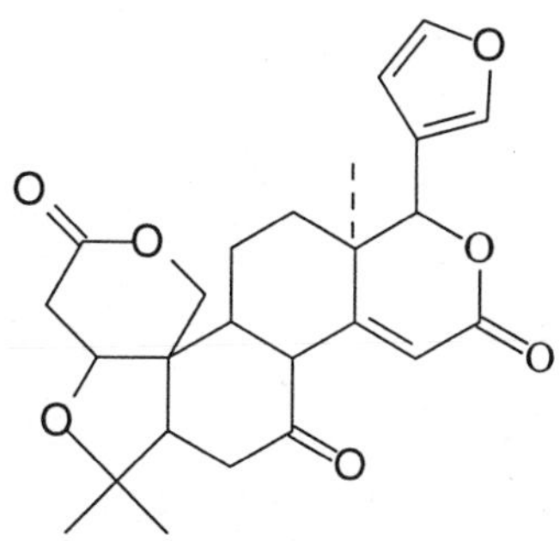

Deoxylimonin (**8**)

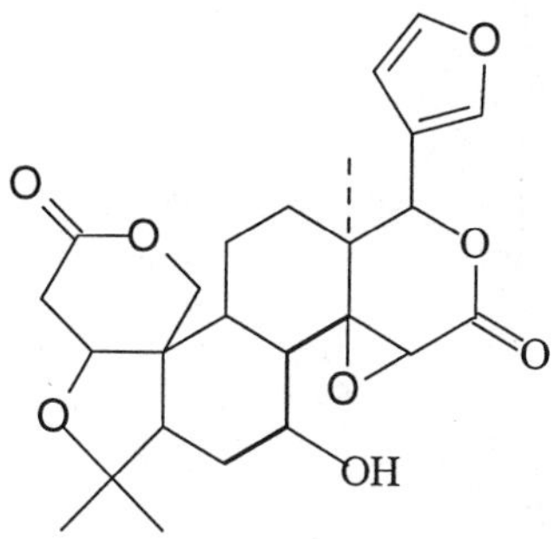

Limonol (**9**)

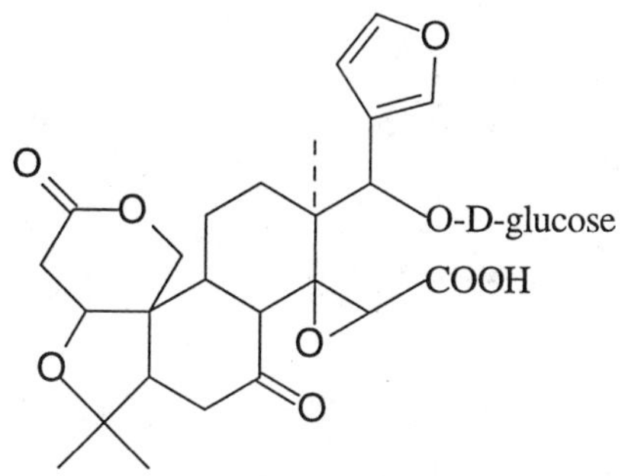

Limonin 17-b-D-glucopyranoside (**10**)

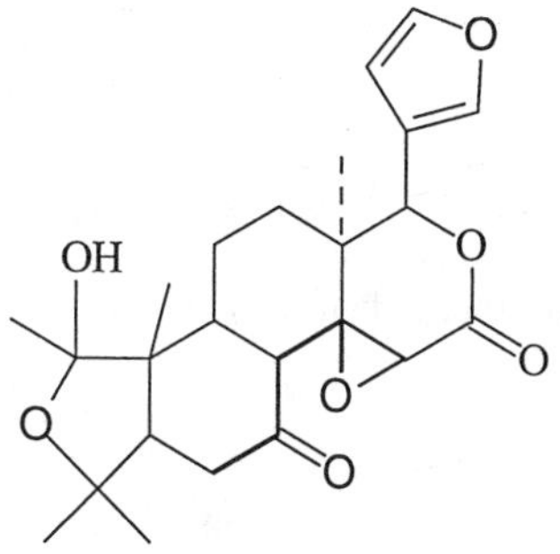

Ichangensin (**11**)

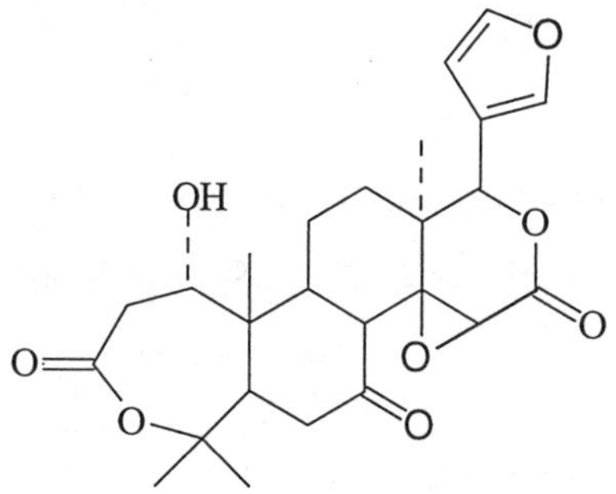

Deacetylnomilin (**12**)

Figure 1. *Continued*

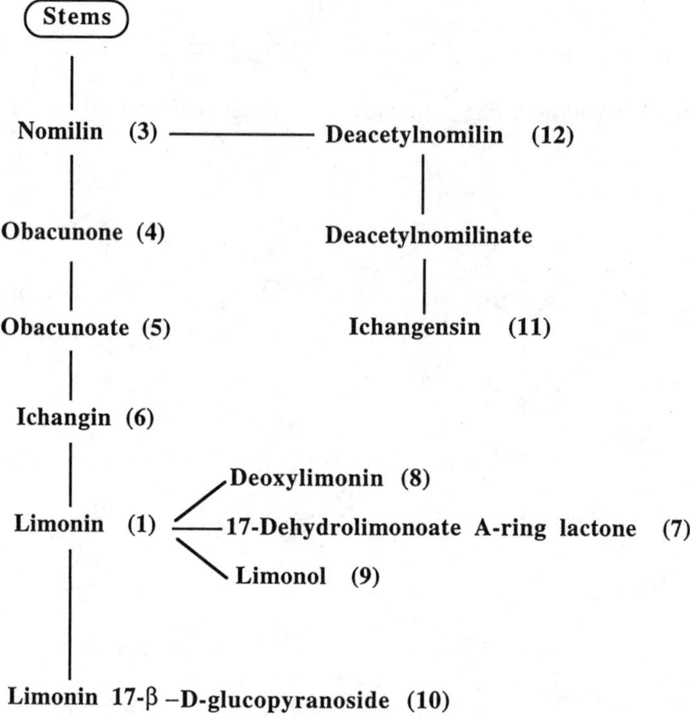

Figure 2. Biosynthetic pathways of limonoids in *Citrus*

enzyme, limonin D-ring lactone hydrolase (2, 3). Generally, limonin bitterness develops several hours after the juice is extracted at room temperature, or overnight if kept in a refrigerator.

Limonin bitterness: importance to the citrus industry

While limonin (**1**) bitterness is not a problem in intact, undamaged fruit, it has long been recognized as a problem in juices. Delayed bitterness lowers the value of certain citrus juices and must be reduced by costly blending or adsorption techniques. A very limited amount of juice has been debittered by adsorption methods on a commercial scale, but none of adsorbents are specific for limonin and each has drawbacks (4-6). Processed citrus products such as sections, vesicles, pieces, etc. are also susceptible to limonin bitterness. These non-juice forms cannot be debittered by adsorption technology or dilution by blending. The harvested fresh fruit must have low levels of the bitter limonoids to begin with for the production of those processed products. It would be certainly better for the production of juices. One method that could be employed is to use modern molecular biology techniques to add new genes to commercial citrus species to reduce the levels of the bitter limonoids during fruit development and maturation. Several laboratories around the world are now using standardized methodology to successfully create transgenic citrus plants (7-11). We would like to utilize this methodology to create transgenic citrus trees that have fruits free from limonin.

Biosynthesis and metabolism of limonin and its closely related limonoids

Thirty-six limonoid aglycones have been isolated from *Citrus* and its hybrids. These aglycones are present in intact fruit tissues as open D-ring forms such as limonoate A-ring lactone (**2**), while in seeds they are present in both open and closed forms. It appears that these citrus limonoids are biosynthesized through at least four different limonoid pathways: the limonin pathway, the ichangensin pathway, the calamin pathway and the 7-acetate limonoid pathway (12). The products of the limonin pathway are found in all citrus species. The products of the other three pathways are only found in certain species of citrus and its close relatives.

Radioactive tracer work has shown that nomilin (**3**) is biosynthesized by the terpenoid pathway from acetate in the phloem region of stems and translocated to other parts of plants such as leaves, fruit tissues and seeds (13,14). At those locations, nomilin is further biosynthesized to other limonoids. Limonoid biosynthesis occurs at each location independently, thus the composition of limonoids in fruit tissues, seeds and leaves are different from each of the others. Limonin is biosynthesized from nomilin via obacunone (**4**), obacunoate (**5**) and ichangin (**6**) (15-17) (Fig. 2)

Limonin appears to be the predominant end product of the limonin pathway. However, some limonin is further metabolized to 17-dehydrolimonoate A-ring lactone (**7**), deoxylimonin (**8**), limonol (**9**) and other compounds (18-21). These metabolites are all nonbitter and only occur in seeds in low amounts.

In fruit tissues and seeds, the limonoid aglycones are converted to 17-β-D-glucopyranoside derivatives such as limonin 17-β-D-glucopyranoside (**10**) during late stages of fruit growth and maturation (22-24). Limonoid glucosides are present only in mature fruit tissues and seeds. The enzyme responsible for this conversion is limonoid UDP-D-glucose transferase (25). This enzyme activity appears in fruit tissues and seeds at the onset of fruit maturation and its activity is retained throughout the ripening process. Twenty such limonoid glucosides have been isolated from *Citrus* and they are all nonbitter tasting. This natural debittering process takes place only in mature fruit tissues and seeds.

Unlike most other citrus species in which limonin is predominant (over 50% of total limonoids) *Citrus ichangensis* accumulates ichangensin (**11**) as the predominant limonoid in its fruit tissues and seeds (26). Fruit tissues contain 50 times as much ichangensin as limonin. They also contain relatively high concentrations of deacetylnomilin (**12**).

Radioactive tracer work has shown that ichangensin is biosynthesized from nomilin via deacetylnomilin and deacetylnomilinate (27). The fact that 55% of the recovered radioactivity fed as nomilin appeared as deacetylnomilin suggests that deacetylnomilin is a direct metabolite of nomilin in *C. ichangensis*. Concentrations of deacetylnomilin and deacetylnomilinate in this species are much higher than those of other *Citrus*. The enzyme activity responsible for the conversion of nomilin to deacetylnomilin is unique to this species and its hybrids. The metabolites of this pathway, ichangensin, deacetylnomilin and deacetylnomilinate are all nonbitter.

Genetic manipulation

The most promising approach for solving the limonin bitterness problem at the present time is through the use of genetic engineering techniques. A gene which codes for an enzyme which either converts bitter limonin to a nonbitter product or diverts the limonin biosynthetic pathway to prevent limonin from being synthesized, could be inserted into the genome of the citrus varieties which have a limonin bitterness problem. As techniques have been developed for inserting genes into citrus species, this approach has a good chance of success provided the proper genetic material is selected.

Specific target enzymes Based on the metabolic and catabolic pathways of limonin in *Citrus*, several enzymes have been identified for possible genetic manipulation. Specifically, we are currently looking at three enzyme systems: 1) limonoate dehydrogenase, 2) limonin UDP-D-glucose transferase and 3) nomilin deacetylase.

Limonoate dehydrogenase This enzyme catalyzes the conversion of limonoate A-ring lactone (**2**) to nonbitter 17-dehydrolimonoate A-ring lactone (**7**). The presence of this enzyme was first demonstrated in a soil bacterium, *Arthrobacter globiformis* (28), later several other species of soil bacteria were shown to possess this enzyme activity. This dehydrogenase activity is also present in *Citrus*. The metabolite, 17-dehydrolimonoate A-ring lactone, has been isolated from various citrus tissues (19), and the enzyme activity has been also demonstrated by radioactive tracer work (18). As both the activity of the dehydrogenase and the concentrations of 17-dehydrolimonoate A-ring lactone are very low in citrus tissues, it would be very difficult to isolate the enzyme from *Citrus*. However, high levels of the enzyme can be induced in cells of *Arthrobacter globiformis* by growing the organism with limonoids as the carbon source. The enzyme has been partially isolated and characterized from this source (18).

Insertion of a gene of limonoate dehydrogenase into navel orange genome might reduce the concentration of limonoate A-ring lactone, the precursor of limonin, by converting it to 17-dehydrolimonoate A-ring lactone which can not be converted to limonin in prepared juice.

Limonoate dehydrogenase was purified from cell-free extracts of *Arthrobacter globiformis* by ammonium sulfate fractionation, Blue dye ligand affinity column chromatography and DEAE ion exchange HPLC (29). These steps provided a 428-fold increase in purity over the cell-free extract. The largest increase in purification was obtained by biospecific elution of the dehydrogenase from the Cibacron Blue dye ligand column. SDS gel electrophoresis revealed a single protein band having a M_r of 31000. Peptide analysis of the 31 kD band indicated the following N-terminal amino acid sequence: ^{1}Met-Pro-Phe-Asn- Arg-^{6}Leu-Glu-Asp-Glu-Val-Ala-^{12}Ile-Val-Val-Gly-Ala. A cDNA library has been prepared from this bacterium and the gene for this enzyme is being screened with cDNA probes built from the amino acid sequence information.

Limonoid UDP-D-glucose transferase Limonin bitterness is a problem in juices extracted from early-season to mid-season winter fruit, but not a problem in juice extracted from the late season fruit. As the fruit ripens, the concentration of limonoate A-ring lactone (**2**) decreases (23,24). This natural limonin debittering process has been known for over a century, but the mechanism of this metabolism was not understood until the recent discovery that limonoid glucosides are present in mature fruit tissues and seeds. We observed that in navel orange the initial appearance of limonin 17-β-D-glucopyranoside (limonin glucoside)(**10**) and the sudden decrease of limonoate A-ring lactone concentrations occur simultaneously. This suggested that limonoate A-ring lactone is being converted to limonin glucoside. Radioactive tracer work confirmed this hypothesis, showing that mature citrus fruit tissues are capable of converting aglycones to their corresponding glucoside derivatives (30). Twenty limonoid glucosides have been identified in citrus species and their hybrids.

The enzyme responsible for this conversion has been identified as limonoid UDP-D-glucose transferase (25). Insertion of copies of this gene with altered expression controls, such as constitutive expression, may result in preventing the accumulation of significant levels of any limonoate A-ring lactone at time of harvest.

The enzyme has been partially purified from albedo tissues of navel oranges by ammonium sulfate fractionation, uridine 5'-diphosphoglucuronic acid (insolubilized on cross-linked 4% beaded) agarose affinity chromatography and DEAE-ion exchange HPLC. SDS gel electrophoresis revealed one major and a few very minor protein bands. The major band has a M_r of 58000 (on a 12% gel) and has the transferase activity. Radio active tracer work showed that this single enzyme appears to catalyze the glucosidation of all limonoid aglycones. The N-Terminal amino acid sequence of the enzyme has been determined. Also, the enzyme was broken down to segments by lysine-C proteinase and the amino acid sequences of three of its segments have been determined. A cDNA library of the navel orange genome is being prepared and it will be screened with probes prepared from the amino acid sequence information.

Nomilin deacetylase As previously discussed, *Citrus ichangensis* and its hybrids possess another limonoid biosynthetic pathway from nomilin (**3**) to deacetylnomilin (**12**) (26, 27, 31). The conversion of nomilin to deacetylnomilin is catalyzed by an enzyme, nomilin deacetylesterase. The existence of nomilin deacetylase in *C. ichangensis* has been demonstrated by radioactive tracer work (27). This enzyme is apparently only present in *C. ichangensis* and its hybrids. The enzyme catalyzes the hydrolysis of the acetyl group at the C1 position on the A-ring of nomilin to leave a hydroxyl group forming deacetylnomilin which is tasteless. By contrast in most citrus fruit the acetyl group at the C1 position on the A-ring is eliminated by the enzyme nomilin acetyl-lyase creating a double bond between the C1 and C2 positions in the A-ring forming obacunone (**4**) which is further converted through a couple of more steps to limonin. The fruit of *C. ichangensis* is hard to grow in the United States. It may be that the enzyme can be obtained from ichangensis hybrids such as Yuzu, Sudachi, or Ichang lemon. If the gene for nomilin deacetylase could be isolated from *C. ichangensis* and inserted with a constitutive promoter into commercial cultivars, it is probable that in these genetically engineered plants nomilin could be converted to deacetylnomilin. Accumulation of the nonbitter deacetylnomilin instead of limonoate A-ring lactone would also reduce delayed bitterness.

Other possible enzymes. Limonol (**9**) is a nonbitter limonoid (21). This compound is a minor limonoid and the pathway from limonin to limonol has not been investigated. This enzyme gene may be useful as specific target enzyme gene, but we assume that the activity is very low and it may not be practical to isolate the enzyme from citrus tissues. Deoxylimonin (**8**) is also a nonbitter limonoid present in *Citrus*. (20). The conversion of limonin to deoxylimonin requires a multi-enzyme system. Therefore, it is not practical to utilize this pathway for genetic engineering manipulation.

Summary

We presented here a novel genetic engineering approach for eliminating the problem of delayed bitterness in citrus juices caused by limonin (**1**). The target enzymes of our genetic engineering efforts, limonoate dehydrogenase (LDH) and limonoid UDP-D-glucose transferase (Gtase) both act on the immediate precursor to limonin, limonoate A-ring lactone. Chemical modification of this precursor to 17-dehydrolimonoate A-ring lactone (**7**) by LDH or to limonin 17-β-D-glucopyranoside (**10**) by Gtase result in the formation of nonbitter metabolites and block the production of limonin. These enzymes have been purified, limited amino acid sequences determined, and the genes encoding these proteins are being identified and cloned.

There are several advantages to a genetic engineering solution to the limonin bitterness problem. First, it deals specifically with impeding the formation of limonin in citrus juices and does not alter other desirable juice constituents. In contrast, treatment of juices with adsorption methods reduces limonin as well as other juice constituents, some of which may be desirable. Second, from regulatory and consumer viewpoints, this in situ debittering process during fruit growth, development and ripening is all "natural" and should be readily acceptable to both groups. Finally, this technology is applicable to both the production of juices and non-juice forms of processed products. In contrast, adsorption methods are applicable to only juice production.

References

1. Albach, R. F; Redman, G. H. *Phytochemistry* **1969**, *8*, 127-143.
2. Maier, V. P.; Beverly, G. D. *J. Food Sci.* **1968**, *33*, 488-492.
3. Maier, V. P.; Hasegawa, S.; Hera, E. *Phytochemistry* **1969**, *8*, 405-407.
4. Johnson, R. L.; Chandler, B. V. *J. Sci. Food Agric.* **1985**, *36*, 480-484.
5. Shaw, P. E.; Wilson, C. W. III. *J. Agric. Food Chem.* **1985**, *50*, 1205-1207.
6. Puri, A. *Bitterness in Foods and Beverages*. Elsevier, **1992**, pp 309-90.
7. Kobayashi, S.; Uchimiya, H. *Japanese J. Genetics* **1989**, *64*, 91-97.
8. Hidaka, T.; Omura, M.; Ugaki, M.; Tomiyama, M.; Kato, A.; Ohshima, M.; Motoyoshi, F. *Japanese J. Breeding* **1990**, *40*, 199-107.
9. Vardi, A.; Bleichman, S.; Aviv, D. *Plant Sci.* **1990**, *69*, 199-206
10. Moore, G. A.; Jacobs, C. C.; Neidigh, J. L.; Lawrence, S. D.; Cline, K. *Plant Cell Reports* **1992**, *11*, 238-242.
11. Pena, L.; Cervera, M.; Juarez, J.; Ortega, C.; Pina, J. A.; Duran-Vila, N.; Navarro, L. *Plant Sci.* **1995**, *104*, 183-191.
12. Hasegawa, S.; Herman, Z. *Secondary-Metabolite Biosynthesis and Metabolism*, Plenum Press, New York, NY, **1992**, pp. 305-317.

13. Hasegawa, S.; Herman, Z.; Orme, E. D.; Ou, P. *Phytochemistry* **1986**, *25*, 2783-2785.
14. Ou, P.; Hasegawa, S.; Herman, Z.; Fong, H. C. *Phytochemistry* **1988**, *27*, 115-118.
15. Hasegawa, S.; Hoagland, J. E. *Phytochemistry* **1977**, *16*, 469-471.
16. Hasegawa, S.; Herman, Z. *Phytochemistry* **1985**, *24*, 1973-1974
17. Herman, Z.; Hasegawa, S. *Phytochemistry* **1985**, *24*, 2911-2913
18. Hasegawa, S.; Maier, V. P.; Bennett, R. D. *Phytochemistry* **1974**, *13*, 103-105.
19. Hsu, A. C.; Hasegawa, S.; Maier, V. P.; Bennett, R. D. *Phytochemistry* **1973**, *12*, 563-567.
20. Dreyer, D. L. *J. Org. Chem.* **1965**, *30*, 749-751.
21. Bennett, R. D.; Hasegawa, S. *Phytochemistry* **1982**, *21*, 2349-2354.
22. Hasegawa, S.; Bennett, R. D.; Herman, Z.; Fong, C.H.; Ou, P. *Phytochemistry* **1989**, *28*: 1717-1720.
23. Hasegawa, S.; Ou, P.; Fong, C. H.; Herman, Z.; Coggins, C. W., Jr.; Atkin, D. R. *J. Agric. Food Chem.* **1991**, *39*, 262-265.
24. Fong, C. H.; Hasegawa, S.; Herman, Z.; Ou, P. *J. Sci. Food Agric*, **1991**, *54*, 393-398.
25. Suhayda, C. G.; Hasegawa, S. *Plant Physiol.* **1994**, *105*, 127.
26. Bennett, R. D.; Herman, Z.; Hasegawa, S. *Phytochemistry* **1988**, *27*, 1543-1545.
27. Herman, Z.; Hasegawa, S.; Fong, C. H.; Ou, P. *J. Agric. Food Chem.* **1989**, 37, 850-851.
28. Hasegawa, S.; Bennett, R. D.; Maier, V. P.; King, A. D., Jr. *J. Agric. Food Chem.* **1972**, *20*, 1031-1034.
29. Suhayda, C. G.; Omura, M.; Hasegawa, S. *Phytochemistry* **1995**, *40*, 17-20.
30. Herman, Z.; Fong, C. H.; Hasegawa, S. *Phytochemistry* **1991**, *30*, 1487-1488.
31. Ozaki, Y.; Miyake, M.; Maeda, H.; Ifuku, Y.; Bennett, R. D.; Herman, Z.; Fong; C. H.; Hasegawa, S. *Phytochemistry* **1991**, *30*, 2659-2661.

Chapter 8

Effect of Amide Content on Thermal Generation of Maillard Flavor in Enzymatically Hydrolyzed Wheat Protein

Qinyun Chen and Chi-Tang Ho

Department of Food Science, Cook College, New Jersey Agricultural Experiment Station, Rutgers University, New Brunswick, NJ 08903

Wheat gluten was hydrolyzed using fungal protease to obtain wheat gluten hydrolysate (WGH), and the resulting hydrolysate was deamidated to produce deamidated wheat gluten hydrolysate (DWGH). Acid-wheat gluten hydrolysate (AWGH) was prepared by hydrolyzing wheat gluten under strong acidic conditions. The three hydrolysates were reacted with glucose in closed systems and the volatile compounds that were generated were isolated and identified. About 11 times the amount of volatile compounds was generated in the DWGH-G system than was generated in the WGH-G system. By comparing the volatiles generated in these three systems, a clear trend was found that the deamidation reaction significantly increased the formation of aroma compounds.

Hydrolysis of proteins has been used to improve the quality of food for 100 years by adding the resulting mixture of amino acids and peptides. An older example is soy sauce, made by the action of mold enzymes in steamed soybeans and wheat. This enzymatic process causes the cleavage of proteins at various sites on the protein molecule and results in an increase of amino acids, oligopeptides or polypeptides. Although some short chain peptides or polypeptides resulting from the hydrolysis of meat or milk protein have been reported in food systems, the role of such hydrolysates as precursors in the generation of flavor compounds has not been investigated. Wheat flour is a major cereal that is rich in protein content. Therefore, enzymatically hydrolyzed wheat gluten would be a good substance to be used to improve the flavor quality of many foods.

Hydrolyzed vegetable protein (HVP), traditionally produced by acid hydrolysis, has been used to produce various types of flavors by the Maillard reaction in the food industry. Hydrochloric acid is commonly used in the production of protein hydrolysates because it works fast and yields a product with a highly accceptable savory profile (*1*). However, in recent years, there has been some concern about the safety of HVP due to the presence of chloropropanols in HVP. These chloropropanols are formed by the interaction of

0097–6156/96/0637–0088$15.00/0

hydrochloric acid with glycerol-containing lipids and may have harmful effects on humans if present in high concentrations (*1*).

Studies on the protein deamidation in food systems has recently been of great interest to the food industry. This is due to the fact that the deamidation reaction changes functional properties of protein such as solubility, emulsion property, and foaming ability (2). Deamidation is a reaction that results in the loss of the amide group of glutamine and asparagine residues in protein molecules, and further results in the release of ammonia. The ammonia released from protein deamidation will interact with other reactants such as reducing sugars when heated via the Maillard reaction possibly affecting both flavor and color attributes of the final product. With these points in mind, the major focus of this work was on the effect of amide content on the generation of aroma compounds via a thermal reaction using Wheat Gluten Hydrolysate (WGH), Deamidated Wheat Gluten Hydrolysate (DWGH), and Acid-Hydrolyzed Wheat Gluten (AWGH) as flavor precursors.

MATERIALS AND METHODS

Preparation of Wheat Gluten Hydrolysate (WGH) and Acid-Hydrolyzed Wheat Gluten (AWGH). Wheat gluten was purchased from Sigma Chemical Company (MO). Fungal Protease 500,000 was purchased from Gist-Brocades Food Ingredients, Inc. (PA). 10 g of wheat gluten and 1 g of fungal protease were weighed into a flask and 100 ml of distilled water was added. The flask was put in an incubator shaker (New Brunswick Scientific Company, Inc., NJ) at 180 RPM at 55°C for 2 hrs. The hydrolyzed mixture was then filtered to obtain the filtrate. The filtrate was then freeze-dried to obtain the WGH. The degree of hydrolysis as measured by the ninhydrin method was 11.5%. It means that the major components in the WGH are oligopeptides.

3 g of wheat gluten was weighed into a reaction flask to which 100 ml of 6 N HCl solution was added. The flask was tightly capped and put in an oven at 130°C for 24 h. The reaction mixture was filtered to obtain the filtrate. The solvents of the filtrate were evaporated with a rotary evaporator under vacuum. The residue was dissolved with 90 ml of distilled water, and the pH was adjusted to 7 with 1 N sodium hydroxide solution and then diluted to 100 ml with distilled water. The resulting solution was then freeze-dried to obtain the AWGH.

Preparation of Deamidated Wheat Gluten Hydrolysate (DWGH). 30 ml of hydrolyzed wheat gluten solution (WGH) was contained in a reaction bottle to which 120 ml of water was added. The pH of the solution was adjusted to 9. The bottle was capped and put in an oven at 110°C for 3 h to carry out the deamidation reaction. After deamidation, the bottle was cooled to room temperature. The pH of the deamidated solution was adjusted to 12, and then the solution was degassed overnight under vacuum in order to remove ammonia. The pH of the sample solution was then adjusted to 7 and the sample was freeze-dried to obtain the DWGH.

Maillard Reaction and Volatile Extraction. Three equal amounts (3 g of wheat gluten) of WGH, DWGH, and AWGH, were separately reacted with 0.9 g of glucose at 155 °C in three reaction vessels for two hours. The volatile compounds were extracted with methylene chloride, and concentrated with a Kuderna-Danish concentrator to the appropriate concentrate of solutions. The three volatile solutions were analyzed by gas chromatography (GC) and GC-mass spectrometry. The volatile compounds were quantified by using dodecane as an external standard.

RESULTS AND DISCUSSION

In the WGH, DWGH, and AWGH model systems we identified and quantified 55, 51, and 53 volatile compounds, respectively (Table I). The major volatiles identified were furans, pyrazines, aldehydes, alcohols, ketones, pyrrolizines. These volatile compounds were mainly derived from Strecker degradation, sugar degradation, lipid degradation, and the further interactions of these degradation products.

Strecker Aldehydes. The four Strecker aldehydes identified were 2-methylpropanal, 2-methylbutanal, 3-methylbutanal, and phenylacetaldehydes, which were derived from their corresponding free amino acids: valine, isoleucine, leucine and phenylalanine, respectively. Wheat gluten contains 6.1-8% leucine, 3.3-4.5% isoleucine, and 5.3-6.2% valine. Strecker aldehydes were the major volatile compounds generated in WGH-G, DWGH-G and AWGH-G model systems. The amount of Strecker aldehydes generated in the AWGH-G and DWGH-G systems were much higher than from WGH-G system.

Compared to the AWGH which contained mainly free amino acids, DWGH consisted of mainly peptides and small amounts of free amino acids. However, the DWGH still generated a significant quantity of Strecker aldehydes when heated with glucose. These results are consistent with the observation of Rizzi (*3*). Rizzi heated fructose with different peptides, tripeptides, and mixtures of their corresponding amino acids. It was found that dipeptides containing valine and leucine produced significant amounts of the Strecker aldehydes, 2-methylpropanal and 3-methylbutanal, despite the blocked amino group or carboxyl group. These experimental results indicated the Strecker degradation of peptides will take place in spite of their peptide bonding.

Compared with 2-methylpropanal, 3-methylbutanal, and 2-methylbutanal, two other aldehydes, acetaldehyde and formaldehyde, were not detected in the model systems, but their corresponding substituted flavor compounds were identified. Acetaldehyde, having a low boiling point of 21°C, might either be too volatile to be detected or be very active and completely react with other components. Several heterocyclic compounds with ethyl-substituents were identified in the model systems.

Alkylpyrazines. Besides the sugar-derived carbonyls, Strecker aldehydes, furans, and furanones were identified from the model systems (WGH-G, DWGH-G, and AWGH-G). The major volatile compounds identified were alkyl-substituted pyrazines. Table I lists seventeen alkyl-substituted pyrazines identified in the WGH-G, DWGH-G, and AWGH-G model systems.

Compared with pyrazines generated from the WGH-G system, a significant amount of pyrazines were generated in the DWGH-G and AWGH-G systems. Although the nitrogen content of the wheat gluten hydrolysate was greater than the deamidated wheat gluten hydrolysate, there were 17 times more pyrazines generated in the DWGH-G system than the WGH-G system.

Table I. Volatile Compounds Generated from the Thermal Reaction of WGH, DWGH and AWGH with Glucose

Compounds		Amount (ppm)		
		WGH-G	DWGH-G	AWGH-G
Strecker Aldehydes				
2-methylpropanal		9.8	277.1	351.6
3-methylbutanal		141.2	2301.7	2912.0
2-methylbutanal		39.1	776.5	1043.8
phenylacetaldehyde		1.3	21.2	28.7
	total	190.1	3355.3	4336.1
Alcohols				
2-methylpropanol		0.7	-	9.8
1,2-propanediol		0.6	-	8.4
1,3-butanediol		8.0	14.8	-
2,3-butanediol		9.0	15.3	-
cyclotene		29.5	97.7	249.8
2,3-dihydro-3,5-dihydroxy-6-methyl-4H-pyran-4-one		6.3	-	-
	total	54.1	127.8	268.0
Other Carbonyls				
2,3-butanedione		1.4	-	-
2-butanone		2.2	29.8	31.2
hydroxyacetone		19.4	312.8	243.3
acetoin		11.6	50.9	41.5
methyl propanoate		-	-	20.7
hexanal		3.0	15.3	-
2-hydroxy-3-pentanone		0.6	13.8	13.7
4-chlorobutanoic acid		13.0	47.3	57.8
2-heptanone		4.3	-	-
benzaldehyde		10.1	57.1	63.7
acetophenone		3.2	-	-
2-phenylpropenal		1.3	-	-
4-phenyl-2-butanone		-	24.8	-
2-phenyl-2-butenal		1.3	21.1	-
5-methyl-2-phenyl-2-hexenal		5.6	22.0	54.9
	total	77.0	595.0	526.8
Furans and Furanones				
2-furfural		9.3	31.6	86.9
2-furfuryl alcohol		15.0	54.7	41.1
2-methyltetrahydrofuran-3-one		2.2	18.7	44.3

Table I. Continued

Compounds	Amount (ppm)		
	WGH-G	DWGH-G	AWGH-G
2-acetylfuran	0.9	11.0	25.8
5-methyl-2-furfural	2.6	29.5	69.5
5-hydroxymethyl-2-furfural	35.8	85.3	431.2
2-phenylfuran	6.4	45.7	107.2
dihydro-5-propyl-2(3H)-furanone	8.9	28.4	-
total	81.1	304.9	806.0
Pyrazines			
pyrazine	27.7	183.6	251.1
methylpyrazine	30.5	234.6	239.8
2,5-dimethylpyrazine	15.6	723.3	391.1
ethylpyrazine	2.5	14.1	18.0
2,3-dimethylpyrazine	1.5	14.4	13.0
vinylpyrazine	1.9	4.1	10.7
2-methyl-5-ethylpyrazine	-	13.7	19.6
trimethylpyrazine	4.7	166.9	100.7
2-methyl-5-ethylpyrazine	-	13.7	19.6
2-methyl-5-vinylpyrazine	-	-	21.1
2,5-dimethyl-3-ethylpyrazine	-	61.0	54.2
2,3-dimethyl-5-ethylpyrazine	-	-	21.3
2-methyl-5-propenylpyrazine	-	9.1	19.5
2-methyl-5-propylpyrazine	-	9.1	13.7
2-methyl-3-propylpyrazine	-	-	7.4
2,5-dimethyl-3-propylpyrazine	1.0	66.3	37.0
2,5-dimethyl-3-isobutylpyrazine	1.2	34.9	21.5
total	86.6	1543.6	1265.9
Other N-containing Compounds			
2-acetylpyridine	1.8	14.4	30.1
pyrrole-2-carboxaldehyde	3.1	-	-
2-acetylpyrrole	0.9	19.6	24.6
2,5-dioxo-3-methylpiperazine	4.5	-	24.0
indole	3.7	-	-
5-acetyl-2,3-dihydro-1H-pyrrolizine	11.2	141.6	339.9
5,6,7,8-tetrahydro-3-methylquinoline	4.7	26.9	74.1
5-propionyl-2,3-dihydro-1H-pyrrolizine	-	8.4	52.5
5-acetyl-6-methyl-2,3-dihydro-1H-pyrrolizine	-	-	47.1
2-acetyl-pyrido(3,4-d)imidazole	1.3	35.3	40.7
total	31.2	236.2	633.0

Table I. Continued

Compounds		Amount (ppm)		
		WGH-G	DWGH-G	AWGH-G
S-containing Compounds				
dimethyldisulfide		1.8	21.8	16.9
thiophene-2-carboxaldehyde		2.0	11.3	35.0
thiophene-3-carboxaldehyde		3.7	-	-
2-acetylthiophene		1.8	9.6	56.3
2-formyl-5-methylthiophene		2.5	-	-
2-acetylthiazole		5.0	8.8	100.7
	total	16.8	51.5	208.9

Furans and Furanones. There were eight furans and furanones identified in the WGH-G, DWGH-G, and AWGH-G model systems. 5-Hydroxymethyl-2-furfural, derived from the thermal degradation and caramelization of glucose in the Maillard reaction, was a dominant product for all model systems. The amount of cyclic oxygen-containing compounds generated in the AWGH-G system was much higher when compared to the other two systems. This may be due to the fact that amino acids catalyze the degradation of glucose, and that the AWGH contains the highest amount of amino acids.

Pyrroles and Pyridines. In the three model systems, pyrroles were found in a minute amount, and 2-acetylpyrrole and pyrrole-2-carboxaldehyde were the only pyrroles detected. 2-Acetylpyridine was found in a minute quantity in the three model systems. The sensory properties of 2-acetylpyridine have been associated with popcorn, bread, cracker and coffee aromas (*4*).

S-Containing Compounds. Four thiophene derivatives were identified in the WGH-G, DWGH-G, and AWGH-G systems; however two of them were not found in the DWGH and WGH-HCl systems. This is probably because of a partial desulfuration reaction and the removal of hydrogen sulfide during either the deamination reaction or during acid hydrolysis.

Thiazoles are somewhat unique among the Maillard products because they contain both a nitrogen and a sulfur atom in the same ring. Even though wheat gluten contains about 2% cysteine, only 2-acetylthiazole was identified in the model systems. This was, again, probably due to the lower abundance of hydrogen sulfide as a result of the deamidation reaction and acid-hydrolysis, which remove both ammonia and hydrogen sulfide.

Dimethyldisulfide was the only non-cyclic sulfur compound identified in the WGH-G, DWGH-G, and AWGH-G model systems resulting from the condensation of the two molecules of methanethiol, derived from methionine.

Some Amino Acid-Specific Compounds. As listed in Table I, there were three pyrrolizines and one pyridoimidazole identified in the model systems. Pyrrolizines are a very important class of volatile compounds generated by the reaction of proline with reducing sugars. 2-Acetyl-pyrido(3,4-d)imidazole was another amino acid-specific Maillard reaction product identified in the model systems, which was derived from the reaction of histidine with dicarbonyls.

Proline is the second most abundant amino acid (13-18%) contained in wheat gluten, and plays a very important role in flavor formation during food processing. A great deal of work has been carried out by Tressl et al. (*5*) on the volatile components generated in proline-specific Maillard reactions. The most abundant proline-specific Maillard reaction products are 2,3-dihydro-1H-pyrrolizines.

The Maillard reaction of WGH-G, DWGH-G, and AWGH-G produced three 2,3-dihydro-1H-pyrrolizines. The most abundant pyrrolizine was 5-acetyl-2,3-dihydro-1H-pyrrolizine (5-ADHP). 5-Acetyl-2,3-dihydro-1H-pyrrolizine was identified in all three systems. 5-Propionyl-2,3-dihydro-1H-pyrrolizine was found in the DWGH-G and AWGH-G systems; while 5-acetyl-6-methyl-2,3-dihydro-1H-pyrrolizine was only identified in the AWGH-G system. It is very interesting to note that the amount of pyrrolizines generated were very different in each of the three model systems. The yields of pyrrolizines generated in the Maillard reactions were in this order: AWGH-G >> DWGH-G >> WGH-G. The main structural difference of proline between WGH, DWGH, and AWGH was that while the proline in the AWGH was free form, most of the proline in WGH and DWGH existed as residues of peptides. The free proline reacted with α-dicarbonyls much faster than proline-containing peptides to form iminium ions which act as key compounds for proline specific compounds. The reason that only small amounts of pyrrolizines were generated in the DWGH-G and WGH-G systems was that most of proline-containing peptides had more than three residues which were sterically hindered to react with α-dicarbonyls to form reactive intermediates.

Histidine-specific Maillard reaction products have not been investigated to any appreciable extent, probably due to the fact that they do not contribute any characteristic flavors to cooked foods. Gi et al. (*6*) identified 2-acetyl-pyrido(3,4-d)imidazole and 2-acetyl-pyrido(3,4-d)imidazole by the reaction of histidine with glucose at roasting and autoclaving conditions. While the wheat gluten contained relatively high amounts of histidine (up to 1.8-3.2%) DWGH contained 0.78% of free histidine. It is interesting to find that the yields of 2-acetyl-pyrido(3,4-d)imidazole found in the DWGH-G and AWGH-G systems were 35.3 and 40.7 ppm, respectively. Again, the results indicate the fact that the peptides which contain histidine will react with glucose to form the peptide Schiff bases. The peptide Schiff bases undergo a rearrangement reaction to promote the hydrolysis of the peptide bonds leading to the formation of free histidine. Histidine then undergoes further reactions leading to the formation of pyridoimidazoles. The yield of pyridoimidazole found in the WGH-G system was 1.3 ppm which was much less than that found in the DWGH-G and AWGH-G systems (35 & 40 ppm), respectively.

Effects of Deamidation on Flavor Formation

The deamidation reaction resulted in the release of ammonia from the residues of glutamine and asparagine in the wheat gluten hydrolysate. The ammonia released during the Maillard reaction will participate in the formation of heterocyclic compounds such as pyrazines, and also participate in pigment formation (*7*).

The results listed in Table I demonstrate that the reaction of DWGH with glucose produces much greater amounts of volatile compounds than is produced by the interaction of WGH with glucose. About 17 times the amount of Strecker aldehydes, 8 times the amount of other carbonyls, 2.5 times the amount of hydrocarbons, 3.7 times the amount of furans, 18 times the amount of pyrazines, 4 times the amount of other N-containing compounds, and 3 times the amount of S-containing compounds were produced in the DWGH-G system than in the WGH-G system.

An examination of the difference in the amount of volatile compounds generated in the WGH-G and DWGH-G systems indicated that ammonia, a reactive species, played a very important role in the Maillard reaction. Shibamoto (*8*) found that ammonia participated in the formation of N-containing volatile compounds in the Maillard reaction by reacting ammonia and hydrogen sulfide with glucose. Shibamoto (*9*) reacted ammonia with rhamnose, and identified 65 volatile compounds which included pyrazines, pyrroles, and imidazoles. Another study was carried out by Hwang (*10*) to investigate the participation of amide groups of glutamine in the pyrazine formation by the Maillard reaction. The results indicated that the relative contribution of amide nitrogen to pyrazine formation was greater than the contribution of α-amino nitrogen. Other studies were carried out to demonstrate that ammonia reacted with sugars or sugar degradation products to form melanoidins. According to Kato (*11*) melanoidin is thought to consist mainly of a repeating aromatic moiety. Izzo (*12*) investigated the effects of residual amide content on the aroma generation and formation of brown color by comparing the reactions of deamidated and undeamidated wheat gluten with glucose in model systems. The experiments demonstrated that the deamidation reaction of wheat gluten greatly decreased the pigment formation in the Maillard reaction when the deamidation degree was higher than 20%.

It was observed that the reacted mixture of WGH-G contained large amounts of dark colored products while the mixture from the reaction of DWGH with glucose contained only very small amounts of colored products. Investigation of this phenomena and the formation of volatile compounds from the WGH-G and DWGH-G systems demonstrated that the deamidation reaction of wheat gluten hydrolysates would increase the formation of aroma compounds and decrease the formation of brown color.

The fact that ammonia could enhance the generation of brown color and suppress pyrazine formation was investigated (*12*). It was thought to be attributed to the reactive nature of the chemistry of the ammonia molecule. The ammonia released from the amide groups of proteins or peptides during the Maillard reaction interacts with glucose leading to the formation of an unstable intermediate which when rearranged produces a primary amine. The primary amine, being quite reactive itself, then undergoes further reactions with another molecule of glucose or another carbonyl compound. The end result is a polymerization of the glucose in the early stages of the Maillard reaction resulting in a subsequent shift in chemical events away from flavor formation. The traditional mechanism for flavor formation could have been suppressed so that the net result was the formation of pigments with little formation of volatiles.

An examination of the quantitative data of aroma formation in the heated WGH-G or DWGH-G samples provides a rather clear trend. Even though the ammonia released from the amides of proteins or peptides during the Maillard reaction could participate in both the volatiles formation and pigment formation by reacting with reducing sugars, it probably interacted with reducing sugars to generate colored polymers.

Literature Cited

1. Nagodawithana, T. W. *Savory Flavors*. Esteekay Associates, Inc.: Milwaukee; **1995**, pp. 401-434.
2. Hamada, J. S. *CRC Crit. Rew. Food Sci. Nutri.* **1994**, *34*, 283-292.
3. Rizzi, G. P. In *Thermal Generation of Aroma*; Parliment, T. H.; McGorrin, R. J.; Ho, C.-T., Eds. ACS Symp. Ser. 409; American Chemical Society: Washington, D. C. 1989, pp. 172-181.
4. Maga, J. A. *J. Agric. Food Chem.* **1981**, *29*, 895-898.
5. Tressl, R.; Helak, B.; Martin, N.; Kersten, E. In *Thermal Generation of Aroma*; Parliment, T. H.; McGorrin, R. J.; Ho, C.-T., Eds. ACS Symp. Ser. 409; American Chemical Society: Washington, D. C. 1989, pp. 156-171.
6. Gi, U.; Baltes, W. In *Thermally Generated Flavors.* Parliment, T. H.; Morello, M. J.; McGorrin, R. J., Eds. ACS Symp. Ser. 543; American Chemical Society: Washington, D. C. 1994, pp. 263-269.
7. Izzo, H.; Ho, C.-T. *J. Agric. Food Chem.* **1993**, *41*, 2364-2367.
8. Shibamoto, T. *J. Agric. Food Chem.* **1977**, *25*, 206-208.
9. Shibamoto, T.; Bernhard, R. A. *J. Agric. Food Chem.* **1978**, *26*, 183-187.
10. Hwang, H. I.; Hartman, T. G.; Rosen, R. T.; Ho, C.-T. *J. Agric. Food Chem.* **1993**, *41*, 2112-2115.
11. Kato, H.; Tsuchida, H. *Prog. Food Nutr. Sci.* **1981**, *5*, 147-156.
12. Izzo, H.; Ho, C.-T. *J. Agric. Food Chem.* **1991**, *39*, 2245-2248.

Chapter 9

Production of Volatile Compounds in Suspension Cell Cultures of *Coriandrum satibrum* L. and *Levisticum officinales*

Hsia-Fen Hsu[1] and Jui-Sen Yang[2,3]

[1]**Food Industry Research and Development Institute, Hsinchu, Taiwan, Republic of China**
[2]**Institute of Marine Biology, National Taiwan Ocean University, Keelung, Taiwan, Republic of China**

Cell cultures of coriander (*Coriander satibrum* L.) and lovage (*Levisticum officinales*) were investigated for the production of compounds useful in medicine and/or flavor. The production of lovage volatile compounds was only achieved in association with the organization of embryoids in the culture. However, 21 and 23 volatile components were identified in the culture media of coriander Cs0 and Cs1, respectively, 12 days after inoculation. Coriander and lovage cells were mixed to stimulate the production of secondary metabolites. Further investigation of cell growth and the metabolite production in mixed cultures is needed.

Lovage (*Levisticum officinalis*) and coriander (*Coriandrum satibrum* L.), Umbelliferae, have been used traditionally as medicine and spice in the Orient. Dried roots of lovage were used for medicine and as a food ingredient without any extraction process. The most important components of lovage for medicine and spice are the volatile compounds lingustilide and butylidene phthalide (*1*). However, it has been difficult to quantify the amount of the dried roots needed for pharmaceutical performance because lovage roots from different sources contained varying amounts of the useful components. Lovages of good quality require delicate culture conditions such as optimum temperature and proper soils. It is not easy to control the quality of natural lovages for medicine.

Coriander, a small herb, was and is used fresh as an important ingredient in Oriental cuisine for its strong, attractive aroma. It was not convenient to use fresh coriander in the food industry because mass production of coriander requires much land for growing it and it is a labor-intensive crop. Postharvest handling of coriander is also difficult.

[3]Corresponding author

0097–6156/96/0637–0097$15.00/0

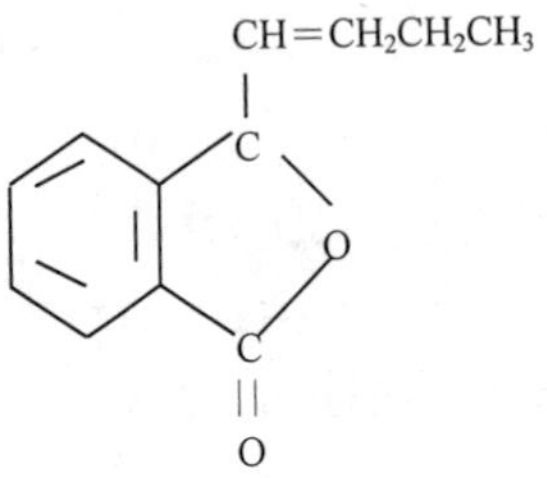

(I) butylidene phthalide

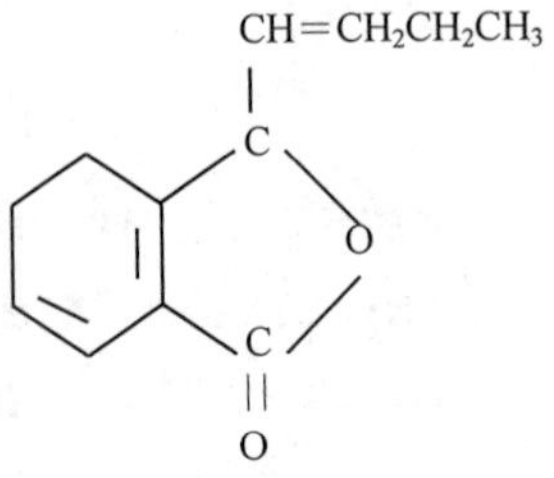

(II) ligustilide

Biotechnology is very interesting to the flavor industry because it has the potential for mass production of important flavoring materials at relatively low costs (*2*). Plant cell culture is an aspect of biotechnology which has been given prominence along with other aspects of this rapidly developing technology. Recently, plant cell culture has become a technique for producing secondary metabolites (*3, 4*) as well as for breeding. In plant cell culture it is essential to establish a stable, high-metabolite-producing cell line to enhance production efficiency (*5*). The composition of the culture medium is very important for cell induction, cell growth and also metabolite formation in plant tissue and cell cultures (*3, 4, 6*).

The aim of the present work is to initiate cell cultures of lovage and coriander to study cell growth and volatile component production. Mixing of cultures of lovage and coriander cells was initiated for studying the stimulation effect of two kinds of cells on metabolite production.

Materials and Methods

Callus cells were induced from petioles of lovage (*Levisticum officinalis*) and coriander (*Coriandrum satibrum* L.) on Murashige and Skoog (MS) medium (*7*) with 20 μM kinetin, 10μM naphthaleneacetic acid (NAA), 2%(w/v) glucose and 0.2%(w/v) phytagel. The medium was adjusted to pH 5.7 prior to autoclaving at 121°C for 15 minutes. The aggregated cells were selected from callus by growth rate, transfered into liquid media and subcultured every three weeks. The liquid medium was prepared without phytagel. The cell lines from lovage and coriander were named YL and Cs, respectively.

Cells cultured in liquid media were separated from the media by filtering them through a stainless steel sieve with a pore size of 1 mm^2. Cells filtered by the steel sieve were then inoculated into various media. The inoculum consisted of 2g fresh weight of the cells per 200 mL medium in a 500 mL flask. Various media on the mineral base of MS medium were supplemented with growth regulators, vitamins, amino acids, carbohydrates and Umbelliferae vegetable juice. Effects of the additives to the media on cell growth and volatile production were studied.

All cultures were carried out at 25°C under 16 hours/day illumination (250 lux). The cell suspension was agitated on a gyratory shaker with a speed of 100 rpm. The cell growth rate was calculated in gram fresh cell weight per gram inoculated cell per day after 21 days.

Two hundred grams of the cultured cells were suspended in 200 mL distilled water and blended for 5 min with a homogenizer. Volatile compounds of the cells were collected with a Likens-Nickerson apparatus (*8*) for 2 hours with an extracting solvent (70 ml of pentane and ether 1:1). The extracts were dried with anhydrous Na_2SO_4 and concentrated with a spinning- band distillation apparatus. A Shimadzu GC-8A gas chromatograph with a Chrompack fused silica capillary column CP-Wax 52CB (0.32 mm X 50 m) was used with hydrogen as carrier gas (flow rate of 1.2 mL/min) in combination with a flame ionization detector (FID). Injector temperature was 250°C. The column was maintained at 50°C for 10 min and then heated to 200°C at a rate of 2°C/min. The final temperature was held for 40 min. A three-microliter aliquot of reference standard solution (2.22 mg n-nonane in 1.5 ml of 1:1 pentane:ether) was added to extracts of the cells as an internal standard. The volatile compounds were further examined using a Hewlett Packard 5985B GC-MS. Operational parameters were as follows: carrier gas, He; ion source temperature, 200°C; electron energy, 70eV; electron multiplier voltage, 2200V. The volatile concentrates from various cultures were compared with each other and with those from natural plants.

In our previous studies the volatile compounds of lovage, as well as lingustilide alone, inhibited the sprouting of *Apium graveolens* L. var. dulce(Mill) pers and *Brassica chinensis* L. (unpublished data). It was suggested that one of the secondary metabolites, lingustilide, produced by lovage, has some role in inhibiting the growth of other plants which grew in the vicinity of lovages. Therefore, in the present work, a mixed culture for inducing the inhibition components, such as lingustilide, was investigated. Lovage cells were cultured together with the coriander cells in a flask. The effect of the mixed culture on the production of the secondary metabolites from lovage cells was investigated.

Results and discussion

The growth rate of L. *officinalis* cells in MS media with 20 μM kinetin and 10 μM NAA increased with an increase in the concentration of ammonium ion, but decreased above the concentration of 40 mM (Figure 1a). The medium with 25 mM nitrate was good for cell growth although the change of nitrate concentration in the range of 4-40 mM did not have any significant effect on cell growth rate (Figure 1b). The optimum concentration of phosphate for cell growth rate was 25 mM (Figure 1c). The color of the cells cultured in liquid media with ammonium, nitrate and mixed-nitrogen source was white, pale - yellow, and yellow, respectively. The cell line named YL-006 was selected from lovage callus for its fast growth .

The addition of some amino acids, such as tyrosine, cysteine, lysine, glutamate and phenylalanine, into the media had no significant effect on cell growth or flavor production in lovage cell YL-006 although 100 μM glycine was generally added into the media (Figure 2). However, volatile compounds were released from YL-006 cells

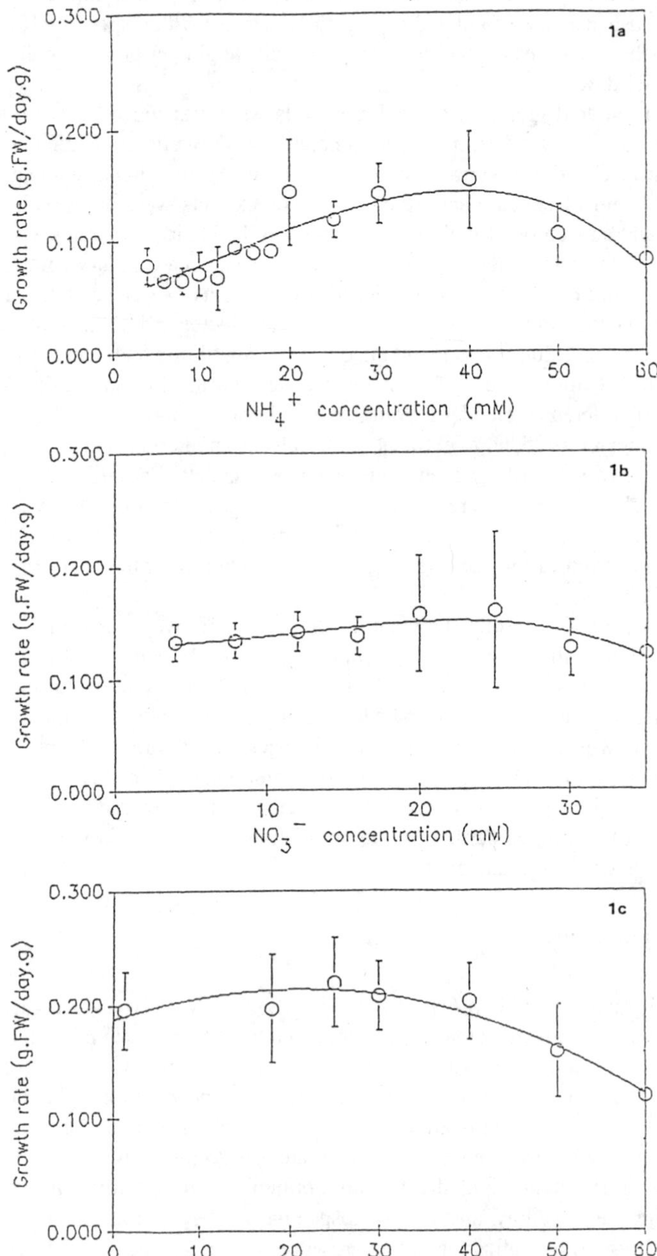

Figure 1. Effects of ammonium (a, top), nitrate (b, middle), and phosphate (c, bottom) concentration on the growth rate of lovage cells.

into the mixed-nitrogen media containing 20 μM kinetin and 10 μM NAA. The flavor compounds were preliminarily identified by gas chromatography. The major components of lovage flavor were not detected in suspension cell cultures although minor compounds appeared.

The embryogenic calli (embryoids) were induced one month after the cells were transferred to the medium with 20 μM kinetin and 0.75 μM NAA. By sensory evaluation lovage flavor was found in the culture media in which embryoids developed and grew. Synthesis of volatile components in lovage was achieved only in association with the organization of embryoids in the lovage culture. There have been many reports of certain compounds produced in association with organization. For example, the synthesis of special metabolites in *Atropa belladonna* and *Lithospermum erythrorhizon* was associated only with the organization of root structures (*9, 10*). Organization seemed to play an important part in the induction of lovage volatile components.

The Cs cells were induced from the petioles of coriander in MS medium with 1.0 μM kinetin and 2.5 μM 2,4-D. The cell lines, Cs0 and Cs1, were selected in the medium with 0.75 μM NAA and 20 μM kinetin. The cells of Cs0 were pale-yellow, aggregated in a diameter of 2 mm and had a doubling time of 6-8 days in growth. The Csl cells were dark-yellow, aggregated in a diameter of 0.5-1 mm and had a doubling time of 4-5 days. During incubation the cells grew fast for two weeks (Figure 3). Two to four days after inoculation flavor components were detected in the media. The amount of the flavor components increased during culture period and after 10-12 days of culturing, many coriander flavor components appeared. Compared with the volatile extract from coriander plant, the extract from culture medium contained several important components of coriander flavor (Figure 4 & Table I). Twenty-one and 23 volatile components were identified by GC-MS in Cs0 and Cs1 culture media, respectively (Table I). Because lovage, celery and coriander are plants belonging to the Umbelliferae family, the volatile extracts of coriander culture media were compared to the volatile composition of lovage and celery. The culture media of Cs0 contained celery volatile components, hexanal and benzaldehyde, and coriander volatile components, octanol and decenal. On the other hand, the media of Cs1 contained celery volatile components, pentyl benzene and alpha-terpineol, and lovage compound, alpha-terpineol. There are also many reports where secondary products are produced in high amounts in undifferentiated cells (*6, 11, 12*). Secondary metabolite synthesis in coriander cells differed from the synthesis in lovage cells.

The possibility of large scale coriander cell culture will be considered for commercial production. The purification process of the volatile components from coriander cell culture media will be a challenge.

The mixed culture of lovage and coriander cells was studied in this work. The mixed culture in the ratio of Cs/YL cells from one to five was investigated for total growth rate. The lowest growth rate occurred in the Cs/YL: 3/1 culture although there was no significant difference among the cultures with various mixed ratios (Figure 5). It was suggested that the highest inhibition in two cell lines occurred in the mixed culture with Cs/YL: 3/1. It was postulated that the Cs/YL: 3/1 culture might produce the highest

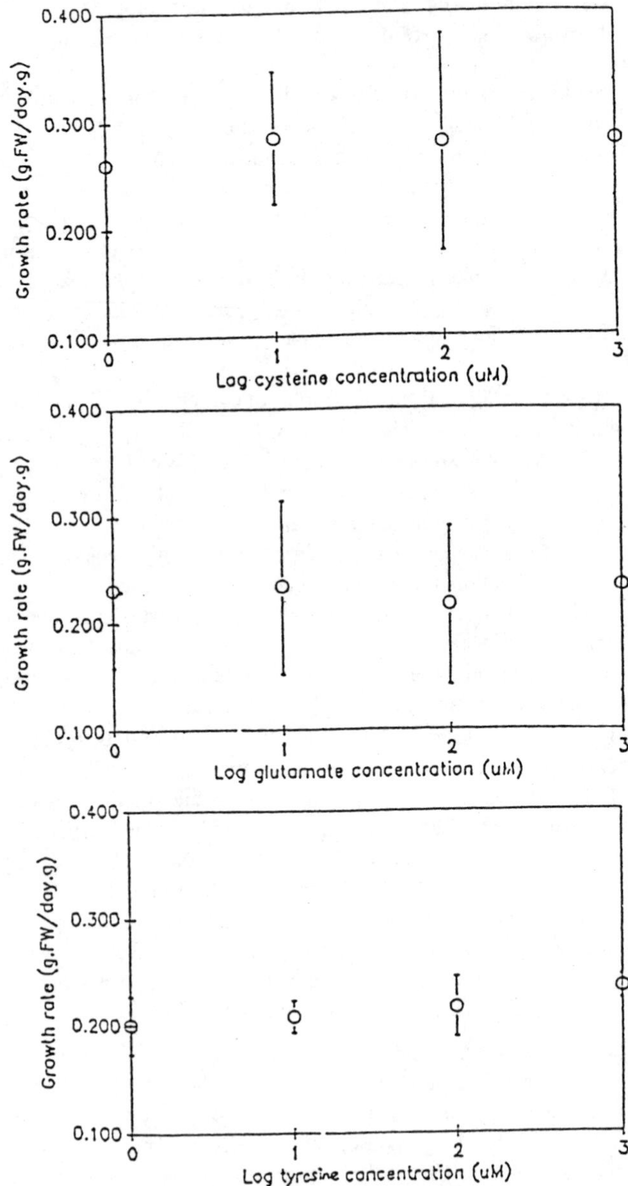

Figure 2. Effects of the concentrations of cysteine, lysine, glutamate, phenylalanine, tyrosine and glycine in culture media on the growth rate of lovage cells.

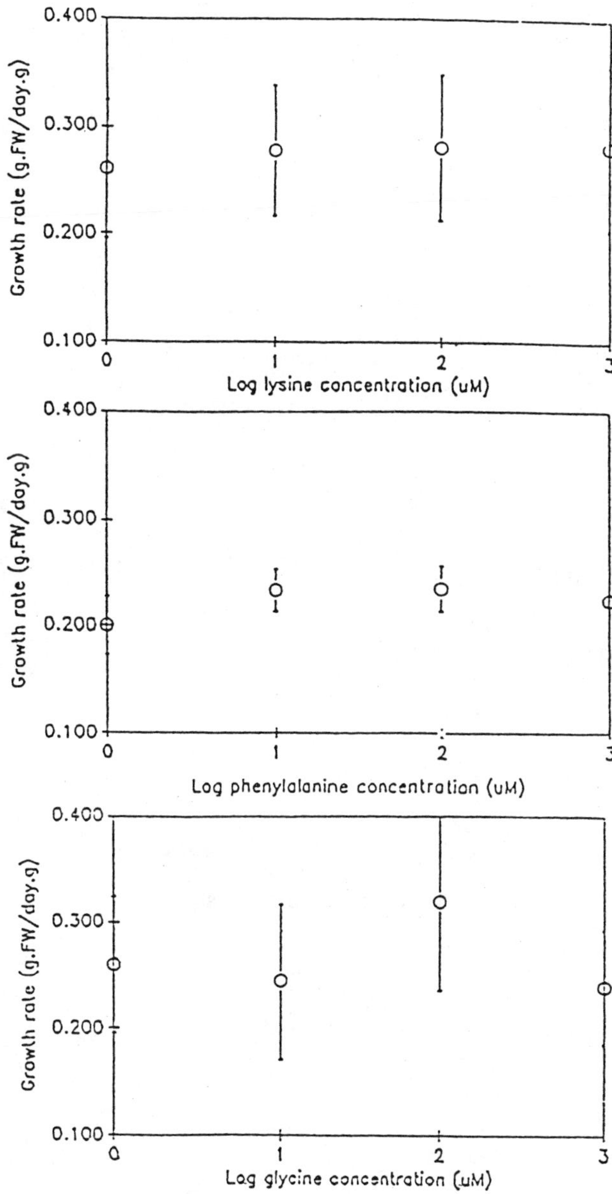

Figure 2. *Continued*

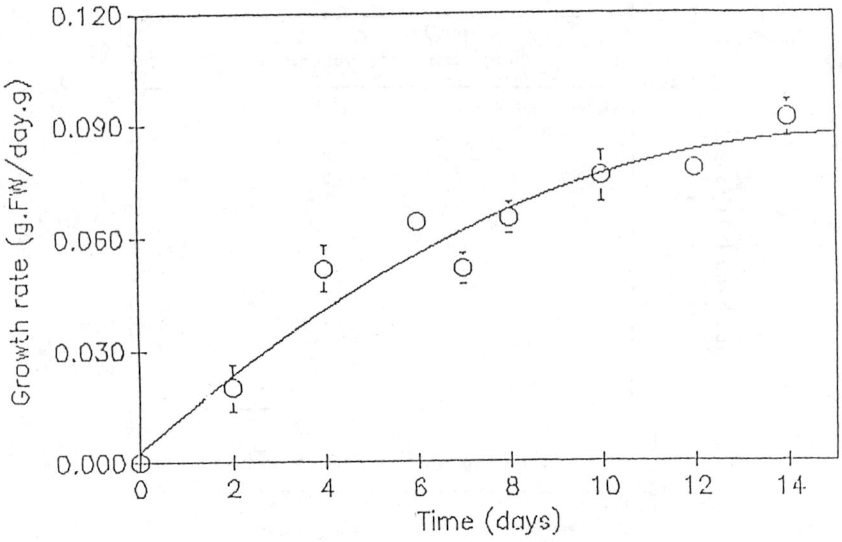

Figure 3. Growth curve of coriander cells.

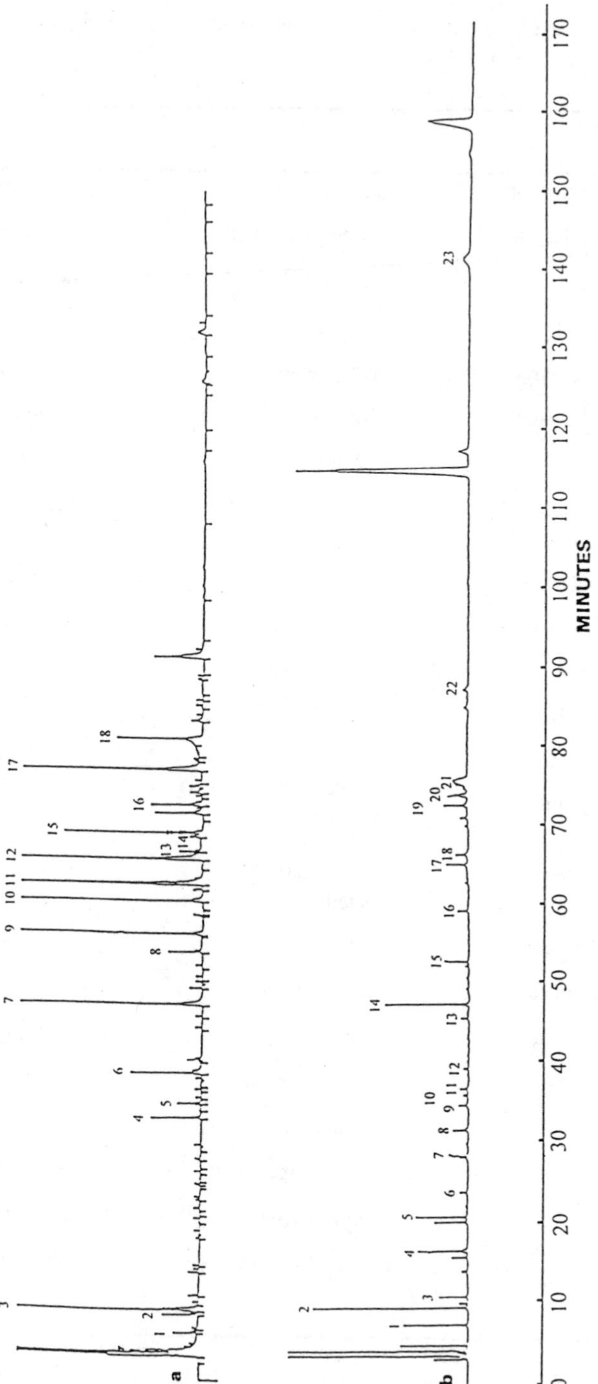

Figure 4. Gas chromatograms of the volatile extracts from (a) coriander plants, and (b) culture medium of coriander.

Table I. Volatile constituents of extracts from culture media of Coriander Cs0 and Cs1 cell lines and plants

peak No.	Plants	Cs0	Csl
1	ethyl formate	ethyl acetate	ethyl formate
2	ethyl acetate	isobutyl aldehyde	ethyl acetate
3	nonane	ethyl alcohol	nonane
4	octanol	2-butanol	isobutyl alcohol
5	cis-hexenyl acetate	1-propanol	(unknown)
6	cis-3-hexen-1-ol	2-methyl-1-propanol	3-methyl-1-butanol
7	decanal	n-butanol	acetoin
8	undecanal	3-hydroxy-2-butanone	pentylbenzene
9	nonanol	furfural	acetic acid
10	dodecanal	(unknown)	furfural
11	undecanol	thujyl alcohol	2-ethyl-1-hexanol
12	trans-2-decenol	trans-2-hexenal	2-methoxy-1-propanol
13	tridecanal	3-cyclohexene-1-methanol	nonanal
14	1-dodecene	3,7-dimethyl-1,6-octadiene	hexadecane
15	2-dodecenal	neral	propylene glycol
16	tetra-decanal	benzylalcohol	ethylene glycol
17	neophytadiene	n-hexanol	furfuryl alcohol
18	trans-2-tridecenal	n-amyl formate	alpha-terpineol
19		1-octanol	tetradecanal
20		2-ethylpiperidine	isovaleric acid
21		benzaldehyde	2,6-octadien-1-ol
22		trans-2-decenal	phenethyl alcohol
23			butyl phthalide

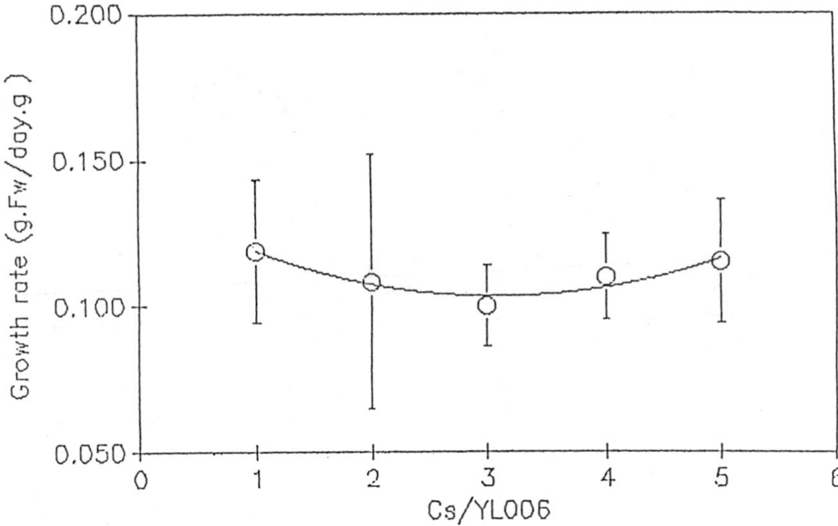

Figure 5. Growth rate of the cells in the mixing culture of coriander and lovage YL-006.

amount of secondary metabolites, such as lingustilide, for an inhibitory effect. Unfortunately, in the mixed cultures coriander cells grew better than lovage cells. Coriander flavor components were produced in mixed cultures but lovage flavor components did not appear. However, cell growth in mixed cultures still remains speculative. Further studies of growth rates of both types of cells and mixed cultures as well as further development of media and culture conditions for the mixed cultures are needed.

Literature Cited

1. Duke, J.A. *Handbook of Medicinal Herbs*; CRC Press, Boca Raton, **1985**; p.391.
2. Wasserman, B.P.; Montville, T.J. In *Biotechnology Challenges for the Flavor and Food Industry*; Lindsay, R.C.; Willis, B.J.; Ed.; Elsevier Applied Science; London and New York, 1989, pp 1-11.
3. Fujita, Y.; Hara, Y.; Ogino, Y.; Suga, C. *Plant Cell Rep.* **1981**, *1*, 59-60.
4. Fujita, Y.; Hara, Y.; Suga, C.; Morimoto, T. *Plant Cell Rep.* **1981**1, *1*, 61-63.
5. Zhong, J.J. *Plant Tissue Cult and Biotech.* **1995**, *1* (2), 75-80.

6. Yamata, Y.; Fujita, Y. In *Handbook of Plant Cell Culture: Techniques for Propagation and Breeding*; Evans, D.A.; Sharp, W.R.; Ammirato, P.V.; Yamata, Y.; Ed.; Macmillan, New York, 1983, pp 717-728.
7. Murashige, Y.; Skoog, *E. Physiol. Plant.* **1962**, *15*, 493-497.
8. Romer, G.; Renner, E. *Z. Lebensm Unters. Forsch.* **1974**, *156*, 329-335.
9. Raj Bhandary, S.B.; Collin, H.A.; Thomas, E.; Street, H.E. *Ann. Bot.* **1969**, *33*, 647-56.
10. Sim, S.J.; Chang, H.H. *Biotech. Lett.* **1993**, *15* (2), 145-150.
11. Brodelius, P. *Hereditas Suppl*. **1985**, *3*, 73-81.
12. Furuya, T.; Yoshikawa, T.; Orihara, Y.; Oda, H. *Planta Med.* **1983**, *48*, 83-87.

Chapter 10

In Vitro Tailoring of Tomatoes To Meet Process and Fresh-Market Standards

M. L. Weaver[1], H. Timm[2], and J. K. Lassegues[1]

[1]Western Regional Research Center, Agricultural Research Service, U.S. Department of Agriculture, 800 Buchanan Street, Albany, CA 94710
[2]Department of Vegetable Crops, University of California, Davis, CA 95610

Expansion of genetic variability, by providing accessability to hidden genetic traits, has been accomplished by development of somaclonal/feedback-inhibition technology. This technology allows processors and breeders to access new genetic traits to improve the fresh-market quality of tomatoes, to minimize the degree of processing and to develop new high-quality products. Through selection of specific chemical and stress environments that challenge the feed-back-inhibition system of plants, over 300 new genetic variants were created with total solids contents of 7.0% to 15.0%. The increase in total-solids content is primarily due to an increase in soluble-solids content. The soluble solids/total solids increased from about 81% to 90% as total solids increased from about 8% to 13%. Insoluble solids decreased from about 22% to 10%. Reducing sugar/soluble-solids, however, remains constant at about 63.5%, over the same range of total solids. Fructose/glucose ranged from 1.25 to 1.13. Other traits of commercial importance: high juice consistency; firmness; different flavors; extended shelf life of 2 to 5 weeks; and different fruit shapes and sizes have been created and isolated in *in-vitro* created variants. To improve the potential of high-solids variants for commercial exploitation other technologies were developed: pollen screening for disease-resistance and environmental-stress tolerance, phytotoxin screening for disease resistance; and nonenzymatic release of pollen protoplasts for direct transfer of genetic traits.

Raw material quality is the key to the development of high-quality fresh-market tomatoes, and to increasing the efficiency and minimizing the degree of processing. It is virtually impossible to minimize processing unless raw-material composition can be standardized and controlled. Creating these desirable changes in raw-

0097–6156/96/0637–0109$15.00/0

material composition requires an expansion of genetic variability. Standard breeding technology, however, often does not allow accessibility to important quality, cultural, and biotic- and abiotic-stress traits even though considerable genetic variability may already exist in the explant tissue. Access to these valuable hidden genetic traits can now be made available through new biotechnology procedures. Induced and spontaneous alterations have been uncovered through somaclonal variations (*1,2*) resulting in several novel and agriculturally useful genetic variants (*3-6*). Somaclonal variation is the result of preexisting genetic traits and cell-culture induced variations. Genes can be uncovered, or modified, by specific chemicals in a culture medium, or they can be expressed in response to biotic or environmental stresses applied to the whole plant, or explants, prior to, or after, cultures have been made. This technology will allow processors to actively participate in the creation and development of raw materials more specifically amendable to meet their particular processing standards.

Creation of High-Solids Content, Tissue-Culture Induced Tomato Variants

A modified somaclonal/feedback-inhibition challenge technology was used to control and direct the *in-vitro* creation of tomato variants. Small leaves taken from 4- to 6-week old plants of well accepted tomato cultivars were sterilized in a 7% Clorox solution for 10 to 15 minutes, and then rinsed 3 times in sterile, glass-distilled water. Leaf rachis were prepared by trimming leaflets to about 0.6 cm on each side of the midrib, cutting off the leaf tip and leaving a small piece of the petiole. The midrib was then cut in half and each cut half was placed with the cut side down on to the culture medium (*7*) in magenta tubes. The medium consisted of MS salts (Gibco); nicotinic acid (1.0 mg/l); thiamine (1.0 mg/l); pyridoxine-HCl (1.0 mg/l); BA (2.5 mg/l); IAA (1.2 mg/l); sucrose (3.0%); myoinositol (100 mg/l); Bacto agar (0.8%). The pH of the mixture was adjusted to 5.8 prior to adding agar. Fifteen ml of medium was dispensed into magenta tubes and autoclaved at 121°C for 25 minutes. To challenge the explant, selected metabolites and antimetabolites of the sugar pathway, various enzyme inhibitors and inorganic chemicals in ascending concentrations were tested. Individual plant were then exposed to heat stress. A specific protocol is now being prepared for patent consideration and cannot be fully described at this time. As a starting point, a concentration that just started to inhibit the formation of shoots was determined. Then a new challenge medium was made starting at this concentration with further increases in the concentration by small increments until all but 1 or 2 of the explants failed to produce shoots. The few explants with shoots were kept in culture until rooted, then transplanted in coil into 10 cm pots until they reached the 3-true-leaf stage, then retransplanted into 10 to 20 liter pots and grown in a greenhouse at 25°C until fruits matured. Fruits were photographed whole and cut in half. A drop of juice was then squeezed onto a differential refractometer (Atago PR-1, NSG Precision Cells, Inc., 195G Central Ave., Farmingdale, N.Y. 11735) to determine the % soluble solids. Fruits with a reading of less than 7.0% were discarded. Seed of fruits with more than 7.0% was saved and stored. Plants generated from these seeds were grown to maturity and then fruit evaluated for soluble-solids content as before. This process was repeated for several generations to test fruit stability. When found to be stable, new explants were made and rechallenged for the creation of new variants. Juice of variants with soluble-solids contents 7.0% were evaluated for total-solids content using a CEM microwave oven (CEM AVC-80, CEM Corporation, 3100 Smith Farm Road, Mathews, N.C. 28105). The difference between total solids and soluble solids was expressed as insoluble-solids content.

Evaluation of High-Solids Tomato Variants

Over 300 high-solids, tissue culture variants were created with total-solids contents from 7.0 to 15.0%. As the total solids increased, soluble solids also increased and insoluble solids decreased (Table I).

Table I. Change in Soluble-Solids/Total Solids and Insoluble Solids/Total Solids as Tomato Variants Increase in Total-Solids Content

Total Solids	*Soluble Solids (%)*	*Insoluble Solids (%)*	*Sol. Solids (%) / Tot. Solids (%)*	*Insol. Solids (%) / Total Solids (%)*
7.8	6.3	1.5	80.8	19.2
8.8	7.3	1.5	83.0	17.0
9.7	8.2	1.5	84.5	15.5
10.8	9.2	1.6	85.2	14.8
11.5	10.2	1.3	88.7	11.3
12.7	11.4	1.3	89.8	10.2

As the soluble-solids content increased from 6.3% to 11.4%, soluble-solids/total-solids increased from about 81% to 90%, and the insoluble-solids/total-solids decreased from about 19% to 10%. Although a high soluble-solids content can result in an increase in the pounds of tomato paste, or sauce, per ton of tomatoes, this does not always mean a big advantage for processors that produce high-consistency products. Often the recovery is greater from thick-juice cultivars that express the highest insoluble-solids content.

High-Solids Content and Reducing-Sugar Content in Tomato Variants

Solids content continues to be a major selection factor in tomato breeding evaluations. As solids content increases, so does the sugar content. Using a Waters HPLC system equipped with Millenium 3.1 software, the reducing-sugar contents of hundreds of high-solids, tissue-culture variants were determined. Twenty ml of tomato juice was extracted in boiling 80% ethanol for 1 h in a crude-fiber extracter. The extract was centrifuged at 600 rpm for 15 min, then filtered twice through a 0.22μm filter to assure all sediment was removed. Five ml of the clarified extract was then run through a Sep Pak-C18 column. Triplicate 10μl samples were then injected into the HPLC. The parameters used were: flow rate of 3.8 cm^3/min at a column temperature of 35°C; attenuation 4X; run time 15 min using a Waters 510 pump; detector was a Waters 410 differential refractometer. A carbohydrate 3.9 X 300 mm column was used for separation. The mobile phase was acetonitrile/water (85:15, v/v). Only curves for fructose and glucose were detected and peak height was used for quantitation. The sum of peak heights of fructose and glucose, determined from standard curves, was used to calculate the percentage of reducing sugars in the total soluble solids. While soluble solids increased as a percentage of total solids (Table I), the reducing-sugar content, as a percentage of soluble solids, tended to remain fairly constant in variants from 7.0% to over 11% soluble solids (Table II).

Table II. Relationship of Reducing Sugar Content to Soluble-Solid Content in High-Solids Tomato Variants

Total Soluble Solids (%)	*Reducing Sugar / Total Soluble Solids (%)*	*Fructose (%)*	*Glucose (%)*	*Fructose / Glucose*
7.0	62.5	2.5	2.0	1.25
8.4	61.9	2.8	2.4	1.17
9.6	62.5	3.2	2.8	1.14
10.0	64.0	3.4	3.0	1.13
11.6	64.7	4.0	3.5	1.14

The average reducing sugar/soluble solids content was 63.5% The average ratio of fructose/glucose was 1.17. Both values compare with those of standard commercial tomato cultivars (*8*).

Other Important Genetic Variants Created. A wide range of fruit sizes and shapes varying from 15-g cherry tomatoes to large 500-g fresh-market variants and 65 to 125 g processing types in elongated, round, and oval shapes were created. Variants with almost solid interiors and with very small locule areas were evaluated for dehydration as dices and slices. Dried products maintained a better shape, and rehydrated to a more natural piece form, than did those from commercial low-solids cultivars. Most variants were extremely firm when harvested fully ripe and could be held at room temperature (21°-25°C) for 2 to 5 weeks without softening and shriveling. With few exceptions, variants had pH values at commercially acceptable levels of 4.0 to 4.4. High-solids variants were very sweet with several distinctly different flavors, especially in cherry types. Juice consistency of some variants was so thick it had to be scraped off the pulper and could not be poured from the beaker. High-solids cherry-tomato variants created from high-solids variants had a firm processing-type interior, not the typical soft interior of most commercial cherry tomatoes.

Field Trial of High-Solids Variants. In cooperation with tomato breeders from industry, several high-solids variants were grown in a commercial field and their fruit was compared with a high-solids commercial cultivar, UC204C (Table III). Fruit of most hybrid crosses among high-solids variants were as large or larger than the commercial standard. The pH of all variants was at, or below, 4.4 and all were lower than that of the commercial standard. All hybrids were higher in soluble solids than the commercial standard. It is usually accepted that as fruit size increases solids content decreases. In this study however, the largest fruit had the highest solids content, 45% more than the standard tomato cultivar. Also, there is usually a negative relationship between high-solids content and juice consistency. Bostwick readings indicated that with fruit of many high-solids variants this relationship may not hold true. While all hybrids had more soluble solids than the standard, all but three had better fruit consistency (lower Bostwick readings) than the control.

Table III. Field Evaluation of Hybrids Made From U.S.D.A. High-Solids Hybrids

Hybrid	*Fruits /lb*	*pH*	*Brix (%)*	*Bostwick (cm/30 sec)*
1	6.6	4.40	7.7	13.5
2	8.0	4.33	7.9	20.5
3	6.6	4.26	7.8	15.0
4	7.3	4.41	7.4	12.0
5	5.3	4.38	9.3	24.0
6	6.2	4.30	7.2	17.0
7	10.0	4.34	7.1	24.0
8	7.3	4.36	8.3	14.5
9	9.0	4.30	7.2	15.0
10	7.3	4.38	7.0	15.0
UC204C (Avg. of 3 trials)	7.2	4.60	6.4	20.60

Other Technologies to Improve the Commercial Potential of High-Solids Tomatoes

To assist in enhancement of the commercial potential of high-solids variants, work on other important genetic traits is needed. Other somaclonal variations, (both desirable and undesirable) can occur along with high-solids variations. Desirable cultural traits such as: optimal foliar cover; plant size; and determinate and indeterminate growth are now being incorporated into high-solids variants. These high-solids variants will also have to be screened for environmental stresses and disease resistance. With modern advances in biotechnology now used to uncover hidden genetic traits, and to transfer these traits to plants of the same and other diverse species, rapid and accurate screening strategies must be developed to identify stress response in new variants.

Pollen Screening for Heat Tolerance. Since reproductive organs are very sensitive to diseases and environmental stresses (*9,10*), pollen activity was chosen as an appropriate screening tool. Pollen germination and tube growth are representative of the growth of the male gametophyte and have been correlated with the growth and function of the sporophyte (*11,12*). Pollen screening has the added advantage of being able to treat hundreds of pollen grains in a single petri dish and reading results in 1 to 3 hours. Pollen can be exposed to many stress factors using little space and low-cost equipment, when compared to expensive greenhouse facilities, or utilizing field studies. Under controlled laboratory conditions pollen can easily be germinated and tube growth initiated in simple and defined media of any time of the year. Pollen function has been used to screen for salt tolerance (*13*), metal tolerance (*14*), and high-temperature tolerance (*15,16*). Using a heat-tolerance, pollen-screening test developed at WRRC (*16*), tomato hybrids created by crossing the heat-tolerant (HT) Grivorski cultivar with U.S.D.A. high-solids variants were exposed to 50 C for 2 h. Data showed (Table IV) that high-temperature tolerance was present in all hybrids in varying degrees.

Table IV. Pollen Screening for Heat Tolerance of High-Solids Tomatoes Created from Grivorski and Exposed Two Hours at 50° C.

Tomato Selection	*Germination (%)*	*Pollen Tube Length (μm)*
Grivorski	78	2950
Hybrid 1	80	3050
Hybrid 2	53	1500
Hybrid 3	82	3100
Hybrid 4	22	240
79N45	0	0

A tomato selection, 79N45, obtained from the Department of Vegetable Crops, University of California, Davis, showed no pollen viability after two hours at 50° C, which is the normal response of pollen of most commercial tomatoes.

Pollen Screening for Disease Resistance Using Phytotoxins. A liquid shake culture was made using a 4:1 (V-8 juice: glass distilled water) medium. The pH was adjusted to 5.8, and the medium sterilized at 121°C for 15 min. Two hundred and fifty ml of medium in 500 ml Erlenmeyer flasks was inoculated with Fusarium and Verticillium races supplied by Dr. Ken Kimball, Moran Seed Co., Davis, CA, and shaken for 7 days at 24°C. The medium was then centrifuged at 1000 rpm to separate out the mass of spores and filaments. The filtrate was run through a 0.45μm filter and 0.25 ml was applied by pipette to several locations on a solid-agar pollen-germination media (*16*) and held overnight. Pollen from different tomato variants and hybrids was shaken onto the areas where toxin had been applied. After 3-6 h, at 100% relative humidity, the areas of agar medium where pollen was applied were removed by cork borer and placed on a microscope slide. A drop of phyloxin-methylgreen dye (*16*) was applied to the agar. A cover glass was placed on the agar and the slide was placed on a heated hot plate. As soon as the agar began to melt the slide was quickly removed. The slide was then microscopically evaluated to determine the percentage of pollen germination and to measure the length of pollen tube growth.

Table V. Pollen Screening of High-Solids Variants and Hybrids Made Among High-Solids Variants for Disease Resistance

Tomato Selection	*Fusarium oxysporum Pollen Germination (%)*	*Verticillum albo-atrium Pollen Tube Length (μm)*
Variant 1	65	2
Variant 2	5	35
Variant 3	72	80
Variant 4	82	85
Hybrid 1	15	78
Hybrid 2	5	10
Hybrid 3	67	25
Ace	8	15
Ace 55	76	82

In cooperation with industry tomato breeders, pollen-screening tests for disease resistance were initiated using commercial cultivars and breeding selections with known disease resistance. These tests indicated that a relationship existed between pollen-screening tests and standard root-inoculation tests. High-solids variants, or hybrids, were not evaluated at that time because their disease resistance was not known. Based on these studies, preliminary pollen-screening tests were conducted on high-solids tomato variants and hybrids among them (Table V). The data showed that there were differences in pollen function of both variants and hybrids in response to toxins from both Fusarium and Verticillium cultures. Pollen of two cultivars, 'Ace' known to be susceptible and 'Ace55' resistant to both disease organisms, showed a positive relationship between pollen function and known disease resistance. As soon as storage tests on the filament/spore mass are completed the mass will be ground and used for root inoculations of seedlings of the same variants and hybrids to determne their disease response and compare it with that of pollen function response to the phytotoxin.

Other Methodologies Developed to Create and Commercialize New High-Quality Tomato Variants. When this study to tailor raw material to improve the quality of tomatoes was initiated, it was believed that pollen protoplasts could be used to accept new genetic traits to create and transfer genes from one plant source to another. Pollen is basically a protoplast covered with a thick cell wall. Once the cell wall is removed, DNA can enter the protoplast by direct uptake (*19*), or by electroporation into the pollen tube (*20*). The obstacle to using pollen protoplasts is the cell wall which is composed of sporopollenin, one of the most impervious natural materials known. This material cannot be disintegrated by any of the common cell-wall degrading enzymes without causing protoplasm disfunction. While searching for alternative methods of protoplast release, it was found (*18*) that viable pollen protoplasts of beans could be removed in less than five minutes using dilute salt solutions. By using a solidified agar medium with different mineral contents, protoplasts can be released from the pollen of tomato, pepper, zucchini, peas, alfalfa, cucumber (*21*). Since pollen tubes are basically extended protoplasts they can also be used to take up DNA. By varying the concentration and type of mineral, primarily calcium, in the protoplast-release medium, it was possible to stop the growth of the pollen tube just as it emerged from the cell wall creating a partial protoplast. By holding the partial protoplast in a DNA solution followed by pollination of the stigma with the partial protoplast, or by insertion of the DNA-impregnated partial protoplast directly into the ovary some fertilization occurred, but as yet the transfer of genetic traits has not been definitely confirmed. The procedures are continually being modified.

Literature Cited

1. Evans, D.A.; Sharp; W.R.; Medina-Filko, H.P. *Amer. J. Bot.* **1984**, *71*, 759-774.
2. Larkin, P.S.; Scowcraft, W.R. *Theor. Appl. Genet.* **1981**, *60*, 197-214.
3. Brettelli, R.I.; Ingram, D.S. *Biol. Rev.* **1979**, *54*, 329-345.
4. Heinz, D.J. In *Mutation breeding of vegetatively propagated perennial crops;* GAO/AEP: Vienna, **1972**, pp. 53-59.
5. Chaleff, R.S. *Science* **1983**, *219*, 676-682.
6. Nickel, L.G. *Crop Sci.* **1977**, *17*, 717-719.
7. McCormick, S.; Niedermeyer, J.; Fry, J.; Barnason, A.; Horsch, R.; Fraley, R. *Plant Cell Reports* **1986**, *5*, 81-84.
8. Kadar, A.A.; Morris, L.L.; Stevens, M.A.; Albright-Holen, M. *J. Amer. Soc. Hort. Sci.* **1978**, *103*, 6-13.

9. El Ahmadi, A.B.; Stevens, M.A. *J. Amer. Soc. Hort. Sci.* **1979**, *104*, 691-696.
10. Sugizama, T.; Iwahor, S.; Takahashi, K. *Acta. Hort.* **1966**, *4*, 63-69.
11. Tanksley, S.D.; Zamir, D.; Rick, C.M. *Science* **1981**, *213*, 453-455.
12. Mulcahy, D.L. *Science* **1971**, *171*, 1155-1156.
13. Sacher, R.F.; Mulcahy, D.L. *Plant Physiol.* **1981**, *67*, 67-96.
14. Searcy, K.B.; Mulcahy, D.L. *Amer. J. Bot.* **1985**, *72*, 1695-1699.
15. Weaver, M. L.; Timm, H.; Silibernagel, M. J.; Burke, D. W. *HortScience* **1985**, *110*, 797-799.
16. Weaver, M. L.; Timm, H. *HortScience* **1989**, *24,* 493-495.
17. Owczarak, A. *Stain Technology* **1952**, *27*, 249-251.
18. Weaver, M. L.; Timm H.; Breda, V.; Gaffield, W. *J. Amer. Soc. Hort. Sci.* **1990**, *115*, 640-643.
19. Kanji, O. O.; Gamborg, I.; Miller, R. A. *Can. J. Bot.* **1972**, pp. 2077-2080.
20. Href, A.; Abdul-Baki, J. A.; Saunders, B. F.; Pillarelli, E.A. *Plant Sci.* **1980**, *7*, 181-190.
21. Weaver, M. L. 1992. Methods to obtain intact viable protoplasts from pollen grains. U.S. Patent #5, 169, 777.

Enzyme and Microbial Transformations

Chapter 11

Generation of Flavors by Microorganisms and Enzymes: An Overview

Karl-Heinz Engel and Irmgard Roling

Lehrstuhl für Allgemeine Lebensmitteltechnologie, Technische Universität München, D–85350 Freising-Weihenstephan, Germany

The generation of individual flavor compounds or complex flavor mixtures by the use of microorganisms and enzymes, the topic of subsequent symposium contributions, is summarized. The two basic approaches, *de novo*-synthesis in the course of microbial fermentations and biotransformations of suitable precursors are outlined. The increased compositional and structural knowledge of the substrates/precursors needed and the tailor-made design of the microorganisms/enzymes employed are presented as bases for strategies to optimize the biogeneration of flavors.

Fermentation is one of the original traditional biotechnological methods for preservation of foods. This primary purpose is frequently accompanied by the formation of typical flavors. For centuries, the microorganisms and enzymes involved have been employed almost unwittingly. Modern biotechnology makes use of the increasing knowledge of the underlying scientific principles and is starting to exploit the advantages offered by biocatalysts in a more specific way. The incorporation of biotechnological steps in the manufacture of high fructose corn syrup, the production of sweeteners, organic acids, vitamins or amino acids has been well established (*1*). Due to the advances in microbial fermentation and enzyme technology, individual flavor compounds or complex flavor mixtures, examples of low-volume but high-value products, are increasingly becoming targets for production on an industrial scale (*2*). The exploitation of such techniques is especially attractive, because flavors or flavor compounds obtained via biotechnology are considered to be natural by regulatory authorities in many countries, as long as certain conditions, such as the natural origin of the raw material, have been met. Biotechnological processes leading to flavor production can be divided into two major groups: *de novo*-synthesis in the course of microbial fermentation (*3*) and biotransformations/bioconversions of suitable precursors either by microorganisms or by enzymes (*4,5*). Recent research developments in these areas are described in the subsequent symposium contributions.

0097–6156/96/0637–0120$15.00/0

Non-volatile precursors. An essential prerequisite for optimum flavor generation by microorganisms or enzymes is detailed knowledge of the composition and the availability of the substrates/precursors needed. Flavor precursors have been objects of intensive studies (*6,7*). A class for which tremendous progress in knowledge of composition and distribution has been achieved recently is the glycoconjugated precursors. Structures of numerous non-volatile, flavorless glycosides of monoterpenes, *nor*isoprenoids, and shikimic acid metabolites present in fruits, wines, and some vegetable products have been elucidated (*8*). Aroma liberation can result from either acid- or enzyme-catalyzed hydrolysis. The application of suitable hydrolytic enzymes and enzyme preparations, respectively, has been reported (*9*). Knowledge acquired of the precursors available in certain fruits, combined with the possibilities offered by modern enzyme technology, e.g. the tailoring of specific biocatalysts, will open new dimensions to influence the release of bound aroma compounds.

Another class of compounds well known as non-volatile flavor precursors are unsaturated fatty acids. The lipoxygenase-catalyzed biogeneration of aroma active C_6 and C_9 aldehydes and alcohols from C_{18} polyunsaturated fatty acids is an important mechanism well studied in plant systems (*10,11*). An analogous process, the oxygenase-catalyzed conversion of fatty acids to oxylipins by diverse marine life, such as algae, has emerged as an exciting new trend (*12*). The biogeneration of flavor compounds from these structurally unique oxylipins might reveal a source of unexplored metabolic activity.

Biotransformations/Bioconversions. The efficiency of microbial *de novo* synthesis of aroma compounds can be increased by offering suitable precursors which serve as starting points for biotransformations (single step reactions) or bioconversions (multi-step reactions) yielding the desired products. Natural sources containing the required precursors in high amounts can be used directly as substrates. An exemplary approach is the fermentative production of (R)-γ-decalactone from castor oil by *Candida lipolytica*. This process makes use of the fact that ricinoleic acid, the precursor metabolized via β-oxidation, represents 90% of the triglyceride fatty acids of castor oil (*13*). Alternatively, single compounds isolated from abundant natural sources, e.g. terpenes from essential oils or the above mentioned C_6 aldehydes and alcohols from plant tissues, can be subjected to highly specific microbially catalyzed reaction sequences.

Enzymes. The use of enzymes is an integral part of many important processes in food production. Hydrolytic enzymes especially are employed on an industrial scale, mainly because no costly regeneration of cofactors is required, in contrast to oxidoreductases. The release of specific fatty acid profiles by lipases in the course of cheese manufacture, or the cleavage by proteases of peptide fragments in protein hydrolyzates that otherwise will cause bitterness are examples for the impact of enzyme-catalyzed reactions on the final flavor of foods (*5*).

The outstanding features of enzymatic reactions, e.g. high substrate specificity even in complex matrices, high reaction specificity, mild reaction conditions and reduction in waste product formation, are also of importance in the synthesis of single flavor compounds. Two additional factors have boosted the application of enzyme technology in the synthesis of flavor substances: (i) the stability of enzymes (lipases, proteases) in organic solvents allows the catalysis of reactions which are not feasible in aqueous medium, e.g. esterifications, transesterifications and lactonizations, thus providing access to a broad spectrum of important volatiles (*5*); and (ii) the increasing knowledge of the influence of absolute configuration on the flavor properties of chiral compounds and analytical progress in the determination of naturally occurring enantiomeric compositions, which have increased the need for biocatalyzed reactions resulting in the "correct" enantiomer. Enzyme-catalyzed biotransformations of prochiral substrates as well as kinetic resolutions of racemic precursors can also be applied (*14,15*).

Recombinant DNA techniques. Mutagenesis and selection techniques based on classical bacteriological and genetic methods are common procedures to optimize and standardize microorganisms used in food fermentations. Recombinant DNA techniques offer the potential of altering the properties of microorganisms more precisely in terms of production efficiency, product quality, safety, and diversity (*16*). A broad spectrum of recombinant microorganisms is available for industrial and agricultural applications (*17*). Recombinant DNA techniques are applied in such traditional areas as sake and beer brewing (*18*). A flavor-related example is the reduction of the amount of diacetyl, one of the major off-flavors in beer. The construction of a brewer's yeast containing a bacterial acetolactate carboxylase gene has been described. This yeast has the ability to convert acetolactate, the precursor of diacetyl, to acetoin which has no impact on beer flavor (*19,20*). Due to public controversy about recombinant DNA techniques, the flavor industry has been reluctant to make use of genetically modified organisms, but such methods will definitely become more common in the future.

An area at the forefront of commercial applications of genetic engineering is the production of enzymes from genetically modified organisms. The milk-clotting protease, chymosin, has been the first food ingredient produced via recombinant DNA techniques to be cleared for food use (*21*). The use of designed enzymes adjusted to specific process requirements will also provide new possibilities in the field of flavors.

Literature Cited

1. Cheetham, P.S.J. *Chemistry & Industry* **1995**, 265-268.
2. Berger, R.G. *Aroma Biotechnology*, Springer Verlag: Berlin, Germany, 1995.
3. Mizutani, S.; Hasegawa, T. *Perfumer & Flavorist* **1990**, *15*, 265-268.
4. Gatfield, I.L. *Perfumer & Flavorist* **1995**, *20*, 5-14.
5. Christen, P.; Lopez-Munguia, A. *Food Biotechnology* **1994**, *8*, 167-190.
6. *Flavor Precursors: Thermal and Enzymatic Conversions*; Teranishi, R.; Takeoka, G.R.; Güntert, M., Eds.; ACS Symposium Series 490; American Chemical Society: Washington, D.C., 1992.
7. *Progress in Flavour Precursor Studies*; Schreier, P.; Winterhalter, P., Eds.; Allured: Carol Stream, Illinois, 1993.

8. Williams, P.J.; Sefton, M.A.; Marinos, V.A. In *Recent Developments in Flavor and Fragrance Chemistry;* Hopp, R.; Mori, K., Eds.; VCH Verlagsgesellschaft: Weinheim, Germany, 1993.
9. Gunata, Z,; Dugelay, J.; Sapis, J.C.; Baumes, R.; Bayonove, C. In *Progress in Flavour Precursor Studies*; Schreier, P.; Winterhalter, P., Eds.; Allured: Carol Stream, Illinois, 1993, pp 219-234.
10. Hatanaka, A. *Phytochemistry* **1993**, *34*, 1201-1218.
11. Winterhalter, P.; Schreier, P. In *Flavor Science*; Acree, T.E.; Teranishi, R., Eds.; American Chemical Society: Washington, D.C., 1993, pp 225-258.
12. Gerwick, W.H. *Biochim. Biophys. Acta* **1994**, *1211*, 243-255.
13. Gatfield, I.L., Sommer, H. In *Recent Developments in Flavor and Fragrance Chemistry;* Hopp, R.; Mori, K., Eds.; VCH Verlagsgesellschaft: Weinheim, Germany, 1993, pp 291-304.
14. Schreier, P. In *Progress in Flavour Precursor Studies*; Schreier, P.; Winterhalter, P., Eds.; Allured: Carol Stream, Illinois, 1993, pp 45-61.
15. Engel, K.-H. In *Flavor Precursors: Thermal and Enzymatic Conversions*; Teranishi, R.; Takeoka, G.R.; Güntert, M., Eds.; ACS Symposium Series 490; American Chemical Society: Washington, D.C., 1992, pp 21-31.
16. Geisen. R.; Stander, L.; Leistner, L. *Food Biotechnol.* **1990**, *4*, 497-504.
17. *Recombinant Microbes for Industrial and Agricultural Applications*; Muraoka, Y.; Imanaka, I.; Eds.; Marcel Dekker: New York, New York, 1993.
18. *Biotechnology & Genetic Engineering Reviews*; Tombs, M.P., Ed.; Intercept: Andover, 1991.
19. Vogel, J.; Wackerbauer, K.; Stahl, U. In *Genetically Modified Foods: Safety Aspects*; Engel, K.-H.; Takeoka, G.R.; Teranishi, R., Eds.; ACS Symposium Series 605; Americal Chemical Society: Washington, D.C., 1995, pp 160-170.
20. Takahashi, R.; Kawasaki, M.; Sone, H.; Yamano, S. In *Genetically Modified Foods: Safety Aspects*; Engel, K.-H.; Takeoka, G.R.; Teranishi, R., Eds.; ACS Symposium Series 605; Americal Chemical Society: Washington, D.C., 1995, pp 171-180.
21. Flamm E.L. *Bio/Technology* **1994**, *12*, 152-155.

Chapter 12

Sensory Analysis and Quantitative Determination of Grape Glycosides

The Contribution of These Data to Winemaking and Viticulture

Patrick J. Williams and I. Leigh Francis

Australian Wine Research Institute, P.O. Box 197, and Cooperative Research Centre for Viticulture, P.O. Box 145, Glen Osmond, South Australia 5064, Australia

The contribution that glycosidically-bound volatiles make to varietal wine aroma was determined by sensory descriptive analyses. Sensory studies were made on grape glycosides a) hydrolyzed under accelerated conditions i) *in vitro* and back added to wine, and ii) *in situ* in wine; b) hydrolyzed under conditions of natural aging. These studies confirmed the role of glycoside hydrolysis in the expression of varietal wine aroma. The sensory data have lent support to the development of a method for quantifying total glycosides in wine grapes, thus giving an indication of juice 'richness'. This assay, made through a determination of the glycosyl-glucose, offers the possibility of an objective pre-harvest measure of grape, and hence potential wine quality.

Recognition of the presence of glycosidically-conjugated flavor precursors in fruits of all major horticultural classes has, in the last few years, stimulated much interest in these compounds. Advances in research on glycosidic flavor precursors of plant-derived foods has been the subject of several recent reviews *(1-4)*.

Grapes and wines were among the earliest products to be investigated in this field with research centering initially on monoterpene glycosides; this research helped in elucidating the role of monoterpenes as flavor compounds of floral grape varieties *(5)*. The subsequent recognition of glycosides of C_{13} norisoprenoid compounds and of shikimic acid-derived metabolites as precursors of non-floral grape flavor, was a later development *(6)*. Further aspects of the involvement of glycosides in the flavor of grapes and wines have been recently discussed *(7-9)*.

New developments in glycoside research of grapes and wines are: a) the use of formal sensory descriptive analysis to investigate the precise role of glycosides in flavor expression, and b) the possibility of grape and wine quality evaluation through the quantification of glycosides. These developments are discussed here.

0097–6156/96/0637–0124$15.00/0

Sensory Studies on Grape Glycosides Hydrolyzed *in vitro* and Back-Added to Wine.

The first studies in this series determined the sensory effects of accelerated hydrolysis of grape glycosides, *in vitro*, for three white wine varieties, ie Semillon, Sauvignon Blanc and Chardonnay *(10)* and the black grape, Shiraz *(11)*. The primary aim of these works was to determine, by sensory analysis, whether the isolated grape glycosides on hydrolysis and back addition to a wine medium, produced an aroma that was related to the aroma of wines made from the varieties. It was found that glycosides from each of the varieties, on acid hydrolysis gave aromas that had sensory properties common to wines made from the same juices. While the hydrolysates from the three white varieties showed several shared aroma attributes each was, nevertheless, distinctive. By direct comparison of the sensory properties of the hydrolysates with wines for three of the varieties *ie* Chardonnay, Semillon *(10)*, and Shiraz*(11)* it was evident that glycosidic flavor precursors contribute to important aroma attributes in these wines. In the case of Shiraz, it was further shown that precursor hydrolysates contain aroma compounds that are important to high quality wines of the variety *(11)*.

Sensory Studies on Glycosides Hydrolyzed *in situ*.

Having established that glycosides can contribute to wine flavor, the next series of experiments examined processing steps that could be used to accelerate glycoside hydrolysis in white winemaking. A high temperature-short time treatment of either juices or wines produced no discernible effect on wine aroma, as measured by duo-trio difference tests. However, storage of wine protected from air at 45°C for several weeks gave wines with aromas clearly different from the controls. Sensory descriptive analysis showed that this thermal treatment produced wines with complex aromas, similar in character to the aroma properties of wines that had been cellar stored for many months *(12)*. Headspace gas chromatographic studies showed that there were significant increases in the amounts of norisoprenoid compounds as a result of the treatment. Headspace analysis also confirmed that the thermal conditions used were sufficient to hydrolyze glycosides and release the same norisoprenoids *(13)*. Furthermore, statistical treatment of the instrumental and sensory data indicated that the concentration of several of the norisoprenoids in the wine headspace was significantly correlated with aroma descriptors from the descriptive analysis study *(14)*. These experiments thus showed that rapid hydrolysis of glycosides *in situ* can be effected by heat treatment, and that this induces desirable flavor changes.

The Effect on White Wine Aroma of Altering the Concentration of Glycosides in Juice.

It could be argued that the conditions of accelerated hydrolysis employed in the *in vitro* and *in situ* hydrolysis experiments described above were far removed from those experienced by a wine during vinification and natural aging. For example, an increased temperature may alter the balance of volatiles released by acid catalyzed hydrolysis compared to the pattern of products that would be obtained at typical cellar

Table I. Summary of juice[a] treatments for the preparation of wines made with and without glycosides

Code	*Treatment*		*Control*	
	Description	Conditions	Description	Conditions
A	Semillon XAD-2 treated	Semillon juice (21 L) passed through a glass column packed with XAD-2 resin (5.0 kg)	Semillon untreated	Semillon juice (21 L) passed through a glass column packed with glass wool.
B	Thompson Seedless with added Semillon glycosidic isolate	After Treatment A the XAD-2 resin was washed with water (3 x 1 L), eluted with ethanol (1.5 L); the glycosidic eluate was evaporated to near dryness in vacuo, extracted with Freon 11, and taken up in 25% v/v aqueous ethanol (240 ml), which was added to Thompson Seedless juice (21 L).	Thompson Seedless untreated	Thompson Seedless juice (21 L) to which was added 25% v/v aqueous ethanol (240 ml)
C	Semillon with added Semillon glycosidic isolate	As for treatment B, with the glycosidic eluate in 25% aqueous ethanol (240 mL) added to Semillon juice (21 L).	Semillon untreated	Semillon juice (21 L) to which was added 25% v/v aqueous ethanol (240 ml).

[a]All juices were obtained from a commercial winery and filtered before use.

temperatures. Also, the base wine medium may have influenced the aroma of the samples presented for sensory analysis in the *in vitro* experiment. Importantly, neither experiment examined any possible sensory effects of fermentation on the glycosides. Whilst the effects of fermentation on glycosylated constituents in juice is unknown, studies monitoring the concentration of monoterpene glycosides before and after fermentation, indicate that yeast do not assimilate the glycosides *(8, 15)*.

To explore these issues, small-scale fermentations were carried out with Semillon and Thompson Seedless juices that had been treated so that the glycoside concentration in the juices was altered. Sensory analysis of the samples was performed to determine what effect these treatments had on the aroma of the finished wines.

Experiments to Determine the Aroma of Wines Made With and Without Glycosides.

Small-scale fermentations were carried out with Semillon juice that had been treated using XAD-2 resin so that the glycosidic fraction was either absent, intact, or augmented *(14)*. In addition, a Thompson Seedless juice was fermented with and without added Semillon glycosides. Table I gives a summary of the juice treatments. Experiment A consisted of intact Semillon juice as a control, with a volume of Semillon juice that had been passed through a bed of XAD-2 resin as the treatment. The effectiveness of the XAD-2 resin in removing glycosides from the juice in this experiment is described below. The treatments in experiments B and C were the appropriate juice with a quantity of Semillon glycosidic isolate added, so that the treated Thompson Seedless juice in B contained Semillon glycosides at single strength, and the treated Semillon juice in C contained glycosides at double strength.

For each experiment, the controls and treated juices were divided into three replicate lots for fermentation. Malolactic fermentation did not proceed in any of the treatment or control samples except for one replicate for treatment B which showed a malic acid decrease of 20%. After fermentation, the samples were bottled and cellar stored at 18±4 °C.

Difference Testing at Different Stages of Cellar Storage. To determine whether any flavor change for the wines had occurred as a result of the juice treatments, duo-trio aroma difference tests were done between control replicates and treatment replicates after 6, 12, 20, and 27 months cellar storage. The first three series of comparisons were done between randomly selected pairs of control and treatment replicates.

The results of the difference tests indicated that after 12 months aging in the bottle there was no significant difference in aroma between wines made from the control juices and wines made from the treated juices. Testing of the wines after 20 months storage, and also after 27 months storage – when 45 out of a possible 48 combinations of pairs of control and treatment samples were compared – showed that for each of the three experiments there were significant differences in aroma between the control and treated wines.

To assess the effect of the juice treatments on wine flavor, sensory descriptive analysis was undertaken by a trained panel of 11 judges, scoring a subset of the samples from these experiments in duplicate. There was insufficient volume of sample to allow all of the treatments to be assessed, and so only two wines were analysed, ie the treatment wine from experiment A (made from Semillon juice that had been treated by XAD-2 resin to remove the glycosides), and the treatment wine from experiment C (made from Semillon juice with added glycoside isolate) (see Table I).

From Figure 1, which shows a plot of the mean descriptive analysis data for the two wines, it is clear that differences between the two were considerable. The wine that had been fermented from juice treated so as to have its glycosides removed, was rated as lower in all attributes than the augmented sample, except for the attribute grassy. In particular, the attributes lime, honey, oak, and toasty were significantly enhanced for the latter sample. Importantly, these were characterizing features shown in the descriptive analysis studies of Semillon hydrolyzed glycosides in the *in vitro* and *in situ* studies and also of bottled-aged Semillon wines *(14)*. A significant difference between these samples was also observed for the floral attribute, however judge inconsistency for this attribute, as measured by an analysis of variance, meant that the validity of this conclusion was doubtful.

This work demonstrated that increasing or decreasing the concentration of glycosides of juice affects wine aroma after an aging period. Because the hydrolysis of glycosides occurred under natural conditions in this study, the sensory effects that were observed are likely to be the same as those that would occur in a Semillon wine during conventional vinification and storage. In the earlier *in vitro* and *in situ* studies it was possible that, due to differences in activation energy among the glycosides and their aglycons, hydrolysis at elevated temperatures may have produced volatile compounds in proportions different from that obtained in hydrolysis at cellar temperature, which may have been largely responsible for sensory effects observed. From this study, in which the variable of accelerated hydrolysis is removed, it is evident that the glycosides alone are major contributors to aroma. It is further evident that a period of cellar storage provides time for slow acid-catalysed hydrolysis of the glycosides to occur resulting in the release of volatiles and the expression of varietal aroma. Observation of this delayed expression also suggest that the grape glycosides that release aroma volatiles apparently pass through the fermentation unchanged, without being either degraded or hydrolyzed by the fermentation process.

Quantification of the Glycosylated Secondary Metabolites Through a Determination of the Glycosyl-Glucose (G-G).

In formally establishing the link between the aroma characteristics of hydrolyzed grape glycosides and wine aroma, this sensory research has given strong support to the results of earlier studies into the chemical composition of the volatiles released from the glycosides *(16-20)*. Together these sensory and compositional studies show the importance of glycosylated secondary metabolites of grapes to potential wine flavor, thus to wine quality and, by implication, to grape quality. It is logical, therefore, that an analysis of the glycosylated secondary metabolites, could give an objective measure of quality. This concept means that, in parallel with research that is necessary to investigate individual flavor molecules and to develop methods to quantify each one of these and its precursor(s), some more rapid method of analysis of the total glycosylated flavor precursors is appropriate.

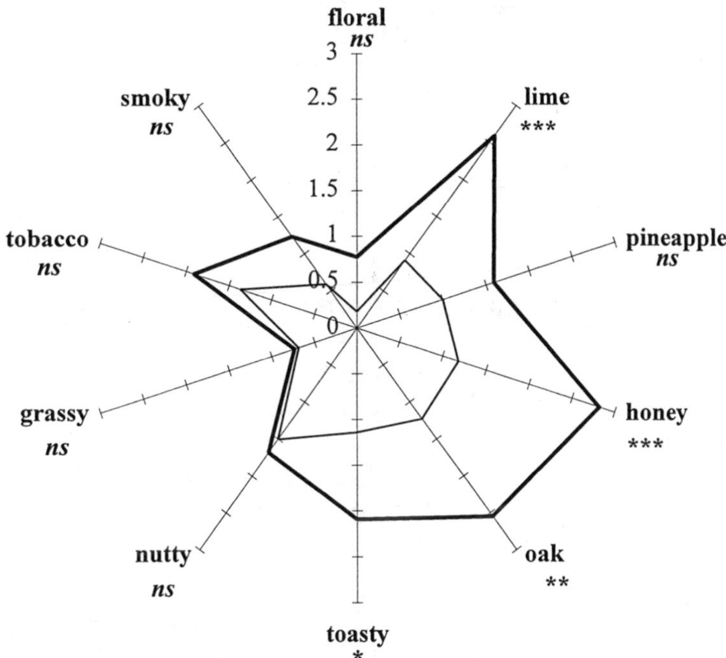

Figure 1. Sensory profile plot of mean aroma intensity ratings (n=11 judges x 2 replicates) for attributes of a wine made from Semillon juice treated with XAD resin ——, and a wine made from Semillon juice with glycoside isolate added ▬▬ . Also shown are significance levels where ns, *, **, *** indicate not significant, and significant at $p<0.05$, 0.01, 0.001 respectively.

When considering the hundreds of glycosidically-bound flavor compounds that are present in grapes, it is evident that all of them are glucosides, with the glucose moiety usually further substituted, ie, to give a disaccharide. Hydrolysis of grape flavor precursors, therefore, yields an equimolar proportion of aglycons and D-glucose. Based on this, a determination of the glycosyl-glucose (G-G) concentration will give an estimation of the total concentration of glucosylated secondary metabolites present in the fruit including, of course, those difficult-to-measure flavor compounds *(21)*. This approach of quantifying the G-G complements an ion chromatographic method for the determination of the monosaccharides released from grape glycosides *(22)*.

A G-G assay has now been developed that accurately and precisely measures the total concentration of glucosides in grapes and wines *(23)* (see Figure 2). Application of the assay has allowed observation of the sharp increase that occurs in G-G with fruit ripening and the range of G-G in juices of a single variety grown in different regions *(23)*. As part of the sensory work described above, the assay was

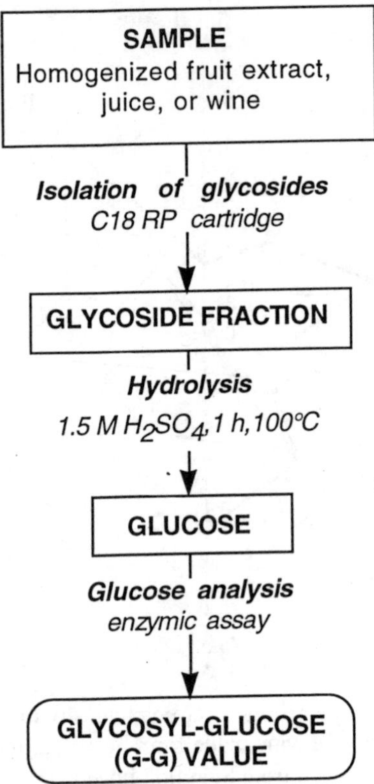

Figure 2. Schematic of the G-G assay

used to determine changes in the glycoside concentration of wines and juices resulting from the treatments applied. A strong negative correlation was found for wines' storage period and glycoside concentration, and a decrease in G-G was also observed as a result of thermal treatment *(14)*. From the experiments described above to determine the effect on white wine aroma of altering the concentration of glycosides in juice, it was found that the untreated Semillon juice contained 310 μM G-G while passage through the column of XAD-2 resin left the treated juice with a G-G concentration of 20 μM. This confirmed that XAD-2 treatment *(24)* removed more that 90% of the juice glycosides under the conditions of this experiment.

The assumption that G-G as a measure of total glucosidically bound secondary metabolite concentration can give an index of quality must yet be tested, but it is logical and rationally based. With regard to the free aroma compounds, it is evident that the concentration of these, ie aglycons, is highly variable *(19, 20)* and, in any case, the presence of free aroma constituents affects the flavor status of the fruit at harvest. The sensory studies discussed above show that hydrolysis of grape glycosides is a major contributor to varietal expression, thus the G-G is a measure of the potential of a batch of fruit to give varietal aroma on maturation of wine.

To date most G-G analyses have been applied in pre-harvest studies on the red winemaking variety Shiraz *(25)*. This work on black grapes has demonstrated the feasibility of looking beneath the total G-G at the red-free G-G which is obtained by subtracting the quite large contribution to glycosyl-glucose coming from anthocyanins *(26)*. Determination of red-free G-G has allowed observation, for the first time, of the sharp development in secondary metabolites in Shiraz fruit in the latter stages of soluble solids accumulation. The rate of formation of these important compounds with degrees Brix, and possibly also the time of maturity at which the development accelerates, are influenced by irrigation as well as by the growing season *(25)*. These characteristics are observable only with this analytical method and no other measure of fruit composition, including colour, reveals these trends.

It is evident that only with extensive application by both the viticulture and wine industries will the practical utility of G-G be tested. A far greater body of G-G data on a range of different varieties, all grown under different conditions is needed. To aid this the assay procedure should be automated to provide the mass of information to test the practical use of the G-G parameter. Post-harvest applications must also be explored with monitoring of G-G of musts through vinification, and of wines undergoing secondary fermentation, and then maturation.

Conclusion.

Results from the sensory studies described here confirm that glycosides in grapes act as a source of varietal aroma which is expressed after a period of wine aging. The impact of continued hydrolysis during prolonged storage is now clear from most recent sensory studies on grape glycosides and wines of the variety Semillon (Francis I.L., Tate M.E., Williams P.J., The Australian Wine Research Institute, submitted for publication).

In establishing the significance of glycosides through sensory analyses, this work has consolidated earlier research into the chemical composition of grape glycosides and laid a foundation for future studies to address some long-standing viticultural and enological challenges. Thus, pre-harvest prediction of grape quality and the suitability of grapes for particular wine styles are just some of the applications that the simple glycosyl-glucose assay could meet. However, further sensory analysis, together with more detailed compositional data, into the nature of glycosidic precursors are clearly required. The identity of compounds responsible for specific flavor attributes of particular grape varieties, and ways for viticulturists and winemakers to manipulate grape flavor, may be forthcoming from such work.

Acknowledgments.

We thank Wies Cynkar and Mariola Kwiatkowski for the G-G analyses and the Orlando Wyndham Group Pty Ltd for generously donating grape juice. The staff and students of the AWRI who participated in the sensory panels are also thanked. Financial support was provided by the Grape and Wine Research and Development Corporation.

Literature Cited.

1. Stahl-Biskup, E.; Intert, F.; Holthuijzen, J.; Stengele, M.; Schultz, G. *Flavour Fragr. J.* **1993**, *8*, 61-80.
2. Williams, P. J. In *Flavor Science: Sensible Principles and Techniques*; Acree, T. E.; Teranishi, R., Eds.; ACS Professional Reference Book Series; American Chemical Society: Washington DC, **1993**; pp 287-303.
3. Williams, P. J.; Sefton, M. A.; Marinos, V. A. In *Recent Developments In Flavor and Fragrance Chemistry, Proceedings of the 3rd International Haarmann and Reimer Symposium;* Hopp, R.; Mori, K., Eds.;VCH: Weinheim **1993**; pp 283-290.
4. Winterhalter, P.; Schreier, P. *Flavour Fragr. J.* **1994**, *9*, 281-287.
5. Strauss, C. R.; Wilson, B.; Gooley, P. R.; Williams, P. J. In *Biogeneration of Aromas;* Parliament, T.H.; Croteau, R., Eds.; ACS Symposium Series No. 317; American Chemical Society: Washington DC, **1986**; pp 222-242.
6. Williams, P. J.; Sefton, M. A.; Wilson, B. In *Flavor Chemistry, Trends and Developments.*; Teranishi, R.; Buttery, R. G.; Shahidi, F., Eds.; ACS Symposium Series No. 388; American Chemical Society: Washington DC, **1989**; pp 35-48.
7. Williams, P. J.; Allen, M. S. In *Analysis of Fruits and Nuts*; Linskens, H. F.; Jackson, J. F., Eds.; Modern Methods of Plant Analysis Vol 18; Springer-Verlag: Berlin, **1995**; pp 37-57.
8. Park, S. K.; Noble, A. C. In *Beer and Wine Production: Analysis, Characterisation and Technological Advances*; Gump, B., Ed.; ACS Symposium Series No. 536; American Chemical Society: Washington DC, **1993**; pp 98-109.
9. Noble, A. C. In *Understanding Natural Flavors*; Piggott, J. R.; Paterson, A., Eds.; Blackie Academic and Professional: Glasgow, **1994**; pp 228-242.
10. Francis, I. L.; Sefton, M. A.; Williams, P. J. *J. Sci. Food Agric.* **1992**, *59*, 511-520.
11. Abbott, N. A.; Coombe, B. G.; Williams, P. J. *Am. J. Enol. Vitic* **1991**, *42*, 167-174.
12. Francis, I. L.; Sefton, M. A.; Williams, P. J. *Am. J. Enol. Vitic* **1994**, *45*, 243-251.
13. Leino, M.; Francis, I. L.; Kallio, H.; Williams, P. J. *Z Lebensm Unters Forsch* **1993**, *197*, 29-33.
14. Francis, I. L. The Role of Glycosidically-Bound Volatile Compounds in White Wine Flavour. PhD Thesis, The University of Adelaide, **1994**.
15. Gunata, Y. Z.; Bayonove, C. L.; Baumes, R. L.; Cordonnier, R. E. *Am. J. Enol. Vitic.* **1986**, *37*, 112-114.
16. Versini, G.; Della Serra, A.; Dell'Eva, M.; Scienza, A.; Rapp, A. In *Bioflavor 87;* Schreier, P. Ed.; de Gruyter: Berlin, **1988**; pp 161-170.
17. Versini, G.; Della Serra, A.; Monetti, A.; De Micheli, L.; Mattivi, F. In *Connaissance aromatique des cépages et qualité des vins;* Bayonove, C.; Crouzet, J.; Flanzy, C.; Martin, J. C.; Sapis, J. C., Eds.; Montpellier, **1993**; pp 12-19.

18. Razungles, A.; Gunata, Z.; Pinatel, S.; Baumes, R.; Bayonove, C. *Sci. Aliments* **1993**, *13*, 59-72.
19. Sefton, M. A.; Francis, I. L.; Williams, P. J. *Am. J. Enol. Vitic.* **1993**, *44*, 359-370.
20. Sefton, M. A.; Francis, I. L.; Williams, P. J. *J. Food Sci.* **1994**, *59*, 142-147.
21. Abbott, N. A.; Williams, P. J.; Coombe, B. C. In *Proceedings of the Eighth Australian Wine Industry Technical Conference*; Stockley, C. S.; Jonnstone, R. S.; Leske, P. A.; Lee, T. H., Eds.; Winetitles: Adelaide, **1993**; pp 72-75.
22. Pastore, P.; Lavagnini, I.; Versini, G. *J. Chromatogr.* **1993**, *634*, 47-56.
23. Williams, P. J.; Cynkar, W.; Francis, I. L.; Gray, J. D.; Iland, P.; Coombe, B. G. *J. Agric. Food Chem.* **1995**, *43*, 121-128.
24. Gunata, Y. Z.; Bayonove, C. L.; Baumes, R. L.; Cordonnier, R. E. *J. Chromatogr.* **1985**, *331*, 83-90.
25. McCarthy, M. G.; Iland, P. G.; Coombe, B. G.; Williams, P. J. In *Proceedings of the Ninth Australian Wine Industry Technical Conference*; Stockley, C. S.; Johnstone, R. S.; Sas, A. N.; Lee, T. H., Eds.; Winetitles: Adelaide, **1996**; pp 141-148.
26. Iland, P. G.; Gawel, R.; McCarthy, M. G.; Botting, D. G.; Giddings, J.; Coombe, B. G.; Williams, P. J. In *Proceedings of the Ninth Australian Wine Industry Technical Conference*; Stockley, C. S.; Johnstone, R. S.; Sas, A. N.; Lee, T. H., Eds.; Winetitles: Adelaide, **1996**; pp 133-140.

Chapter 13

Chimeric β-Glucosidases with Increased Heat Stability

Kiyoshi Hayashi, Ajay Singh, Chika Aoyagi, Atsushi Nakatani, Ken Tokuyasu, and Yutaka Kashiwagi

Enzyme Applications Laboratory, National Food Research Institute, 1–2–1 Kannondai, Tsukuba, Ibaraki 305, Japan

In order to improve thermal stability of the ß-glucosidases from *Cellvibrio gilvus,* chimeric enzymes were constructed. Two homologous domains were found between *C. gilvus* and *Agrobacterium tumefaciens* ß-glucosidase. Since the active center of the enzyme locates at the N-terminal domain, the C-terminal domain, showing 40 % homology, was selected to construct chimeric enzymes. Four chimeric enzymes of CHSTY, CHBSA, CHAGE and CHBSM possessing 8%, 22%, 30% and 39% of the amino acid sequence of *A. tumefaciens* ß-glucosidase at the C-terminal were prepared by shuffling the gene. The properties of four chimeric enzymes to pH and temperature were an admixture of the two parental enzymes. It is especially interesting that the thermal stability of the chimeric enzymes was increased 9-16 °C when compared to that of the *C. gilvus* enzyme. Regarding the substrate specificity of the four chimeric enzymes, they are more similar to *A. tumefaciens* than the *C. gilvus* enzyme. This result suggests that enzyme character can be improved by constructing chimeric enzymes.

Several kinds of enzymes have been used in many industries including food industries because enzymes catalyze the reaction at very moderate conditions and the reaction is very specific. Since enzymes are so useful and important, searching for a new enzyme with required characteristics such as higher heat stability, broader substrate specificity, *etc.* is inevitable. Altering some character of the currently used enzyme is extremely difficult. Immobilization of enzymes can sometimes stabilize enzymes but this is not applicable for all cases (*1*). Chemical modification of the enzyme also does not help so much to improve the enzyme character (*2*). There have been no appropriate methods to improve a character of the enzyme.

The ordinal way to find a new enzyme is to carry out screening for the enzymes, since enormous varieties of enzymes have been created during the long time span of evolution (*3*). During the long process of evolution of 3 to 4 billion years, point mutation and gene shuffling occurred in the genes of enzymes. By point mutation in the gene, only one amino acid will be altered, sometime resulting suitable character for living things. By gene shuffling, a wider region of the gene has been exchanged, resulting in drastic alteration of the enzyme character. Of course most of

0097–6156/96/0637–0134$15.00/0

these changes occur very slowly during evolution. As a result, varieties of enzymes with different amino acid sequences, that is different in character, have been created. Screening for a new enzyme is to find a suitable enzyme which is created naturally during this long process.

Because of the advance in biotechnology, enzyme genes can be manipulated in laboratories. Enzymes possessing one or two different amino acids can easily be produced by point mutation of the gene. However, changing only one or few amino acids is not sufficient to improve the enzyme character, though it is easy to inactivate the enzyme by changing one amino acid in the active site (*4*). Gene shuffling will be a good answer to create a new enzyme of desired character (*5-7*). More than 40,000 thousands of available sequences of the protein can be used for designing chimeric enzymes. In this paper, we examined the improvement of heat stability of a enzyme by preparation of chimeric enzymes by employing heat sensitive ß-glucosidase produced by *Cellvibrio gilvus* (*8*) and a heat resistant one from *Agrobacterium tumefaciens* (*9*).

Preparation of Chimeric ß-glucosidase

ß-glucosidase of *C. gilvus*. Cellobiose rather than glucose accumulated when *C. gilvus* ATCC 13127 was cultivated in a medium containing acid swollen cellulose (*10*). Then, accumulated cellobiose has been considered to be utilized in the cell by hydrolysing it to glucose and glucose-1-phosphate by cellobiose phosphorylase (EC 2.4.1.20) (*11-13*). It was found that this unique accumulation of the disaccharide has been caused by a specificity of ß-glucosidase (EC 3.2.1.21) of this strain; substantially lower activity toward cellobiose than cellotriose, cellotetraose, cellopentaose and cellohexaose (*8*). This ß-glucosidase can be regarded as a suitable enzyme for production of cellobiose which is used as non-dietary sugar. However, heat stability of this enzyme is not high enough for the production of disaccharide.

Homology Analysis. *C. gilvus*' ß-glucosidase which belongs to the BGB group of ß-glucosidases (*14-17*) share conserved regions with ß-glucosidases from different organisms including bacteria, yeast and fungi, as shown in Table I. It was also found that the homologous region can be separated into two domains. Among eleven homologous proteins two enzymes produced by *Butyrivibrio fibrisovens* and *Ruminococcus albus* showed that the two homologous domains are inverted as shown in Fig. 1, indicating that these domains may have independent functions. Highest homology scores in the amino acid sequence were obtained between *C. gilvus* ß-glucosidase and *A. tumefaciens* ß-glucosidase (*9*). In particular, the region from Ala-541 to Pro-811 of ß-glucosidase from *A. tumefaciens* is quite similar (about 40%) to the region from Ala-472 to Pro-741 of ß-glucosidase from *C. gilvus* as shown in Fig. 2.

Importance of C-domain. Since deletion of more than 70 bp fragment from the C-terminal part of *C. gilvus* ß-glucosidase gene resulted in the loss of enzyme activity (T.T. Hoa and K. Hayashi, unpublished), the C-terminal region seems to be important for ß-glucosidase activity. Although, the deletion of about 100 amino acid residues near the C-terminal region of α-amylase gene did not affect enzyme activity (*18*), cyclomaltodextrin glucanotransferases lacking 30 amino acids (*19*) and an endoglucanase lacking 75 amino acids (*20*) from the C-terminal end showed no enzyme activity. Keeping in view the importance of the C-terminal region and the location of estimated catalytic center at Asp-291 in the N-terminal region of *C. gilvus* (*16*), the C-terminal region was selected for the construction of chimeric enzymes.

Table I. Homology with *C.gilvus* ß-glucosidases

Microorganisms	Homology Score
Agrobacterium tumefaciens (Bacterium)	1013
Clostridium thermocellum (Bacterium)	839
Kluyveromyces fragilis (Yeast)	784
Saccharomycopsis fibligera B (Yeast)	710
Saccharomycopsis fibligera A (Yeast)	706
Butyrivibrio fibrisovens (Bacterium)	610
Hansenula anomala (Yeast)	596
Ruminococcus albus (Bacterium)	582
Schizophyllum commune (Fungi)	208
Aspergillus wentii (Fungi)	101

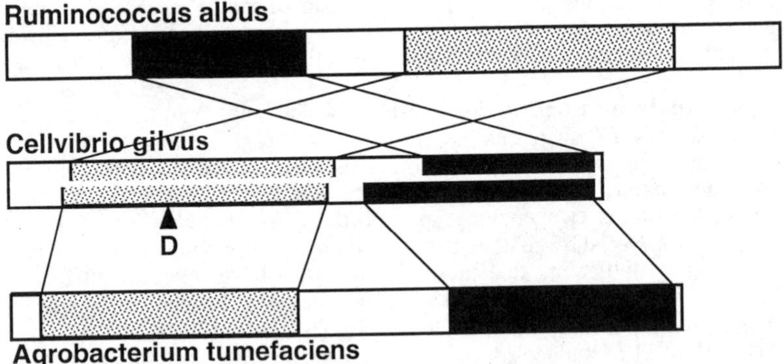

Figure 1. Homologous regions in amino acid sequences of ß-glucosidases from *R. albus, C.gilvus* and *A.tumefaciens*.
Two homologous regions were found in N- terminal (Shaded) and C-terminal (Darkly shaded) regions of *C. Gilvus* ß-glucosidase genes. The two homologous domains were inverted in case of *Ruminococcus albus* ß-glucosidase genes.

```
                                  ——CHBSM—
CG YPVGGIAVKGLLPATWPGPVVYYPSSPLRAIQAQAPNAKVVFDDGRDPARAARVAAGADV  480
                                                ... .** .. . .
AT  AVTLGAARRYRVVVEYEAPKASLDGINICALRFGVEKPLGDAGIAEAVETARKSDIVLL  549
                                 ——CHBSM—

                                      ——CHAGE—
CG ALVFANQWIGEANDAQTLALPDGQEELITSVAGANGRTVVVLQTGGPVTMPWLARVPAVL  540
    .  ...* .*. * ... **. *****..**..* ..*********..****..*.***
AT LVGREGEWDTEGLDLPDMRLPGRQEELIEAVAETNPNVVVVLQTGGPIEMPWLGKVRAVL  608
                                ——CHBSA—       ——CHAGE—
CG EAWYPGTSGGEAIANVLFGAVNPSGHLPATFPQSEQQLPRPKLDGDPKNPELQFAVDYHE  600
    . **** . *.*.*.****.*.*.*.**.***..  .  ..  *. .  *. .  * * *
AT QMWYPGQELGNALADVLFGDVEPAGRLPQTFPKALTD-NSAITDDPSIYPGQDGHVRYAE  667
                                                      ——CHBSA—
CG GAAVGYKWFDLKGHKPLFPFGHGLSYTTFAYSG--LSG-QLKDGRLHVRFKVTNTGNVAG  657
    *  ***.  * .. .****** **.** *....  *** .. .. * *  .***.*. **
AT GIFVGYRHHDTREIEPLFPFGFGLGYTRFTWGAPQLSGTEMGADGLTVTVDVTNIGDRAG  727
                          ——CHSTY—
CG KDVPQVYAAPMSTKWEAP-KRLAAWSKVALLPGETKEVEVAVEPRVLAMFDEKSRTWRRP  717
   .** *.*. . ... * * * * *..*. * **.* .. . ..** ** ** ..  .*
AT SDVVQLYVHSPNARVERPFKELRAFAKLKLAPGATGTAVLKIAPRDLAYFDVEAGRFRAD  787
                                    ——CHSTY—

CG KGKIRLTLAEDASAANATSVTVELPASTLDARGRAR                         752
    **  *..*..* . .* **...**.. .
AT AGKYELIVAASAIDIRA-SVSIHLPVDHVMEP                             818
```

Figure 2. Homology of the amino acid sequences in C-terminal region of *C. Gilvus* and *A. tumefaciens* ß-glucosidases.
Identical and similar amino acid residues are designated by (*) and (.), respectively. Four chimeric enzymes were constructed by shuffling the regions marked by arrow heads.

Construction of Four Chimeric Enzymes. Considering the translation frame and similar regions of both genes, four regions in *C. gilvus* ß-glucosidase gene were selected for substitution with *A. tumefaciens* ß-glucosidase gene. The structures of the four chimeric ß-glucosidase genes are shown in Figure 3. In order to obtain the chimeric enzyme gene of CHSTY where 8% of amino acid sequence are originating from *A. tumefaciens* ß-glucosidase, the region between two AvaI site starting from Pro-759 over stop codon in the plasmid of pcbg1 coding *A. tumefaciens* ß-glucosidase (*9*) were substituted at StyI site in the plasmid of pCG5 coding *C. gilvus* ß-glucosidase (*21*). Three other chimeric enzymes were similarly prepared by substituting the region between SfiI and HinfI sites starting from Asp-660 in pcbg1 at BsaBI site of pCG5 (CHBSA, 22% amino acid sequence originating from *A. tumefaciens* ß-glucosidase), the region between two AvaII sites starting from Ile-594 in pcbg1 at AgeI site of pCG5 (CHAGE, 30% amino acid sequence originating from *A. tumefaciens* ß-glucosidase), and the region between NdeI and HinfI sites starting from Cys-517 in pcbg1 at

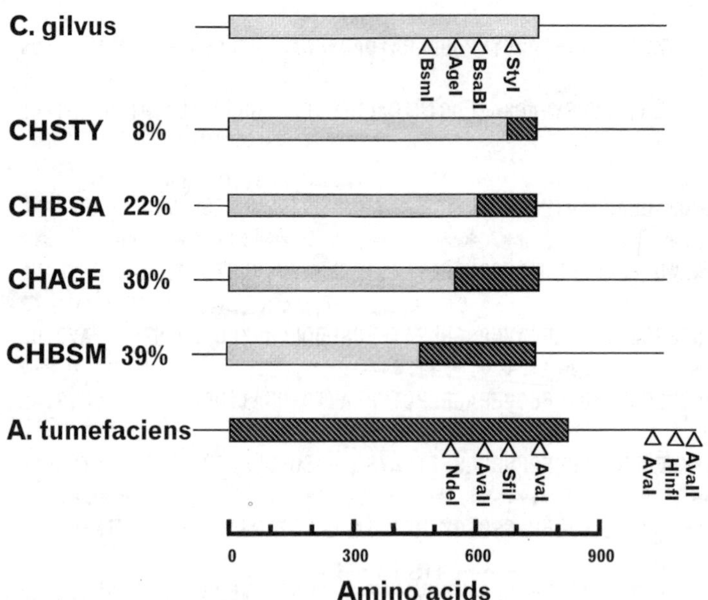

Figure 3. Construction of four chimeric ß-glucosidase genes. Light and dark shadow represent regions derived from *C. gilvus* and *A. tumefaciens*, respectively. Restriction enzymes used for the construction of chimeric enzymes are shown with arrowheads.

BsmI site of pCG5 (CHBSM, 39% amino acid sequence originating from *A. tumefaciens* ß-glucosidase) (*22-23*). The clones expressing chimeric enzymes were isolated on agar plates containing ampicillin (50 µg/ml) and fluorescent substrates of 4-methylumbelliferyl-ß-glucoside (1 mM). Four chimeric plasmids were characterized and confirmed by restriction enzyme digestion.

Characterization of Chimeric ß-glucosidases

Protein Behavior in Column Chromatography. Distinctive differences in the behavior of ß-glucosidases from *C. gilvus* and *A. tumefaciens* have been observed during the purification step using ion-exchange chromatography. Cation-exchange column (SP Sepharose) was used at pH 5.0 to elute ß-glucosidases from *C. gilvus*, whereas anion-exchange column (Q Sepharose) was used to elute *A. tumefaciens'* ß-glucosidases at pH 6.5. All four chimeric ß-glucosidases were found to show similar behavior to that of *C. gilvus*' ß-glucosidase in ion exchange chromatography, indicating that outer surface of the four chimeric enzymes is close to *C. gilvus* enzyme.

Effect of pH on the Chimeric Enzyme Activity. The pH optima for *C. gilvus* and *A. tumefaciens* enzymes were found to be 6.2-6.4 and 7.2-7.4, respectively. Chimeric enzymes showed intermediate profiles of their parents as shown in Figure 4A. The optimum pH of CHSTY, CHBSA, CHAGE and CHBSM enzyme was 6.0, 6.6, 6.8-7.0 and 6.6-7.0, respectively. All the chimeras were stable between pH 4

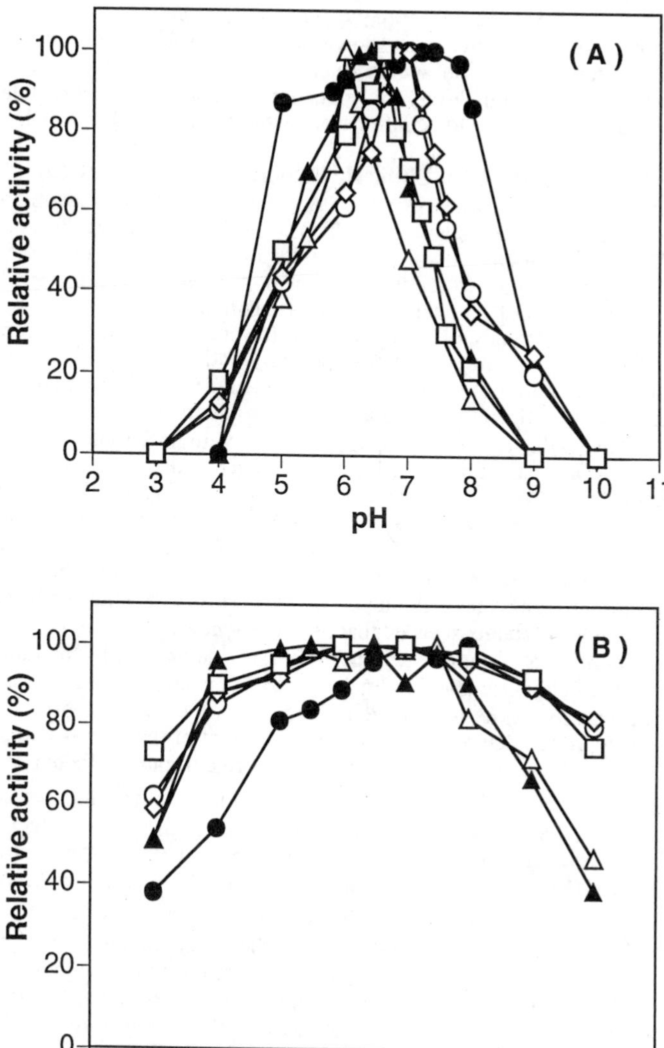

Figure 4. pH-activity (A) and pH stability (B) profiles of four chimeric and two parental ß-glucosidases.
The pH was adjusted with buffers: citrate (pH 3.0-4.0); 2-(N-morpholino) ethane sulphonic acid (pH 5.0-6.8); 3-(N-morpholino) propane sulphonic acid (pH 7.0-8.0); 2-(N-cyclohexylamino) ethane sulphonic acid (pH 9.0-10.0). For pH stability experiments, enzyme was incubated at different pHs for 1 h at 25°C. The residual activities were measured under standard assay condition. *C. gilvus*, ▲; CHSTY, △; CHBSA, □; CHAGE, ◇; CHBSM, ○; *A. tumefaciens*, ●.

and 9, whereas the ß-glucosidases from *C.gilvus* and *A. tumefaciens* were stable at pH 4-8 and pH 5-10, respectively, as shown in Fig. 4B. Substitution of segments in homologous C-terminal region seems to have an influence on pH-activity and stability. In *Bacillus* cyclomaltodextrin glucanotransferase (*19*) and cellulase (*24*), pH-activity profile were found to be influenced by the N- and the C-terminal parts.

Improved Thermal Stability in Chimeric Enzymes. ß-Glucosidase from *C. gilvus* is optimally active at 35°C, whereas that of *A. tumefaciens* exhibit maximum activity at 60°C. With regards to heat stability, ß-glucosidase from *C. gilvus* shows complete activity up to 30°C, retains about 80% of its maximum activity at 35°C, and inactivates completely at 55°C. On the other hand, *A. tumefaciens'* enzyme is stable up to 55°C and even at 65°C, it retains 60% of its maximum activity.

Four chimeric enzymes showed marked differences in their temperature optima (Fig. 5A). The chimeric enzymes were optimally active at 45-50°C, showing an intermediate temperature optimum between *C. gilvus'* and *A. tumefaciens'* enzymes. CHSTY showed the temperature optima of 45 °C with complete loss of activity at 65 °C. CHBSA showed the temperature optima of 45 °C with no activity at 70°C. On the other hand, CHAGE was optimally active at 50°C and exhibited 52% of its maximum activity at 60°C. CHBSM exhibited maximum activity at 50°C and 61% of its maximum activity at 60°C. The temperatures at which 50% loss of the enzyme activities occurred were 41, 47, 50, 55 and 57 and 67°C for *C. gilvus*, CHSTY, CHBSA, CHAGE, CHBSM and *A. Tumefaciens* enzymes, respectively (Fig. 5B). Thus heat stability of chimeric enzymes was increased by 6-16°C as compared to *C. gilvus* enzyme though none of them exceeded that of *A. Tumefaciens* enzyme.

Heat stability may be influenced by only a few amino acid substitutions (*22, 25*). In the case of chimeric isopropylmalate dehydrogenases produced by shuffling genes of *Thermus thermophilus* and *Bacillus subtilis*, the heat stability of chimeric enzymes was approximately proportional to the content of the amino acid sequence from the *T. thermophilus* enzyme (*26*). In general, protein stability increases with the insertion into an α-helix of helix-forming amino acids (alanine, glutamic acid, *etc.*) and decreases with the insertion of helix-breaking amino acids (proline, glycine, *etc.*). The secondary structures of the parental and chimeric enzymes were predicted by Robson's method (*27*). There were similar numbers of helix-breaking but more helix-forming amino acid residues in the α-helix regions of chimeric enzymes than *C. gilvus* enzyme, suggesting that it could be one of the factors influencing heat stability of chimeras. Hydrophobic interaction inside the protein molecule is another important factor in stabilizing protein structure. Hydrophobic cluster analysis (*28,29*) of native and chimeric enzymes revealed that the amino acid substitution from *C. gilvus* to *A. tumefaciens* significantly increased the hydrophobic properties of the chimeric enzymes. These substitutions might be important for heat stability of ß-glucosidase.

Substrate Specificity of Chimeric Enzymes

Difference in Parental Enzymes. ß-Glucosidase from *C. gilvus* has rather strict specificity toward glucose residues in aryl-glycosides (Table II). This enzyme hydrolysed p-nitrophenyl-ß-D-xyloside (pNPXyl) only at 0.81% of the level of p-nitrophenyl-ß-D-glucoside (pNPGlu) and p-nitrophenyl-ß-D-galactoside (pNPGal) at 0.15%. It showed no activity to p-nitrophenyl-ß-D-fucoside (pNPFuc), p-nitrophenyl -ß-D-mannoside (pNPMan), p-nitrophenyl-N-acetyl-ß-D-glucosaminide (pNPGlunAc), p-nitrophenyl-N-acetyl-ß-D-galactosaminide (pNPGalnAc), p-nitro-phenyl-α-D-glucoside (pNP-α-Glu) (Table III). On the other hand, *A. tumefaciens*

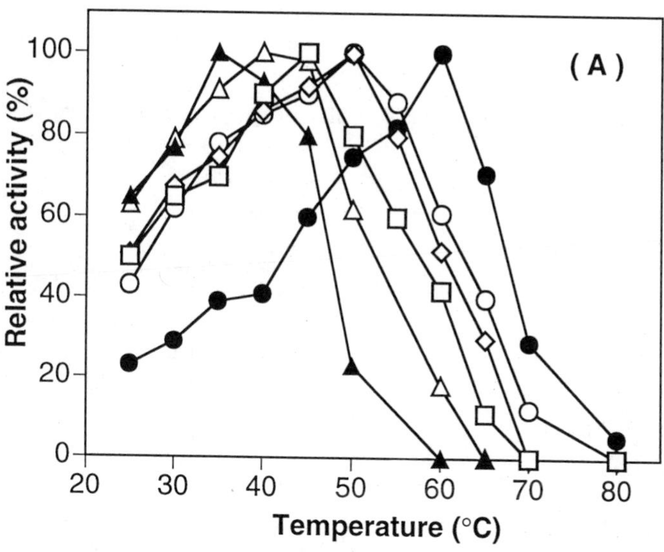

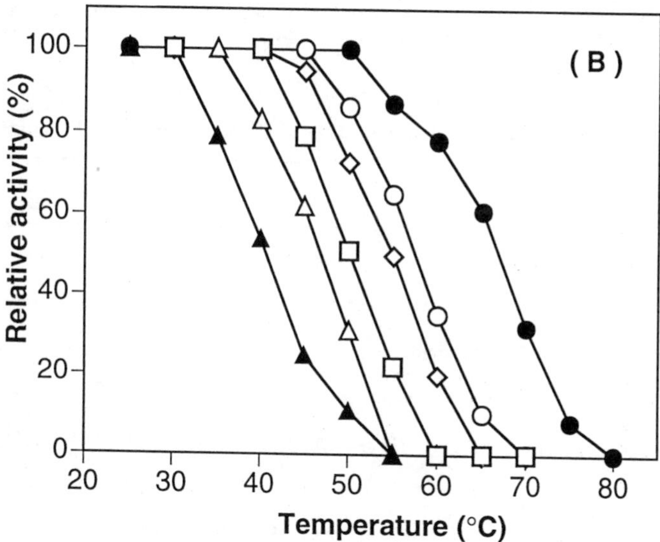

Figure 5. Temperature optima (A) and heat stability (B) profiles of four chimeric ß-glucosidases.
For heat stability experiments, each enzyme at its optimum pH was treated at different temperatures for 1 h. The residual activities were measured under standard assay condition. *C. gilvus,* ▲; CHSTY, △; CHBSA, □; CHAGE, ◇; CHBSM, ○; *A. tumefaciens,* ●.

Table II. Kinetic Parameters of Three Chimeric Enzymes and Two Parental Enzymes

Substrate	CG	CHBSA	CHAGE	CHBSM	AT
p-nitrophenyl-ß-D-glucoside					
Km (mM)	1.806	0.273	0.291	0.270	0.032
Vmax (%)[a]	100	100	100	100	100
p-nitrophenyl-ß-D-xyloside					
Km (mM)	6.261	0.005	0.004	0.004	0.005
Vmax (%)	0.81	7.7	7.2	9.1	14
p-nitrophenyl-ß-D-galactoside					
Km (mM)	10.8	34.4	15.1	16.6	13.1
Vmax (%)	0.15	49	32	43	63
p-nitrophenyl-ß-D-fucoside					
Km (mM)	-	0.277	0.166	0.210	0.123
Vmax (%)	<0.001	11	10	14	20

CG and AT represents *C. gilvus* and *A. tumefaciens*, respectively.
Since partially purified enzymes were used in this assay, measured Vmax values were expressed as relative values to the Vmax values for pNPGlu.

Table III. Relative velocity of Three Chimeric Enzymes and Two Parental Enzymes

Substrate (1 mM)	CG	CHBSA	CHAGE	CHBSM	AT
p-Nitrophenyl-ß-D-glucoside	100	100	100	100	100
p-nitrophenyl-ß-D-mannoside	<0.0020	<0.12	<0.10	<0.070	<0.11
p-nitrophenyl-N-acetyl-ß-D-glucosaminide	<0.00088	<0.19	<0.25	<0.30	0.47
p-nitrophenyl-N-acetyl-ß-D-galactosaminide	<0.002	<0.21	<0.30	<0.26	<0.21
p-nitrophenyl-α-D-glucoside	<0.0063	<0.25	<0.20	<0.17	<0.28
Glu-Glu	44.1	65.4	34.6	49.2	27.6
Glu-Glu-Glu-Glu-Glu-Glu	70.9	45.7	51.4	23.6	<0.58

CG and AT represents *C. gilvus* and *A. tumefaciens*, respectively.
Measured velocities are were expressed as relative values to the velocity values for pNPGlu.

enzyme exhibited 14%, 63%, 20% and 0.47% activity on pNPXyl, pNPGal, pNPFuc and pNPGlunAc, respectively. Affinity of AT to pNPGlu (Km=0.032 mM) and pNPXyl (0.005 mM) is extremely low compared to CG (1.806 and 6.261 mM, respectively). *A. tumefaciens* enzyme specificity toward aryl-glycoside substrates is broader than of *C. gilvus* enzyme, though the former can hardly hydrolyse celloheptaose.

Chimeric Enzymes. Due to low expression of one chimeric enzyme of CHSTY, this enzyme can not be subjected to this experiment. Based on the measurement of pNPGal, AT and CG including chimeric enzymes loosely recognizes the hydroxy residue at C-4 position of ß-D-glucose. Significant differences in relative velocity for hydrolysing pNPGal are observed between CG (0.15%) and AT (63%) including chimeric enzymes (32-49%), though no prominent difference in Km (10.8-34.4 mM). All enzymes including the three chimeric enzymes strictly recognize the hydroxy residue at the C-2 position of ß-D-glucose since pNPMan was not hydrolyzed as shown in Table II and III. The three chimeric enzymes were found to be closer to *A. tumefaciens* enzyme since they showed similar Km and Vmax to that of *A. tumefaciens* enzyme for the substrate of pNPXyl, pNPGal and pNPFuc. However, when pNPGlu is used as substrate, Kms of the three chimeric enzymes (0.270-0.293 mM) were just between the two parental enzymes of *C. gilvus* (1.81 mM) and *A. tumefaciens* (0.032 mM). Furthermore, the relative velocity for hydrolysing cellohexaose shows that chimeric enzymes are closer to *C. gilvus* enzyme than *A. tumefaciens* enzyme (Table III).

While the Km value of the chimeric citrate synthases have similarly been found to be lower than those of the parental enzymes (*30*), substrate affinity decreased by about two fold in active human-yeast chimeric phosphoglycerate kinase engineered by domain interchanges (*31*). However, no significant differences were found between the Km values of parental and chimeric isopropylmalate dehydrogenases (*26*). Replacement of the catalytic base Glu400 by glutamine in *Aspergillus niger* glucoamylase was found to affect both substrate ground-state binding and transition state stabilization (*32*). Km values for maltose and maltoheptaose were 12- and 3- fold higher for the Glu400>Gln mutant, with Kcat values 35- and 60-fold lower, respectively, as compared to those of the wild type enzyme. Similarly in *Aspergillus awamori* glucoamylase mutants, Ser119>Tyr, Gly183>Lys and Ser184>His, slightly higher activity for maltose hydrolysis and lower activity for isomaltose as compared wild was observed by Sierks and Svensson (*33*). The observed increase in selectivity was attributed to the stabilization of the maltose transition-state complex for each enzyme. Modulation of binding energy by mutation could be attributed to modification in hydrogen bonding (*32, 34, 35*). In the case of chimeric enzyme consisting of an amino-terminal domain of phenylalanine dehydrogenase and a carboxy-terminal domain of leucine dehydrogenase, the chimeric enzyme showed a broad substrate specificity in the oxidative deamination, like phenylalanine dehydrogenase. However, it acted much more effectively than phenylalanine dehydrogenase on isoleucine and valine. The substrate specificity of the chimeric enzyme in the reductive amination was an admixture of those of the two parent enzymes (*36*).

Effect of Chimeric Regions on the Enzyme Character. Thus heat stability, pH-activity and substrate specificity were changed distinctly by substituting different segments of *C. gilvus* ß-glucosidase gene with that of *A. tumefaciens*. It is interesting to note that changes in heat stability were more pronounced with the increased size of insertion fragment. Thermal stability was found to increase in the order of CHSTY

<CHBSA<CHAGE<CHBSM. However, no distinct differences in the measured kinetic parameters of Km and Vmax for each substrate were found among the three chimeric enzymes of CHBSA, CHAGE and CHBSM. This results suggested that amino acid residues contributing the thermal stability distribute evenly at least in the C-terminal region of the amino acid sequence of *A. tumefaciens* ß-glucosidase. This region may also play an important role in determining substrate specificity. In case of chimeric mammalian hexokinase, where a catalytic domain locate at C-terminal domain, N-terminal was found to be responsible for interacting with inhibitor (*37*).

In chimeric isopropylmalate dehydrogenase from an extreme thermophile, *Thermus thermophile*, and a mesophile, *Bacillus subtilis*, the stability of each chimeric enzyme was also proportional to the content of the amino acid sequence from the *T. thermophile* enzyme (*26*). The thermal stability of the chimeras was also intermediate between that of the highly labile Type II hexokinase and the relatively stable Type I isozyme (*37*).

Creation of a New Enzyme.

Valuable information on the structure-function relationship has so far been successfully obtained by preparation of chimeric enzymes from two functionally related proteins, sharing extensive sequence similarity (*38,39*). Several different combinations of homologous C-terminal regions of ß-glucosidases from *C. gilvus* and *A. tumefaciens* successfully resulted in the formation of enzymatically active chimeric enzymes and heat stability has improved significantly. Preparation of chimeric enzymes can be a new method for creating new enzymes with desired character. Enzymes with improved properties can be obtained through constructing chimeric enzymes.

Acknowledgements

We are thankful to Drs. L.A. Castle and R.O. Morris, University of Missouri, for useful discussions and providing the recombinant plasmid carrying ß-glucosidase gene from *A. tumefaciens*. Thanks are also due to Drs. H. Taniguchi and S. Sasaki for critical discussion.

References

1. *Enzyme technology;* Lafferty R. M., Ed.; Springer-Verlag: New York, NY, **1983**; pp 193-270.
2. *Enzyme chemistry and molecular biology of amylases and related enzymes*; The amylase research society of Japan, Ed.; CRC Press: Boca Raton, Fl, **1994**; pp 46-67.
3. Watson, J. D.; Hopkins, N. H.; Roberts, J. W.; Steitz, J. A.; Weiner, A. M. *Molecular biology of the gene*; Benjamin/Cummings: Menlo Park, CA, **1988**; pp 1097-1163.
4. Davis, L. G.; Kuehl, W. M.; Bettey, J. F. *Molecular biology*; Applenton & Lange: Norwalk, CN, **1994**, pp 738-744.
5. Wales, M. E.; Wild, J. R. *Meth. Enzymol.* **1991**; Vol. 202, pp 687-706.
6. Onodera, K.; Sakurai, M.; Moriyama, H.; Tanaka, N.; Numata, K.; Oshima, T.; Sato, M.; Katsube, Y. *Prot. Engin.* **1994**, *7*, 453-459.
7. Hjelmstad, R. H.; Morash, S. C.; McMaster, C. R.; Bell, R. M. *J. Biol. Chem.* **1994**, *269*, 20995-21002.
8. Kashiwagi, Y.; Aoyagi, C.; Sasaki, T.; Taniguchi, H. *Agric. Biol. Chem.* **1991**, *55*, 2553-2559.

9. Castle, L.A.; Smith, K.D.; Morris, R.O. *J. Bacteriol.* **1992**, *174*, 1478-1486.
10. Storwick, W.O.; King, K.W. *J. Biol. Chem.* **1960**, *235*, 303-307.
11. Kitaoka, M.; Sasaki, T.; Taniguchi, H. *Biosci. Biotech. Biochem.* **1992**, *56*, 652-655.
12. Tariq, M. A.; Hayashi, K. *Biochem. Biophys. Res. Com.* **1995**, *214*, 568-575.
13. Tariq, M. A.; Hayashi, K. *Carbohyd. Res.* **1995**, *275*, 67-72.
14. Beguin, P. *Ann. Rev. Microbiol.* **1990**, *44*, 219-248.
15. Paavilainen, S.; Hellman, J.; Korpela, T. *Appl. Environ. Microbiol.* **1993**, *59*, 927-932.
16. Kashiwagi, Y.; Aoyagi, C.; Sasaki, T.; Taniguchi, H. *J. Ferment. Bioeng.* **1993**, *75*, 159-165.
17. Singh, A.; Hayashi, K. *Adv. Appl. Microbiol.* **1995**, *40*, 1-44.
18. Yamane, K.; Hirata, Y.; Furusato, T.; Yamazaki, H.; Nakayama, A. *J. Biochem.* **1984**, *96*, 1849-1858.
19. Kaneko, T.; Song, K. B.; Hamamoto, T.; Kudo, T.; Horikoshi, K. *J. Gen. Microbiol.* **1989**, *135*, 3447-3457.
20. Ohmiya, K.; Deguchi, H.; Shimizu, S. *J. Bacteriol.* **1991**, *173*, 636-641.
21. Kashiwagi, K.; Aoyagi, C.; Sasaki, T.; Taniguchi, H. In *Genetics, Biochemistry and Ecology of Lignocellulose Degradation;* Shimada, K.; Ohmia, K.; Kobayashi, Y.; Hoshino, S.; Sakka K.; Karita, S. Eds.; UNI Publishers: Tokyo, **1993**, pp 368-377.
22. Singh, A.; Hayashi, K.; Hoa, T. T.; Kashiwagi, Y.; Tokuyasu, K. *Biochem. J.* **1995**, *305*, 715-719.
23. Singh, A.; Hayashi, K. *J. Biol. Chem.* **1995**, *270*, 21928-21933.
24. Nakamura, A.; Fukumori, F.; Horinouchi, S.; Masaki, H.; Kudo, T.; Uozumi, T.; Horikoshi, K.; Beppu, T. *J. Biol. Chem.* **1991**, *226*, 1579-1583.
25. Nosoh, Y.; Sekiguchi, T. *Trends Biotechnol.* **1990**, *8*, 16-20.
26. Numata, K.; Muro, M.; Akutu, N.; Nosoh, Y.; Yamagishi, A.; Oshima, T. *Protein. Eng.* **1995**, *8*, 39-43.
27. Garnier, J.; Osguthorpe, D. J.; Robson, B. *J. Mol. Biol.* **1978**, *120*, 97-120.
28. Gaborioud, C.; Bissery, V.; Benchetrit, T.; Mornon, J. P. *FEBS Lett.* **1987**, *224*, 149-155.
29. Henrissat, B.; Raimbound, E.; Tran, V.; Mornon, J. P. *CABIOS* 6, **1990**, 3-5.
30. Molgat, G. F.; Donald, L. T.; Duckworth, H. W. *Arch. Biochem. Biophys.* **1992**, *298*, 238-246.
31. Mas, M. T.; Chen, C. Y.; Hitzman, R. A.; Riggs, A. D. *Science* **1986**, *233*, 788-790.
32. Frandsen, T. P.; Dupont, C.; Lehmbeck, J.; Stoffer, B.; Sierks, M.R.; Honzatko, R. B.; Svensson, B. *Biochemistry* **1994**, *33*, 13808-13816.
33. Sierks, M. R.; Svensson, B. *Protein Eng.* **1994**, *7*, 1479-1984.
34. Olsen, K.; Christensen, U.; Sieks, M. R.; Svensson, B. *Biochemistry* **1993**, *32*, 9686-9693.
35. Sierks, M. R.; Svensson, B. *Biochemistry* **1993**, *32*, 1113-1117.
36. Kataoka, K.; Takada, H.; Tanizawa, K.; Yoshimura, T.; Esaki, N.; Ohshima, T.; Soda, K. *J. Biochem. (Tokyo)* **1994**, *116*, 931-936.
37. Tsai, H. T.; Wilson, J. E. *Arch. Biochem. Biophys.* **1995**, *316*, 206-214.
38. Sode, K.; Yoshida, H.; Matsumura, K.; Kikuchi, T.; Watanabe, M.; Yasutake, N.; Ito, S.; Sano, H. *Biochem. Biophys. Res. Com.* **1995**, *211*, 268-273.
39. Newsted, W. J.; Ramjeesingh, M.; Zywulko, M.; Rothstein, S. J.; Shami, E. Y. *Enz. Microbial Technol.* **1995**, *17*, 757-764.

Chapter 14

Biogeneration of Volatile Compounds via Oxylipins in Edible Seaweeds

Tadahiko Kajiwara, Kenji Matsui, and Yoshihiko Akakabe

Department of Biological Chemistry, Yamaguchi University, 1677–1 Yoshida, Yamaguchi 753, Japan

Unsaturated and saturated fatty aldehydes such as (*Z*, *Z*, *Z*)-8, 11, 14-heptadecatrienal, (*Z*, *Z*)-8, 11-heptadecadienal, (*Z*)-8-heptadecenal, (*Z*, *Z*, *Z*)-7, 10, 13-hexadecatrienal, pentadecanal, (*E*, *Z*, *Z*)-2, 4, 7-decatrienal, (*E*, *Z*)-2, 6-nonadienal and (*E*)-2-nonenal have been identified in essential oils from edible seaweeds as characteristic major compounds. The enzymatic formations of the long-chain fatty aldehydes from fatty acids such as linolenic acid, linoleic acid, oleic acid and palmitic acid, respectively, have been demonstrated. Based on enzymatic formation of (2*R*)-hydroxy-palmitic acid and 2-oxo-palmitic acid from palmitic acid during biogeneration of pentadecanal and on incubation experiments of synthetic (2*S*)- or (2*R*)-hydroxy-palmitic acid and 2-oxo-palmitic acid as substrates with crude enzyme solution of *Ulva pertusa*, the biogeneration mechanism of long chain aldehydes *via* oxylipins is discussed.

Many kinds of seaweeds are cultivated on the coast of Japan for use mainly for food and occasionally for chemical algin (*1*). A group of the kelps which belong to *Laminaria* and *Undaria* are generally called "kombu" and "wakame" in Japanese, respectively. After kombu and wakame, a red seaweed *Porphyra* sp. (asakusa-nori) is the most popular seaweed in Japanese foods. Also lots of green seaweeds *Enteromorpha* sp. (ao-nori), *Ulva* sp. (aosa) and *Monostroma* sp. (hitoegusa) are used for food. These seaweeds have been eaten from ancient days and are highly favored for their aromas, tastes, and textures. Seaweeds are a well recognized source of unique natural products, principally terpenoids. Over the last ten years, chemistry and biochemistry of oxylipins (*2*) have been emerged as a new trend field. In recent years volatile components derived from oxylipins in fresh seaweeds have been

NOTE: Dedicated to Professors Masakazu Tatewaki and Tadao Yoshida on the occasion of their academic retirement.

0097–6156/96/0637–0146$15.25/0

explored and various fatty aldehydes have been identified as characteristic flavor compounds of edible seaweeds. This paper will focus on our developments on the identification and biogeneration of the characteristic long-chain fatty aldehydes in Japanese edible seaweeds.

Flavor Compounds from Edible Seaweeds

Katayama reported volatiles of some air-dried green marine algae at the early stages of gas chromatography (GC) development (*3*). In recent years, we have explored volatile compounds in fifty or more species of wet and undecomposed seaweeds in Japan: green seaweeds *Ulva pertusa*, *Monostroma nitidum*, and *Enteromorpha clathrata*; brown seaweeds *Laminaria japonica* and *Undaria pinnatifida*, and red seaweeds *Porphyra tenera* and *Porphyra yezoensis* by GC and GC-mass spectrometry (MS).

U. pertusa was collected along the Yoshimo coast, Yamaguchi southern part of the Japan sea and along the Charatsunai coast, Muroran, Hokkaido northern part of Pacific Ocean. *M. nitidum* and *E. clathrata* were obtained along the Uchiseura and Ryoshihama coast, Mie mid part of the Pacific Ocean, respectively. The steam-distillates of homogenates of each fresh seaweed gave essential oils: *U. pertusa* $5.4 \times 10^{-3}\%$; *M. nitidum* $7.4 \times 10^{-4}\%$; and *E. clathrata* $9.6 \times 10^{-4}\%$. The characteristic odorous oils were analyzed by GC and GC-MS equipped with fused silica capillary columns (SF-96 and DB-1). Among volatile compounds detected in the oils of the green seaweeds, thirty one compounds were newly identified as volatile components of the Ulvales (*4*). As Table I shows, the aldehydes (C_{10}-C_{17}) were characteristic components in the Ulvales oils. The major aldehydes were (Z, Z, Z)-8, 11, 14-heptadecatrienal and pentadecanal. Particularly, the C_{17}-trienal accounted for 35% of the oils of *Ulva* obtained along the Pacific coast of Hokkaido, Muroran and 8-11% along the Sea of Japan coast of Yamaguchi. A homolog of the C_{16}-trienal, (Z, Z, Z)-7, 10, 13-hexadecatrienal, was also found in *U. pertusa*. The characteristic aldehyde of *M. nitidum* and *E. clathrata* was (*E*, Z, Z)-2, 4, 7-decatrienal. The C_{17}-aldehydes and the characteristic C_{10}-trienal were first identified in seaweeds. However, the C_{17}-unsaturated aldehydes have been reported as volatiles of an aqueous cucumber homogenate (*5*) and green leaves of tobacco at flowering time just after topping (*6*). Closely related aldehydes such as (3Z, 6Z)-3, 6, 11-dodecatrienal, (2*E*, 4Z, 7Z)-2, 4, 7, 12-tridecatetraenal, and (3Z, 6Z, 9Z)-3, 6, 9, 14-pentadecatetraenal have been reported to possess "characteristic seaweeds or algae odor" (*7*).

Katayama has identified aldehydes, monoterpenes, and alcohols in the steam-distillate of some dried *Laminaria* sp. However, they were not detected in fresh kelps, except for the secondary alcohol (*8*). With essential oils of the wet and undecomposed edible kelps, *L. angustata*, *L. japonica*, *Kjellmaniella crassifolia*, *Costaria costata*, *Ecklonia crassifolia*, *E. cava* and *U. pinnantifida* along the Sea of Japan, fifty three compounds including alcohols, aldehydes, esters, ketones, hydrocarbons, and carboxylic acids were identified by comparison of Kovats indices and MS data with those of authentic compounds (*9*, *10*). The nor-carotenoids such as β-cyclocitral, β-homocyclocitral, β-ionone, and dihydroactinidiolide, which have been

Table I. Fatty Aldehydes Identified in Essential Oils from Edible Algae

Compounds	Green alga		Brown alga		Red alga	
	UP	*EC*	*LA*	*LJ*	*PT*	*PY*
(8*Z*, 11*Z*, 14*Z*)-Heptadecatrienal	++	++	-	-	-	-
(8*Z*, 11*Z*)-Heptadecadienal	+	-	-	-	-	-
(8*Z*)-Heptadecenal	+	+	-	-	-	-
(7*Z*, 10*Z*, 13*Z*)-Hexadecatrienal	+	-	-	-	-	-
(6*Z*, 9*Z*, 12*Z*)-Pentadecatrienal	+	-	-	-	-	-
n-Pentadecanal	++	++	-	-	+	+
n-Tetradecanal	+	+	-	-	-	+
n-Tridecanal	+	+	+	+	-	+
n-Dodecanal	-	-	+	-	-	-
(2*E*, 4*Z*, 7*Z*)-Decatrienal	+	+	-	-	-	-
(2*E*, 4*Z*)-Decadienal	+	-	+	-	-	+
(2*E*, 4*E*)-Decadienal	+	-	+	+	+	+
(3*Z*, 6*Z*)-Nonadienal	-	-	-	-	+	-
(2*E*, 6*Z*)-Nonadienal	-	-	++	++	++	++
(2*E*, 6*E*)-Nonadienal	-	-	+	+	-	-
(2*E*)-Nonenal	-	-	++	++	++	++
(2*E*, 4*E*)-Octadienal	-	-	-	+	+	+
(2*E*)-Octenal	-	-	+	+	+	+
(2*E*, 4*Z*)-Heptadienal	-	-	-	-	-	+
(2*E*, 4*E*)-Heptadienal	-	-	+	-	+	+
(2*E*)-Heptenal	-	-	+	+	-	+
Heptanal	-	-	+	-	-	+
Benzaldehyde	-	-	-	-	-	+
Hexanal	-	-	+	+	+	+
(2*E*)-Hexenal	-	-	+	+	-	+
(2*E*)-Pentenal	-	-	-	-	+	+

UP: *Ulva pertusa*, *ES*: *Entermoropha clathrata*, *LA*: *Laminaria angustata var. longissima*, *LJ*: *Laminaria japonica*, *PT*: *Porphyra tenera*, *PY*: *Porphyra yezoensis*.
-: not detected, +: detected, ++: major component

reported as flavor components of an edible red seaweed, *P. tenera* (*11*) seem to be important constituents of some brown algae such as *C. costata* and *A. crassifolia*. The sesquiterpene alcohol, cubenol, was detected in the volatile oils of all of the submitted kelps, whereas the stereoisomer, epicubenol was detected only in *L. japonica* and *K. crassifolia* in small amounts. (*E*, *Z*)-2, 6-Nonadienal, (*Z*, *Z*)-3, 6-nonadienal, (*E*)-2-nonenal, and the corresponding alcohols, which are well known as flavor constituents of cucumber (*12-14*), melons (*13*, *15*), and fish (*16*), were found to be the principle odor contributors in some brown seaweeds. The C_9-aldehydes and alcohols particularly were at their highest concentration in *A. crassifolia*. Recently, (*E*, *Z*)-2, 6-nonadienal has been found in *Cymathere triplicata*, a large brown kelp, in Northern Washington (*17*). (*E*)-2-Nonenal and (*E*, *Z*)-2, 6-nonadienal are formed *via* (*Z*)-3-nonenal and (*Z*, *Z*)-3, 6-nonadienal from 9-hydroperoxy-(*E*, *Z*)-10, 12-octadecadienoic acid and 9-hydroperoxy-(*E*, *Z*, *Z*)-10, 12, 15-octadecatrienoic acid in higher plants, respectively (*18*, *19*). However, studies on the biogeneses of the short chain aldehydes in marine algae are under way in our laboratory.

With the great thalli of the red seaweeds, *P. tenera* and P. *yezoensis*, fatty aldehydes such as, *n*-pentadecanal, (*E*, *Z*)-2, 6-nonadienal, and (*E*)-2-nonenal (Table I), nor-carotenoids and sesquiterpene alcohols were identified (*19*). An essential oil of cultivated conchocelis-filaments of the seaweed contained cubenol, phytol, palmitic acid and the long-chain fatty aldehydes such as tetradecanal, pentadecanal, (*Z*, *Z*)-7, 10-hexadecadienal, (*Z*, *Z*)-8, 11-heptadecadienal and (*Z*)-8-heptadecenal. Flament and Ohloff have reported on the identification of more than 100 volatile constituents of dried thalli of *P. tenera*: nor-carotenoids (α-ionone, β-ionone, dihydroactinidiolide etc.) and unsaturated fatty short-chain aldehydes [(*E*, *Z*)-2, 4-decadienal, (*E*, *E*)-2, 4-heptadienal etc.] (*11*). However, these compounds were detected only as minor components in wet and undecomposed conchocelis-filaments.

Biogeneration of Long-Chain Fatty Aldehydes in Seaweeds

Galliard and Matthew (*5*) have reported the biogenesis of C_{15}, C_{14}, C_{13} and C_{12}-saturated fatty aldehydes from palmitic acid in cucumber fruits. However, that of the unsaturated C_{17}-aldehydes such as (*Z*, *Z*, *Z*)-8, 11, 14-heptadecatrienal, (*Z*, *Z*)-8, 11-heptadecadienal and (*Z*)-8-heptadecenal had not been studied so far. Thus, the enzymatic formation of the long-chain aldehydes from unsaturated fatty acids in a green seaweed, *U. pertusa*, was explored.

Enzymatic Formation of Long-Chain Fatty Aldehydes in a Green Seaweed *U. pertusa*. In preliminary experiments, they were found to increase during incubations of unsaturated fatty acids with homogenates of *U. pertusa* fronds. The acetone powder preparations from the homogenates were shown to retain this activity (*20*). Thus, enzyme solutions solubilized from the acetone powders with 0.1% Triton X-100 were used as a source of the enzyme activity. The products, (*Z*)-8-heptadecenal, (*Z*, *Z*)-8, 11-heptadecadienal and (*Z*, *Z*, *Z*)-8, 11, 14-heptadecatrienal and pentadecanal, were identified by GC and GC-MS analysis using authentic specimens

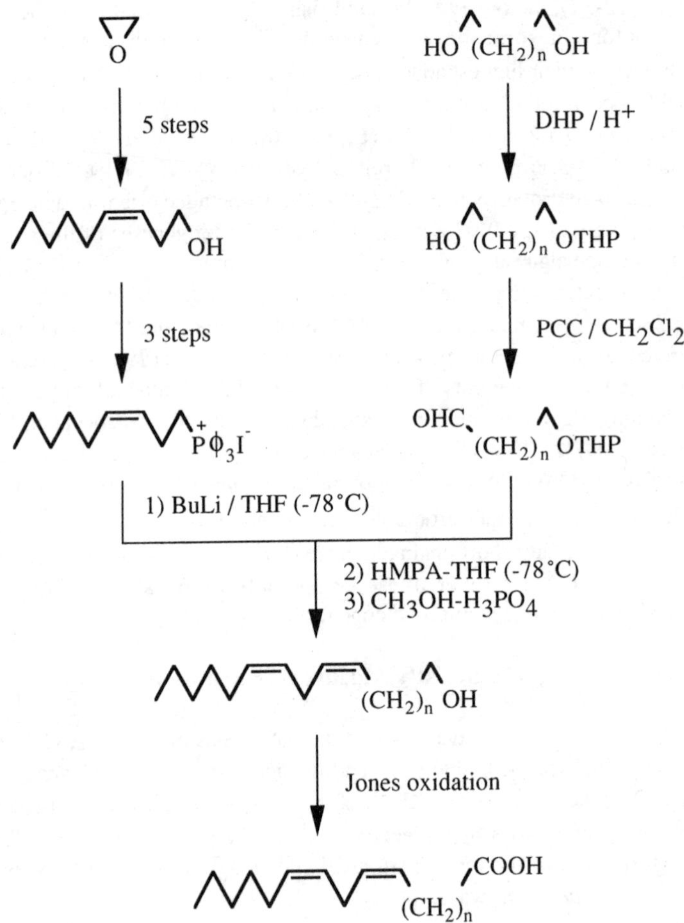

Figure 1. Syntheses of (ω6*Z*, ω9*Z*)-C_{12}-C_{20}-dienoic acids.

prepared by Collins oxidation of the corresponding alcohols, which were synthesized through essentially the same unequivocal routes (Figure 1). The aldehydes formed during the incubations were analyzed by HPLC (Zorbax ODS) of the 2, 4-dinitrophenylhydrazone derivatives quantitatively. The heptadecatrienal increased greatly (over 40 fold), when linoleic acid was incubated with the enzyme solution in phosphate buffer (pH 7.0) at 35°C for 60 min. Also the heptadecatrienal (**1**), heptadecadienal (**2**), heptadecenal (**3**), and pentadecanal (**4**) were formed by incubations with linolenic acid (LNA), linoleic acid (LA), oleic acid (OA) and palmitic acid (PA), respectively (Figure 2). With heat-treated suspensions, the increases were not observed (*20*).

Fronds were homogenized with 50 mM sodium-pyrophosphate buffer (pH 9.0) in a mixer. The homogenate was filtered through four layers of gauze and the filtrate centrifuged at 4,000 *g*. Nearly 60% of the long chain aldehyde-forming activity in the homogenate was found in the 4,000 *g* precipitate. The activity (probably membrane bound) in the precipitate was readily solubilized by an addition of a relatively low concentration (0.2%) of Triton X-100. The solubilized activity showed a maximum activity in the pH range 8.5-9.5.

The substrate specificity for the solubilized long chain aldehyde-forming enzyme was examined using natural C_{18} and C_{20} fatty acids, and some derivatives (alcohol, aldehyde, ester) and synthetic (ω6, ω9)-dienoic acids (Figure 1). Tests of substrate specificity for the unsaturated C_{18} fatty acids, showed that linoleic acid and linolenic acid were the best substrates, γ-linolenic acid, however, was a poor substrate. On the other hand, linoleic acid and the aldehyde isomer could act as a substrate, whereas the methyl ester and alcohol isomers showed no activity for the enyzme. With polyenoic acids such as 20:4 and 20:5, the reactivities were low. For elucidation of the structural requirement of substrate for the aldehyde forming enzyme, an entire series of (ω6Z, ω9Z)-dienoic acids in which the chain length varied from C_{13} to C_{20} (except for linoleic acid), were synthesized by a (Z)-selective Wittig reaction between (Z)-3-nonenyltriphenylphosphonium iodide and the appropriate 2-tetrahydropyranyloxy-alkanal followed by removal of the protective group and Jones oxidation (Figure 1). The synthetic dienoic acids (over 95% purity) were used as substrates for the enzyme after purification by silica gel column chromatography. The C_{15} ~ C_{20}-dienoic acids (n=4~9): (Z, Z)-6, 9-pentadecadienoic acid, (Z, Z)-7, 10-hexadecadienoic acid, (Z, Z)-8, 11-heptadecadienoic acid, (Z, Z)-9, 12-octadecadienoic acid, (Z, Z)-10, 13-nonadecadienoic acid, and (Z, Z)-11, 14-eicosadecadienoic acid, were found to be converted to aldehydes containing one carbon atom less, i.e. (Z, Z)-5, 8-tetradecadienal, (Z, Z)-6, 9-pentadecadienal, (Z, Z)-7, 10-hexadecadienal, (Z, Z)-8, 11-heptadecadienal, (Z, Z)-9, 12-octadecadienal, and (Z, Z)-10, 13-nonadecadienal, respectively (Figure 3). Linoleic acid was the best substrate of all the dienoic acids tested. The reactivity of the acids decreased with both an increase and a decrease in chain length with respect to linoleic acid. C_{13} and C_{14} dienoic acids (n=2~3) act slightly as substrates for the activity. These results suggest that the distance between a (Z, Z)-pentadiene and a carboxy group or a formyl group in the substrate, is an important structural requirement of substrates for long chain aldehyde-forming enzyme

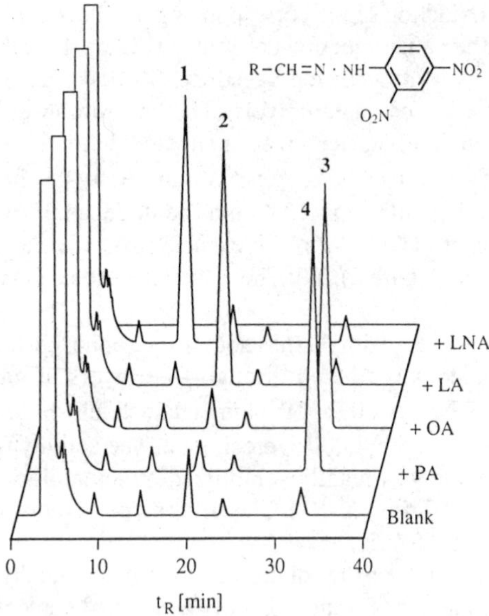

Figure 2. Enzymatic formation of long chain aldehydes in a marine green alga *U. pertusa.*

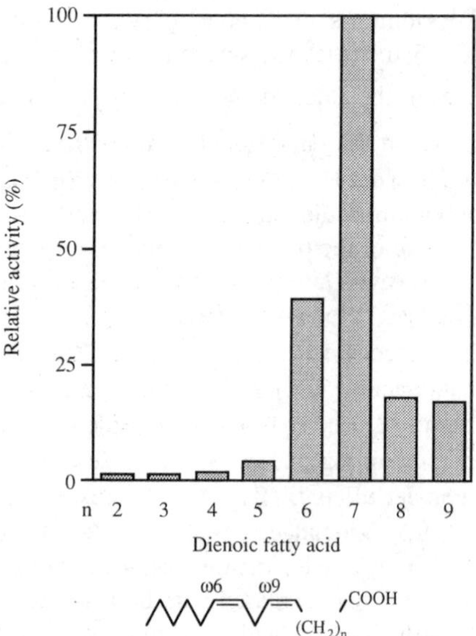

Figure 3. Substrate specificity of a series of (ω6*Z*, ω9*Z*)-dienoic acids for long chain aldehydes-forming enzyme.

in seaweeds. This might partly explain the striking difference in the reactivity between linolenic acid and γ-linolenic acid.

Distribution of Long Chain Aldehyde-Forming Activity in Seaweeds. The long chain fatty aldehyde-forming activity producing(*Z*, *Z*)-8, 11-heptadecadienal from linoleic acids was assayed by quantitative HPLC analysis after conversion of the products to the corresponding 2, 4-dinitrophenylhydrazone derivatives. The distribution of the long chain aldehyde-forming enzyme activity was examined in 25 species of brown seaweeds and 15 species of red seaweeds (Table II). *Dictyota dichotoma*, *Dictyopteris divaricata*, *Gracilaria asiatica* and *Chondria crassicaulis*, showed high activities. In contrast, *Myelophycus simplex*, *L. japonica*, *L. angustata*, *U. pinnatifida*, *Cystoseira hakodatensis*, *Rhodoglossum japonica*, *Schizymenia dubyi*, *Tichocarpus crinitus*, *Neorhodomela larix*, *Rhodomela teres* and *Odonthalia corymbifera* showed slight or no activity. An edible seaweed, *G. asiatica* showed (*Z*, *Z*)-8, 11-heptadecadienal and pentadecanal formation activity of ca. 13 and 16 nmol / mg protein / h from linoleic acid and palmitic acid, respectively. Relative ratios in the activities of the heptadecadienal and pentadecanal formation differed among the various species. These results indicate that the long chain aldehyde-forming activity is distributed among a wide range of green, red and brown seaweeds. The α-oxidation activity that produces pentadecanal from palmitic acid in the same manner as long chain aldehyde-forming enzyme activity had previously been found in terrestrial plants such as in germinated peanut cotyledons (*21*), pea leaves (*22*), potato slices (*23*) and rice seedlings (*24*). Thus, the long chain aldehyde-forming activity probably occurs widely throughout the plant kingdom (*25*).

Purification of Long Chain Fatty Aldehyde-Forming Enzyme and its Properties in *U. pertusa*. An acetone powder or the homogenate from *U. pertusa* fronds was utilized as a source of long chain fatty aldehyde-forming enzyme. A homogenate of *U. pertusa* was centrifuged at 19,000 *g* and the precipitate resuspended with Triton X-100 and glycerol in Na-pyrophosphate buffer (pH9.0). The specific activity of the supernatant (crude enzyme solution; CES) was 3.0-fold higher than that of the homogenate. The CES was directly applied on a DEAE Cellulofine A-500 column; Figure 4 shows a typical elution profile of the activity (*28*). The long chain aldehyde-forming enzyme was adsorbed to the resin and eluted by an increase of NaCl concentration. The enzyme eluted as a shoulder on the main protein peak. With this step, the activity was purified 53-fold compared with the homogenate. The fraction kept at 4°C for 24 hr showed only ca. 50% of the initial enzyme activity.The optimum pH of the enzyme was pH 8.5-9.0. This value was higher compared with those reported in higher plants, i.e. optimum pHs of cucumber fruit and pea leaf are 7.2 and 7.5, respectively (*5*, *26*). Recently, Baardseth et al. (*27*) partially purified the α-oxidation enzyme in cucumber fruits and suggested a similarity between the α-oxidation system in cucumber and that of pea leaf and cotyledons of germinating peanuts by fast protein liquid chromatography using an anion-exchange column. An entire series of saturated fatty acids in which the chain length varied from C_{12} to C_{20}

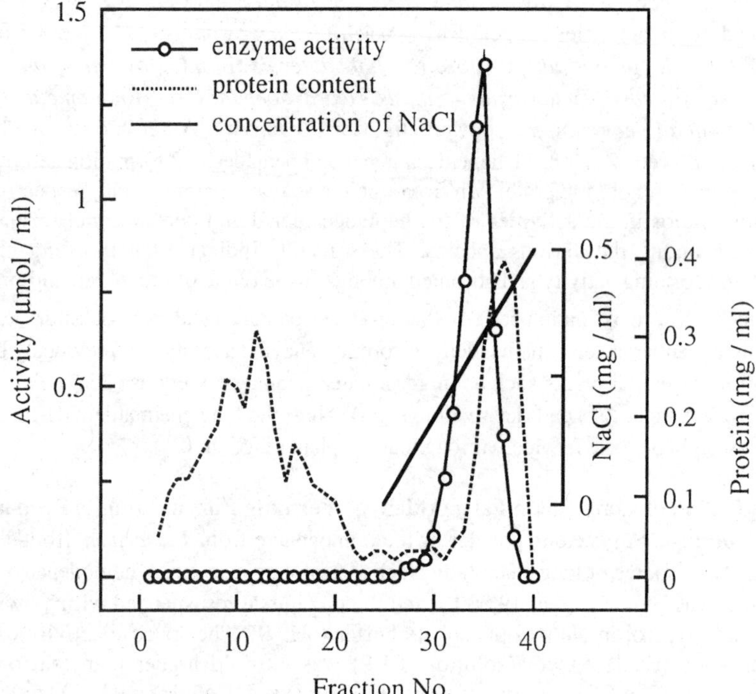

Figure 4. Elution profile of chain aldehyde-forming enzyme in *U. pertusa* from DEAE-Cellulofine A-500 column chromatography.

Table II. Distribution of Long Chain aldehyde-Forming Enzyme activity in marine algae

	Algae	Activity	Algae	Activity
Green algae				
Ulvales	*Ulva pertusa*	+++	*Enteromorpha* sp.	+++
Siphonales	*Codium fragile*	-		
Brown algae				
Dictyotaceae	*Dictyota dichotoma*	+	*Dilophus okamurae*	++
	Dictyopteris divaricata	+	*Dictyopteris prolifera*	+
	Pachydictyon coriaceum	+	*Padina arborescens*	+
Scytosiphonaceae	*Scytosiphon lomentaria*	+	*Endarachne binghamiae*	+
	Colpomenia bullosa	+		
Asperococcaceae	*Myelophycus simplex*	-		
Leathesiaceae	*Leathesia difformis*	+		
Ralfsiaceae	*Analipus japonicus*	+		
Ishigeaceae	*Ishige okamurae*	+		
Laminariaceae	*Laminaria japonica*	-	*Laminaria angustata*	-
Alariaceae	*Alaria crassifolia*	+	*Undaria pinnatifida*	-
Fucaceae	*Pelvetia wrightii*	+		
Cystoseiraceae	*Cystoseira hakodatensis*	-		
Sargassaceae	*Hizikia fusiformis*	+	*Sargassum horneri*	+
	Sargassum thunbergii	+	*Sargassum miyabei*	+
	Sargassum nigrifolium	+	*Sargassum confusum*	+
Red algae				
Gracilariaceae	*Gracilaria asiatica*	++		
Gigartinaceae	*Rhodoglossum japonica*	-	*Chondrus yendoi*	+
	Chondrus ocellatus	+		
Petrocelidaceae	*Mastocarpus pacificus*	+		
Nemastomaceae	*Schizymenia dubyi*	-		
Halymeniaceae	*Grateloupia filicina*	+		
Tichocarpaceae	*Tichocarpus crinitus*	-		
Dumontiaceae	*Neodilsea yendoana*	+		
Rhodomelaceae	*Neorhodomela larix*	-	*Rhodomela teres*	-
	Odonthalia corymbifera	-	*Chondria crassicaulis*	++
Gelidiaceae	*Gelidium elegans*	+		
Bonnemaisoniaceae	*Delisea japonica*	+		

-, not detected; +, detected; ++, strong; +++, very strong.

Table III. Inhibition Effects on Long Chain Aldehyde-Forming Enzyme

Inhibitor	Relative activity (%) *	
	0.1 mM	0.5 mM
Control	100	100
DTT	109	124
GSH	100	110
L-cysteine	112	115
Imidazole	1	2
EDTA	111	105
DEDTC (diethyldithiocarbamate)	93	95
NaCN	62	23
NaN_3	98	85
Glucose oxidase (20 U/ml)	0	
Peroxidase (5 U/ml)	40	

* Enzyme activities are expressed relative in mole number to heptadecadienal obtained from 18:2 with no addition.
S. D. =2%.

was examined. Palmitic acid (16:0) was the best substrate of all acids tested. The reactivity decreased with both an increase and a decrease in chain length with respect to 16:0. Linoleic acid and α-linolenic acid were good substrates. However, γ-linolenic acid was a very poor substrate. These observations were in agreement with the results for the acetone powder preparation of the seaweed (*29*), but not with the substrate specificity of α-oxidation systems in higher plants, e.g. cucumber fruit (*5*, *27*) and peanut cotyledons (*26*).

Table III shows the inhibitory effects of metal ligands, SH-reagents, chelating agents, glucose oxidase and peroxidase on the long chain aldehyde-forming activity. The enzyme activity was inhibited by removal of hydrogen peroxide using peroxidase and nonexistent under aerobic condition with glucose oxidase. Low (0.1 mM) and high (0.5 mM) concentrations of SH-reagents, dithiothreitol (DTT), cysteine and glutathione (GSH) did not inhibit, but stimulated slightly at high concentration. The heavy metal ligand, NaCN, showed inhibition at low concentrations and strong inhibition at high concentrations, whereas chelating agents e.g. ethylenediaminetetraacetic acid (EDTA), showed slight stimulation. When imidazole was added, a marked inhibition has been observed in common with the α-oxidation system in higher plants green leaves and germinating seeds (*5*, *30-32*). The hydrogen peroxide-generating system has been reported to be needed for α-oxidation in germinating seeds (*28*).

Enzymatic Formation of Long-Chain Aldehydes in Conchocelis-Filaments of a Red Seaweed *P. tenera*. The long-chain fatty aldehyde forming enzyme in the preparation from conchocelis-filaments of *P. tenera* enhanced its activity ca 20% by the addition of 0.2% Triton X-100. The pH optimum for linoleic acid was 7.5. Under anaerobic conditions, i. e. in the presence of glucose and glucose oxidase, the enzyme activity was negligible (only 5% of control). Certain metal ligands (such as KCN and NaN_3) showed a strong inhibitory activity (93 and 94%, respectively), whereas EDTA had no effect. These results suggest that the aldehyde-forming enzyme is probably responsible for the generation of an active species of molecular oxygen required to initiate the oxidative process (*5*). The substrate specificity for the enzyme was examined using a series of saturated long-chain fatty acids in which the chain length varied from dodecanoic acid to octadecanoic acid. Palmitic acid was the best substrate of all acids tested. Unsaturated acids such as oleic acid, linoleic acid, α-linolenic acid, γ-linolenic acid were moderate substrates (18:2 > 18:1 > α-18:3 = γ-18:3). These results were in agreement with the substrate-specificity reported for cucumber fruit (*5*) and peanut cotyledons (*26*). With polyenoic acids, such as arachidonic acid (20:4) and eicosapentaenoic acid (20:5), the enzyme showed no activity. The substrate specificity observed for the long chain aldehyde forming enzyme in *P. tenera* was different from that of the activity from thalli of *U. pertusa*. From these results, the substrate specificity for the enzyme activity from the filaments was considered to reflect the composition of volatile aldehydes in the essential oil from the filaments. The long chain aldehyde-forming activity in the thallus stage was much lower (5%) than that in the conchocelis. This fact suggests that the long chain aldehyde-forming enzyme

Table IV. Volatile Compounds Identified in the Essential Oils from Thalli Culture of *U. Pertusa*

Compounds	Peak area (%)	
	Thalli culture	Field fronds
Aldehydes		
(*Z*, *Z*, *Z*)-8, 11, 14-Heptadecatrienal	27.52	24.11
(*Z*, *Z*)-8, 11-Heptadecadienal	2.16	2.16
(*Z*)-8-Heptadecenal	5.27	3.33
(*Z*, *Z*, *Z*)-7, 10, 13-Hexadecatrienal	2.06	0.83
(*Z*, *Z*, *Z*)-6, 9, 12-Pentadecatrienal	-	1.44
n-Pentadecanal	5.60	11.46
n-Tetradecanal	-	1.55
n-Tridecanal	-	1.67
(*E*)-2-Nonenal	-	0.15
(*E*)-2-Octenal	-	0.29
Alcohols		
Phytol	1.55	2.55
Octadecanol	6.44	-
α-Cadinol	2.46	-
Cubenol	1.46	6.93
α-Terpineol	1.66	-
2-Ethylhexanol	6.62	-
Ketones		
6, 10, 14-Trimethylpentadecan-2-one	0.67	0.62
α-Ionone	1.39	0.64
β-Ionone	-	0.35
Lactones		
Dihydroactinidiolide	1.61	0.25
2, 3-Dimethyl-2-nonen-4-olide	-	1.00
Carboxylic acids		
Palmitic acid	2.14	0.60
Myristic acid	1.07	-
Ester		
(*S*)-3-Acetoxy-1, 5-undecadiene	-	0.24
Hydrocarbons		
(*Z*)-7-Heptadecene	10.74	26.79
n-Tetradecane	-	0.10
Limonene	0.89	-
Sulfur compounds		
Benzothiazole	0.49	0.61
Dimethyl trisulfide	1.20	-
Halide		
Chlorobenzene	0.91	-

activity might play a physiological role in the transition from the conchocelis stage to the thallus stage.

Volatile Components and Long Chain Fatty Aldehydes Formation in Thalli Culture of *U. pertusa*.

Long chain fatty aldehydes such as (*Z*)-8-heptadecenal, (*Z*, *Z*)-8, 11-heptadecadienal and (*Z*, *Z*, *Z*)-8, 11, 14-heptadecatrienal have been identified as major volatile compounds in the Ulvaceae essential oils (*4*). They have been shown to be produced enzymatically from unsaturated fatty acids in the field fronds of *U. pertusa* collected from the sea (*20*, *29*). Thus, it remained to establish unequivocally that the enzymatic activity does not derive from the attached bacteria or other epiphytes.

Volatile Compounds in Thalli Culture. A large mass of clean thalli culture, regenerated from protoplasts of *U. pertusa*, was obtained by cultivation in Provasoli's enriched sea water containing antibiotics (*36*). The essential oil was prepared from the thalli culture by a simultaneous distillation extraction (SDE) procedure (*37*). The yield of the essential oil was 8.0 X 10^{-3}% (cf. 4.2 X 10^{-3}% for the field fronds). Twenty-three of the volatile compounds in the oil were identified by GC and GC-MS (Table IV). The major characteristic compounds were long-chain fatty aldehydes as in the oil from the field fronds: mainly *n*-pentadecanal, (*Z*, *Z*, *Z*)-7, 10, 12-hexadecatrienal, (*Z*)-8-heptadecenal, (*Z*, *Z*)-8, 11-heptadecadienal, and (*Z*, *Z*, *Z*)-8, 11, 14-heptadecatrienal. They accounted for ca 40% of the oil and were considered to have the characteristic flavor of seaweeds. Dimethyl trisulfide, 2-ethylhexanol, limonene, α-terpineol, α-cadinol and octadecanol were found as minor components in the thalli culture. These compounds were absent from field fronds (Table IV), although a trace amount of dimethyl trisulfide was present (*38*). The sniffing test showed that the flavor in the culture of *U. pertusa* was similar to that of *Enteromorpha* sp. with the sulfide as a characteristic component (*33*). Thus, dimethyl trisulfide seemed to contribute closely to the flavor of the *Ulva* oil as a top note.

Enzymatic Formation of Long-Chain Fatty Aldehydes in Thalli Culture. Thalli culture or field fronds were homogenized with 50 mM Na-pyrophosphate buffer, pH 9.0 containing 0.2% Triton X-100. The homogenate was filtered through 4 layers of gauze and the filtrate centrifuged at 19,000 *g* for 20 min. Ninety-five per cent of the long chain aldehyde-forming enzyme activity was found in supernatant prepared from a thalli culture of *U. pertusa* (*36*). Enzymatic formation of long-chain fatty aldehydes for the solubilized activity was examined using saturated and unsaturated C_{18}-acids. The products were identified by GC and GC-MS and reversed phase HPLC of their 2, 4-dinitrophenylhydrazone derivatives. The long-chain aldehydes formed from unsaturated fatty acids by the enzymes of the thalli culture were the same as in the field fronds, i.e. (*Z*)-8-heptadecenal from oleic acid, (*Z*, *Z*)-8, 11-heptadecadienal from linoleic acid, (*Z*, *Z*, *Z*)-8, 11, 14-heptadecatrienal from α-linolenic acid, respectively. No activity was observed with stearic acid as the substrate. Oleic acid and linolenic

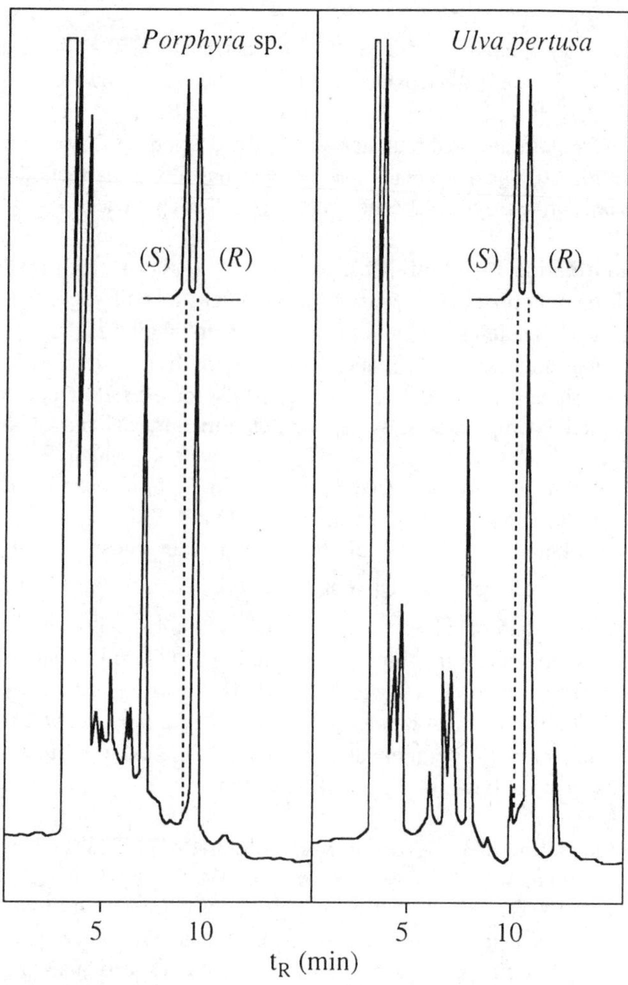

Figure 5. HPLC analysis of MTPA esters of 2-hydroxy-palmitic acid from thalli of *U. pertusa.*

acid were the best substrates. And, α-linolenic acid was a better substrate than γ-linolenic acid. This utilization of oleic acid by the culture preparation was different from that of the field fronds which shows a relatively low reactivity for the C_{18}-monoenoic acid (*20*, *29*).

Examination of the thalli culture on ST3ss and M23 media showed the culture was free of bacteria and epiphytes. Thus, the unsaturated C_{n-1}-aldehydes are produced from Cn-fatty acids by algal enzymes and not by the enzymes of associated organisms (*36*). The clean thalli culture lost its typical foliar thallus morphology and grew as regular and irregular long, tubular shapes (*36*). However, further studies are needed to explain the difference in the minor volatile compounds and in the substrate specificity for the long chain aldehyde-forming enzyme between the culture and the field fronds, i.e. whether or not they are due to the aseptic conditions or the different morphogenetic development of the fronds regenerated from the protoplasts of *U. pertusa*.

Mechanism of Biogeneration of Long Chain Fatty Aldehydes *via* Oxylipins in Seaweeds

In the course of an investigation into the nature of the biosynthetic intermediate of long chain fatty aldehydes, Hitchcock and James proposed that the (*S*)-2-hydroxy-fatty acid serves as an intermediate in the α-oxidation of fatty acids (*40-41*). On the other hand, another mechanism for α-oxidation of fatty acids in peanut cotyledons has been proposed by Shine and Stumpf (*34*), involving hydroperoxy-fatty acids rather than the hydroxy-fatty acids as a transitory intermediate. However, the hydroperoxy intermediate has not been studied so far in marine algae.

Enzymatic Formation of 2-Hydroxy-Palmitic Acid and 2-Oxo-Palmitic Acid from Palmitic Acid in Seaweeds. Fresh fronds of *U. pertusa* were homogenized in phosphate buffer (pH 7.5). The homogenate was extracted and, after methyl-esterification, the methyl esters were separated from the esterified fraction by silica gel chromatography. The 2-hydroxy- and 2-oxo-palmitic acid methylesters were identified in the esterified fraction of the green alga by direct comparison (Rt and MS) with authentic specimens, which were synthesized by O_2-bubbling into the THF solution containing HMPA in presence of LDA (*42*). Both acids were also detected in the red seaweed, *Porphyra* sp., although 2-oxo-phytanic acid has been isolated as a mammalian metabolite of phytanic acid and (*R*)-2-hydroxy-palmitic acid has been detected in higher plants (pea leaf and peanut cotyledons) during the course of investigations on the α-oxidation enzyme system (*40-43, 46*). The enzymatic formation of the 2-hydroxy- and 2-oxo-fatty carboxylic acids was confirmed by using [1-^{13}C] palmitic acid as a substrate for the crude enzyme solution prepared from the green seaweed, *U. pertusa*. The crude fraction containing methyl 2-hydroxy-palmitate was converted to its diastereomeric ester with (*S*)-(-)-MTPA-Cl. The enantiospecificity of the 2-hydroxy acid formation was examined by HPLC analyses. As shown in Figure 5, the MTPA ester of the separated 2-hydroxy-palmitate coincided with the

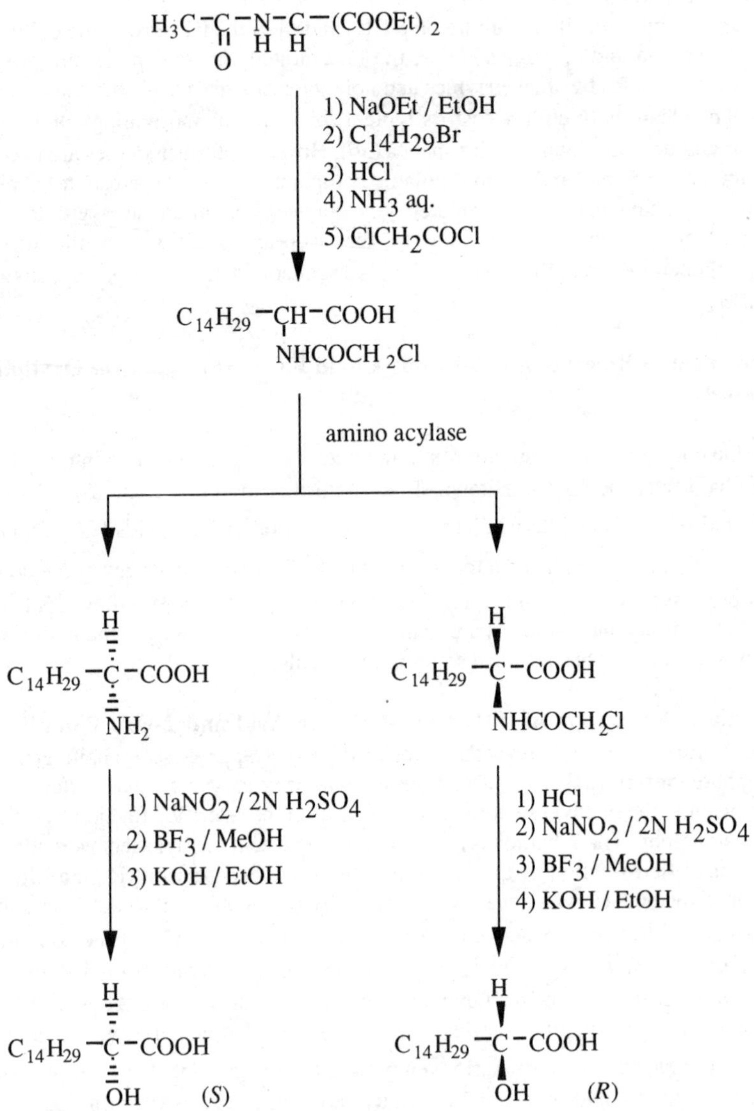

Figure 6. Syntheses of (*S*)-2-hydroxy- and (*R*)-2-hydroxy-palmitic acid.

authentic MTPA ester of (*R*)-2-hydroxy-palmitate, which was synthesized through unequivocal route (Figure 6) on HPLC analyses. The configuration of separated 2-hydroxy palmitate (*4*) was found to be (*R*)-form, and the enantiomeric (*S*)-isomer was not observed (Figure 5). Thus, the enzymatic production of (*R*)-2-hydroxy-palmitic acid was enantiospecific in the green seaweed, *U. pertusa* and the red seaweed, conchocelis filaments of *Porphyra* sp.

Detection of 2-Hydroperoxy-Palmitic Acid during Biogeneration of 2-Hydroxy-Palmitic Acid and Pentadecanal from Palmitic Acid in Seaweeds. If the (*R*)-2- or (*S*)-2-hydroxy-palmitic acid is a transitory intermediate of long chain fatty aldehyde formation, pentadecanal (PD) is expected to increase when the (*R*)-2- or (*S*)-2-hydroxy-palmitic acid as substrates was added to the homogenate of *U. pertusa*. However, the increase of PD was slight or nonexistent (*45*). These results suggest that both the (*S*)- and (*R*)-2-hydroxy-palmitic acid are not intermediates of pentadecanal formation from palmitic acid, but an alternative possible intermediate, 2-hydroperoxy-palmitic acid. Thus, the hydroperoxide was synthesized by inverse addition of the carbanion of palmitic acid to oxygen-saturated ether (*42, 44*). HPLC analyses of the reaction products showed a mixture of 2-hydroxy- and 2-hydroperoxy-palmitic acids (65:35). However, the hydroperoxide was successfully separated from 2-hydroxy-palmitic acid through silica gel column chromatography (CH_2Cl_2 : AcOEt = 10:0 ~ 0:10). The structures of the hydroperoxides were substantiated by ^{13}C-NMR analyses; the chemical shift of 2-hydroxy carbon (69.99 ppm), was lower than that of 2-hydroperoxy carbon (83.45 ppm). The formation of hydroperoxide from palmitic acid in a green seaweed, was explored using authentic 2-hydroperoxy-palmitic acid. Homogenate of a green seaweed, *U. pertusa* was used as an enzyme solution. A reaction mixture of homogenate and palmitic acid was incubated at 25°C for 20 min. Margaric acid (100 nmol) was added to the mixture as an internal standard at the end of the incubation. To the reaction mixture was added an 0.5% AcOEt-acetone (1/1, v/v) solution of ADAM (9-anthryldiazomethane) (50 μl). The ADAM derivative was extracted with diethyl ether / iso-propanol (9/1, v/v) (4 ml X 2). The combined extract solution was centrifuged at 2,000 rpm for 5 min. and was dried over $MgSO_4$. The extract was evaporated *in vacuo* and the residue was dissolved in AcOEt (500 μl). An aliquot (1 μl) was quantitatively analyzed by HPLC with a fluorescence detector (Ex. 365 nm; Em. 412 nm): Wakosil 5C8 column 4.6 mmϕ X 150 mm, mobile phase: CH_3CN / H_2O = 85 / 15, flow rate 1.0 ml / min., pressure 40 kg / cm^2. The retention times of most peaks were compared with those of authentic ADAM derivatives (Figure 7). As shown in Figure 7, peaks **1**, **2**, **3**, **4** and **5** coincided with authentic 2-hydroperoxy-palmitic acid, 2-hydroxy-palmitic acid, 2-oxo-palmitic acid, pentadecanoic acid and palmitic acid, repectively. The peak **1** shifted to peak **2** by addition of $NaBH_4$ to **1**. Thus, the peak **1** was tentatively identified as 2-hydroperoxy palmitic acid in the green seaweed, although it is neccesary to confirm by NMR of the isolate. Based on the above facts, a possible biogeneration pathway of long chain fatty aldehydes *via* oxylipin 2-hydroperoxy-isomers from fatty acid in seaweeds was suggested (Figure 8).

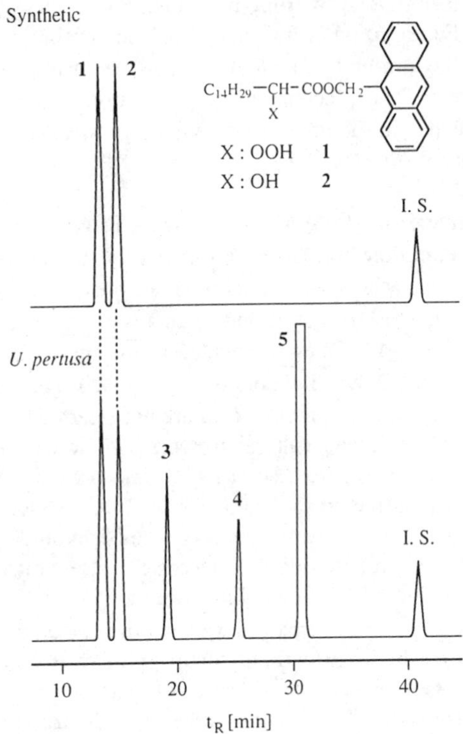

Figure 7. Detection of 2-hydroxy-palmitic acid from *U. pertusa* by HPLC.

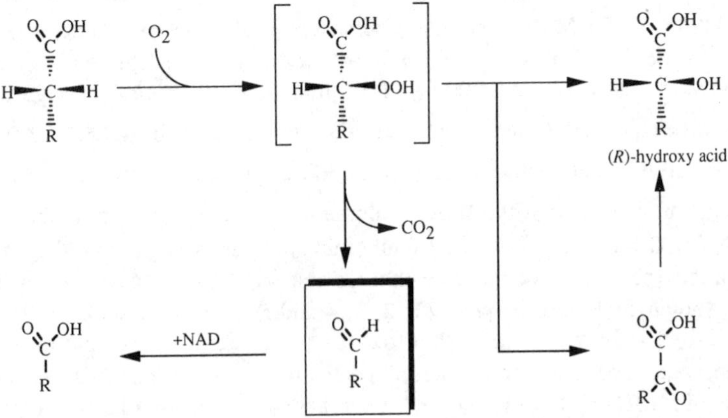

Figure 8. A possible biogeneration mechanism of long chain fatty aldehydes from fatty acids in seaweeds.

Acknowledgements

This work was supported in part by a Grant-in Aid (No. 07660140) from the Ministry of Education, Science and Culture, Japan.

Literature Cited

1. Maeshige, S. In *Biochemistry and Utilization of Marine Algae*; the Japanese Society of Scientific Fisheries Ed.; Koseishya Koseikaku: Japan, 1983; pp 143-153.
2. Gerwick, W. H.; Moghaddam, M.; Hamberg, M. *Arch. Biochem. Biophys.* **1991**, 290, 436-444.
3. Katayama, T. In *Physiology and Biochemistry of Algae*; Lewin, R. A., Ed.; Academic Press: New York, 1962; pp 467-473.
4. Kajiwara, T.; Hatanaka, A.; Kawai, T.; Ishihara, M.; Tuneya, T. *Bull. Japan. Soc. Sci. Fish.* **1987**, 53, 1901.
5. Galliard, T. and Matthew, J. A. *Biochim. Biophys. Acta* **1976**, 424, 26-35.
6. Takagi, Y.; Fujimori, T.; Kaneko, H.; Kato, K. *Agric. Biol. Chem.* **1981**, 45, 769-770.
7. Broekhof, N. L. J. M.; Witteveen J. G.; van der Weerdt, A. J. A. *Recl. Trav. Chim. Pays-Bas* **1986**, 105, 436-442.
8. Katayama, T. *Bull. Japan. Soc. Sci. Fish.* **1961**, 27, 75-84.
9. Kajiwara, T.; Hatanaka, A.; Kawai, T.; Ishihara, M.; Tsuneya, T. *J. Food Sci.* **1988**, 53, 960-962.
10. Kajiwara, T.; Kodama, K.; Hatanaka, A.; Matsui, K. In *Bioactive Volatile Comounds from Plants*; Teranishi, R.; Buttery, R. G.; Sugisawa, H. Eds.; *ACS Sym. Ser.* 525; American Chemical Society: Washington, DC, 1993; pp 103-120.
11. Flament, I.; Ohloff, G. In *Progress in Flavour Research*; Adda, J. Ed.; Elsevier Science Publishers B. V.: Amsterdam, 1984; pp 281-300.
12. Forss, D. A.; Dunstone, E. A.; Ramshaw, E. H.; Stark, W. *J. Food Sci.* **1962**, 27, 90-93.
13. Kemp, T. R.; Knavel, D. E.; Stoltz, L. P. *J. Agric. Food Chem.* **1974**, 22, 717-718.
14. Kemp, T. R. *J. Am. Oil Chem. Soc.* **1975**, 52, 300-302.
15. Yajima, I.; Sakakibara, H.; Ide, J.; Yanai, T.; Hayashi, K. *Agric. Biol. Chem.* **1985**, 49, 3145-3150.
16. Josephson, D. B.; Lindsay, R. C. In *Biogeneration of Aromas*; Parliment, T. H.; Croteau, R. Eds.; *ACS Sym. Ser.* 317; American Chemical Society: Washington, DC, 1986; pp 201-219.
17. Protean, P. J.; Gerwick, W. H. *Tetrahedron Lett.* **1992**, 33, 4393-4396.
18. Fleming, H. P.; Cobb, W. Y.; Etchells, J. L.; Bell, T. A. *J. Food Sci.* **1968**, 33, 572-576.
19. Kajiwara, T.; Kashibe, M.; Matsui, K.; Hatanaka, A. *Phytochemistry* **1990**, 29, 2193-2195.

20. Kajiwara, T.; Yoshikawa, H.; Saruwatari, T.; Hatanaka, A.; Kawai, T.; Ishihara, M.; Tsuneya, T. *Phytochemistry* **1988**, 27, 1643-1645.
21. Markovetz, A. J.; Stump, P. K. *Lipids* **1972**, 7, 159-164.
22. Hitchcok, C.; Mooris, L. J. *Eur. J. Biochem.* **1970**, 17, 39-42.
23. Laties, G. G.; Hoelle, C.; Jacobson, B. S. *Phytochemistry* **1972**, 11, 3403-3411.
24. Kang, M. Y.; Shimomura, S.; Fukui, T. *J. Biochem.* **1986**, 99, 549-559.
25. Kajiwara, T.; Matsui, K.; Hatanaka, A.; Tomoi, T.; Fujimura, T.; Kawai, T. *J. Appl. Phycol.* **1993**, 5, 225-230.
26. Shine, W. E.; Stump, P. K. *Arch. Biochem. Biophys.* **1974**, 162, 147-157.
27. Baardseth, P.; Slinde, E.; Thomassen, M. S. *Biochim. Biophys. Acta* **1987**, 922, 170-176.
28. Kajiwara, T.; Hatanaka, A.; Matsui, K.; Tomoi, T.; Idohara, T. *Phytochemistry* **1944**, 35, 55-57.
29. Kajiwara, T.; Yoshikawa, H.; Matsui, K.; Hatanaka, A.; Kawai, T. *Phytochemistry* **1989**, 28, 407-409.
30. Martin, R. O.; Stumpf, P. K. *J. Biol. Chem.* **1959**, 234, 2548-2554.
31. Hitchcok, C.; James, A. T. *Biochim. Biophys. Acta* **1966**, 116, 413-424.
32. Laties, G. C.; Hoelle, C.; Jacobsen, B. S. *Phytochemistry* **1972**, 11, 3403-3411.
33. Obata, Y.; Igarashi, H. *Bull. Japan. Soc. Sci. Fish.* **1951**, 17, 60.
34. Katayama, T. ; Tomiyama, T. *Bull. Japan. Soc. Sci. Fish.* **1951**, 17, 122.
35. Katayama, T. *Bull. Japan. Soc. Sci. Fish.* **1955**, 21, 420.
36. Fujimura, T.; Kawai, T.; Shiga, M; Kajiwara, T.; Hatanaka, A. *Bull. Japan. Soc. Sci. Fish.* **1989**, 55, 1353-1359.
37. Schultz, T. H.; Flath, R. A.; Mon, T. R.; Eggling, S. B.; Teranishi, R. *J. Agric Food Chem.* **1977**, 25, 446-449.
38. Sugisawa, H. *Fragrances J.* **1988**, 89, 25.
39. Shine, W. E.; Stumpf, P. K. *Arch. Biochem. Biophys.* **1974**, 162, 147-157.
40. Hitchcok, C.; Morris. L. J.; James, A. T. *Eur. J. Biochem.* **1968**, 3, 419-421.
41. Hitchcok, C.; Rose, A. *Biochem. J.* **1971**, 125, 1155-1156.
42. Konen, D. A.; Silbert, L. S.; Pfetter, P. E. *J. Org. Chem.* **1975**, 40, 3253-3258.
43. Vemecg. J.; Draye, J. P. *Biomed. Environ. Mass Spectrom.* **1988**, 15, 345.
44. Adam, W.; Cueto, U. *J. Org. Chem.* **1977**, 42, 28.
45. Kajiwara, T.; Kashibe, M.; Matsui, K.; Hatanaka, A. *Phytochemistry* **1991**, 30, 193-195.
46. Salim-Hanna, M.; Campa, A.; Giuseppe, C. *Photochemistry & Photobiology* **1987**, 45, 849.

Chapter 15

Elimination of Bitterness of Bitter Peptides by Squid Liver Carboxypeptidase

C. Kawabata, T. Komai, and S. Gocho

Technical Research Center, T. Hasegawa Company, Limited, 335-Kariyado, Nakahara-ku, Kawasaki 211, Japan

Molecular and enzymatic properties of serine carboxypeptidase (EC.3.4.1.6.1, CPase Top), isolated and refined from the common squid (*Todarodes pacificus*) liver, were studied. It was found that this enzyme reacts well at the C-terminal position of peptides having hydrophobic amino acids. Because of this property, it was anticipated that this enzyme would have the effect of eliminating bitterness of some peptides. This enzyme was used on bitter peptides prepared by hydrolysis of proteins with pepsin and trypsin. It was found that this CPase Top can eliminate bitter peptides prepared from soy protein and corn gluten.

Carboxypeptidase has been studied by many workers. It has been reported that this enzyme exists in many kinds of various microorganisms, plants, and animals (*1*, *2*, *3*). Because it is quite useful in the food industry, we studied how to produce it inexpensively. We found that this enzyme occurs in the liver of the common squid (*Toardes pacificus*) and determined its characteristic features. We found this enzyme to be useful in eliminating the bitterness of bitter peptides, which is described in this paper.

MATERIALS AND METHODS

Materials and Chemicals. N-acyl peptides were purchased from SIGMA Chemical Co. (St. Louis, MO, U.S.A.), and Peptide Institute Inc. (Osaka, Japan). Soy protein (New Fujipro) was purchased from Fuji Oil Co., Ltd. (Osaka, Japan). Casein (ALACID 720) was purchased from New Zealand Dairy Board (Wellington, New Zealand). Corn gluten (Gluten meal) was purchased from Nihon Shokuhin Kakou Co., Ltd. (Tokyo, Japan). Pepsin (EC.3.23.1)(Pepsin (1:10000), Trypsin (EC.3.4.21.4) (Trypsin Type IX), and L-Glycyl-L-Leucine (Gly-Leu) were purchased from SIGMA Chemical Co. Common squid (*Toardes pacificus*) livers were purchased from Nakamura Gyogyou-bu Co., Ltd. (Aomori, Japan).

0097–6156/96/0637–0167$15.00/0

Preparation of Enzyme. Separation and purification of the enzyme from the liver of the common squid were as follows: Squid livers were mashed and dispersed in five volumes of distilled water, then acidified to pH 4.0. Much oil was eliminated. This was followed by ultrafiltration, salting out with ammonium sulfate (50% saturation), dialyzing and freeze-drying in vacuo, to yield a crude enzyme. A purified enzyme was obtained from this crude enzyme by using column chromatography. Active fractions were separated by cation exchange resin, Mono S (Pharmacia) and further purified by gel-filtration column of Superdex 75 (Pharmacia). The active fractions were collected as the purified enzyme.

Assay of Enzyme Activity. Activity of this enzyme was determined by the increase in ninhydrin color after hydrolysis of Carbobenzoxy-L-glutamyl-L-tyrosine (Z-Glu-Tyr) as substrate at 30°C, pH 3.1. One katal of acid carboxypeptidase activity was defined as the amount of enzyme required to liberate 1 mol of C-terminal amino acid per second.

Preparation of Bitter Peptides. The bitter peptides were prepared by hydrolysis of soy protein or corn gluten with pepsin and hydrolysis of casein with trypsin. In the preparation of bitter peptides from soy protein, a 5% suspension of soy protein was hydrolyzed by pepsin [enzyme/substrate ratio = 1/100 (w/w)] at 37°C at pH 2.0. The water soluble portion was dialyzed and then freeze dried to yield bitter peptides. Bitter peptides from corn gluten and casein were prepared in a similar manner.

Evaluation of Bitterness and Measurement of Liberated Amino Acids. Bitterness was evaluated sensorially by 5 members of a sensory panel by comparing samples to a standard aqueous bitter standard. The standard for bitterness was a Gly-Leu solution, with 1.0% concentration scored as 10 points and 0.1% as 1. Liberated amino acids were measured by the ninhydrin methods after samples were de-proteinized with trichloroacetic acid solution and represented as μmol/ml of Leu.

RESULTS AND DISCUSSION

Enzyme Preparation. The purified enzyme from the squid liver was found to be approximately uniform judging from the SDS-PAGE pattern. This final preparation was purified about 470-fold from squid livers. The molecular weight of this enzyme was estimated at 42,000 by SDS-PAGE and gel filtration. The characteristics of this enzyme were studied with this purified material.

Mode of Action of the Enzyme. Synthetic substrate, Z-Gly-Pro-Leu-Gly, was incubated with this purified enzyme at 30°C, pH 3.1. Measurements of the released amino acids were made with an amino acids analyzer (Hitachi, Model L-8500) throughout the time of incubation. From the reaction of the enzyme on the substrate, Gly at the C-terminus was first released. Next released was Leu, which is in the penultimate position. From the sequential release of amino acids from the C-terminus, it was confirmed that the enzyme is a carboxypeptidase, which was named CPase Top.

Characteristics of CPase Top. With further study of the characteristics of CPase Top, it was found that the optimum temperture was 40°C, with it being stable under 45°C, and that its optimum pH was 4.0, with stability in the range of pH 2-6. The isoelectric point was 6.0. The enzyme activity was almost completely inhibited by *p*-chloromercuribenzoic acid (PCMB), monoiodoacetic acid, diisopropyl fluorophosphate (DFP), $HgCl_2$, and partially inhibited by phenylmethanesulfonyl fluoride (PMSF), *N*-tosyl-L-phenylalanyl chloromethylketone (TPCK), chimostatin, and $CuSO_4$. These results indicate that this enzyme is a member of the serine carboxypeptidase family (EC.3.4.16.1).

Substrate Specificity. The relative hydrolytic activity of CPase Top on various *N*-acyl-peptides was obtained by reacting at 30°C, pH 3.1 for 20 minutes. The relative activities are shown with activity to Z-Glu-Tyr as 100% (Table I). CPase Top shows a preference for peptides having hydrophobic amino acids as Z-Phe-Leu, Z-Tyr-Phe and Z-Tyr-Phe-Leu. When the penultimate amino acid from the C-terminus was a hydrophobic or bulky amino acid as Phe, Tyr, Leu and Glu, the hydrolysis rate of the sustrate was high. When Gly and Pro were penultimate to the C-terminus, activity decreased. Similarly, the same specificity of changing relative activity was obtained for substrates by amino acids penultimate to the C-terminus from CPase produced from *Aspergillus saitoii* (*1*). From this substrate specificity, it was assumed that by releasing a hydrophobic amino acid at the C-terminus of soy peptides and other bitter peptides, this enzyme could eliminate the troublesome bitterness of peptides produced by enzymatic hydrolysis of proteins (*4-8*) in food processing. To prove the above assumption to be true, bitter peptides were prepared from soy protein, casein, and corn gluten, and studies were conducted to eliminate the bitterness of the bitter peptides by using the crude enzyme from squid liver.

Hydrolysis of Bitter Peptides from Soy Protein by CPase Top. Three bitter peptides were incubated with crude CPase Top at 30°C, pH 4.0, with the bitterness and the amount of liberated amino acids being evaluated throughout incubation. When a 1% bitter peptide solution from soy protein was incubated with crude CPase Top (enzyme/substrate ratio = 1 μkat/g), the bitterness diminished as reaction time increased. The bitterness was almost completely eliminated after 15 hr (Fig. 1). The amount of amino acids liberated increased with time of incubation.

Hydrolysis of Bitter Peptides from Casein by CPase Top. When a 0.5% solution of bitter peptides from casein was incubated with crude CPase Top (enzyme/substrate ratio = 2 μkat/g), most of the bitterness was eliminated after 2 hr, but some bitterness remained even after 15 hr (Fig. 2). The liberated amino acids increased with time similarly as with the soy bitter peptides.

Hydrolysis of Bitter Peptides from Corn Gluten by CPase Top. When a 0.5% bitter peptides solution was incubated with crude CPase Top (enzyme/substrate ratio = 2 μkat/g), bitterness diminished similarly in a short time, and was almost undetectable after 15 hr (Fig. 3). Liberated amino acids increased similarly as in the other bitter peptide studies.

Table I Relative Rates of Hydrolysis for CPase Top on Range of Small *N*-Acyl-peptide

Peptide	Relative activity(%)
Z-Phe-Leu[P]	483
Z-Phe-Ala[P]	219
Z-Tyr-Phe-Leu[P]	120
Z-Phe-Tyr[P]	114
Z-Phe-Pro[P]	12
Z-Gly-Pro-Leu-Gly[P]	309
Z-Tyr-Phe[P]	201
Z-Tyr-Glu[P]	188
Z-Phe-Tyr-Leu[P]	172
Z-Glu-Phe[P]	150
Z-Glu-Tyr	(100)
Z-Gly-Val	24
Z-Gly-Phe	21
Z-Gly-Met	19
Z-Gly-Leu	17
Z-Gly-Pro-Leu-Gly-Pro	3
Z-Gly-Pro	0
B-Gly-Lys	0
Z-Pro-Pro[P]	0

P : Partially insoluble at 30℃, pH3.1
Z : Carbobenzoxy-
B : Benzoyl-

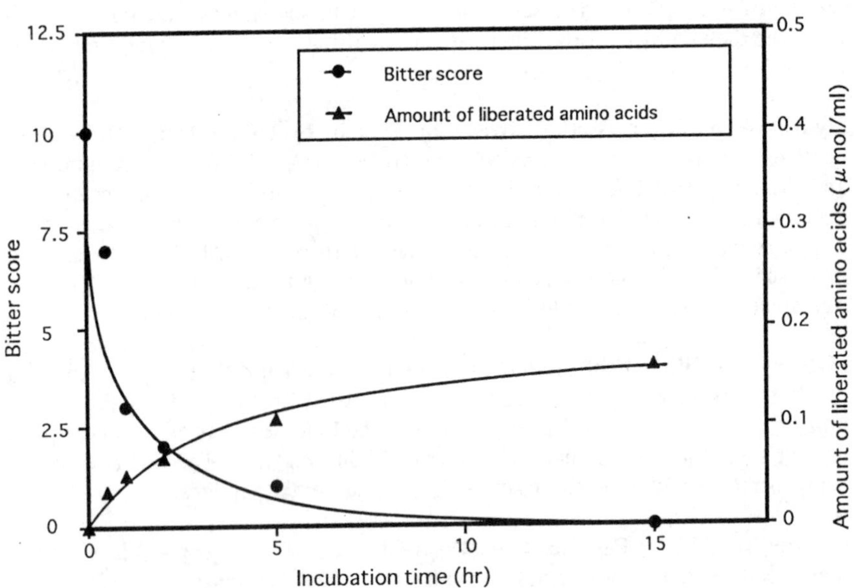

Figure 1. Hydrolysis of bitter peptides from soy protein by CPase Top. Reaction conditions: 30°C, pH 4.0.

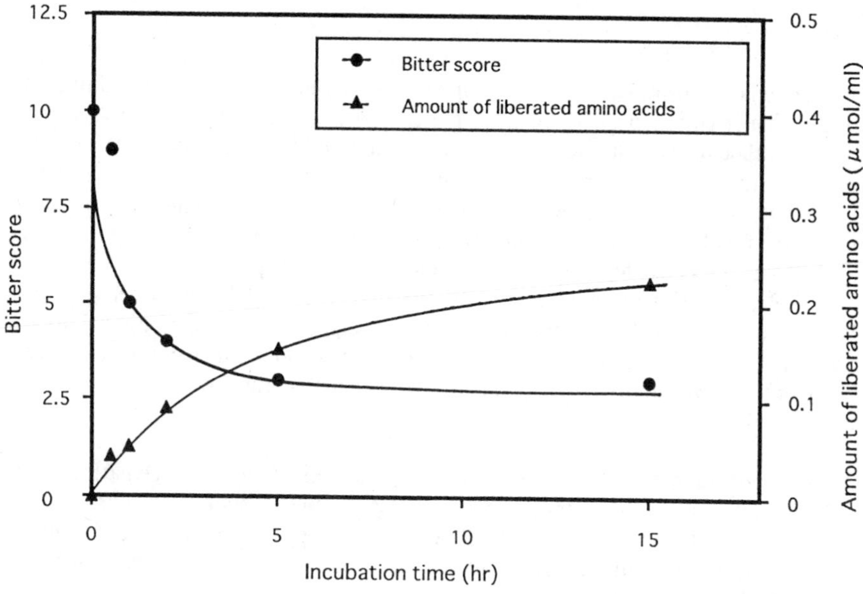

Figure 2. Hydrolysis of bitter peptides from casein by CPase Top. Reaction conditions:, 30°C, pH 4.0.

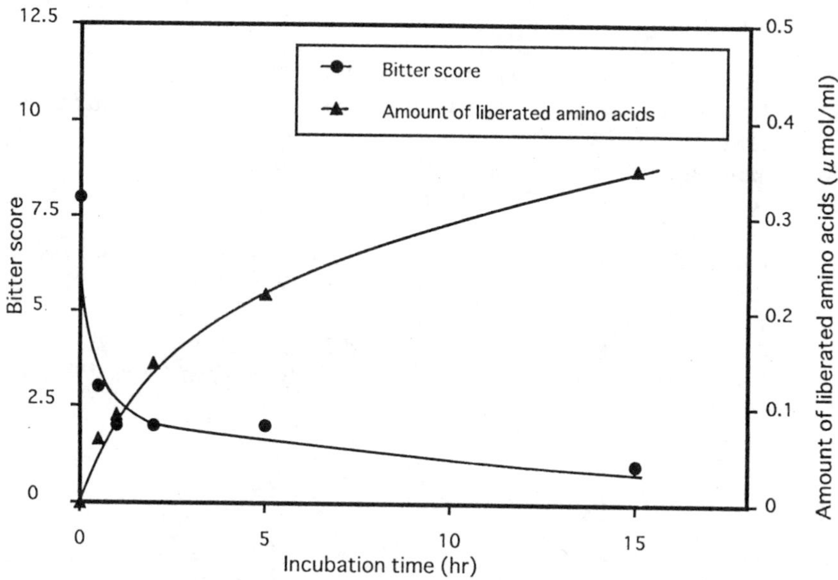

Figure 3. Hydrolysis of bitter peptides from corn gluten by CPase Top. Reaction conditions: 30°C, pH 4.0 .

CONCLUSION

It has been demonstrated that CPase Top enzyme is useful for eliminating bitterness originating from soy protein and corn gluten. With respect to the bitterness from casein, although the bitterness was not completely eliminated, it was effective in reducing most of the bitterness. It is postulated that the bitterness of peptides is due to the hydrophobic amino acids in the C-terminus position (*4*). This assumption is necessary for the elimination of bitterness by liberating a hydrophobic amino acid at the C-terminus by the substrate specificity of CPase Top.

Because of the high effectiveness of CPase Top to eliminate bitterness, and because it is obtained from the edible parts of squids (Japanese traditonal food, Shiokara), the safety of this enzyme is assured. It is hoped that the development of this method for eliminating the bitterness of bitter peptides will be used widely in the food industry.

Acknowledgements. The authors thank Professor Eiji Ichishima, Department of Applied Biological Chemistry, Faculty of Agriculture, Tohoku University, and Professor Souichi Arai, Department of Applied Biological Chemistry, Division of Agriculture and Agricultural Life Science, The University of Tokyo, for their most valuable advice and suggestions.

Literature Cited

1. Ichishima, E. *Biochem. Biophys. Acta* **1972**, *258*, 274-288.
2. Umetsu, H.; Abe, M.; Nakai, T.; Watanabe, S.; Ichishima, E. *Food Chem.* **1981**, *7*, 125-138.
3. Mellors, A. *Arch. Biochem. Biophys.* **1971**, *144*, 281-285.
4. Arai, S.; Yamashita, M.;, Kato, H.; Fujimaki, M. *Agr. Biol. Chem.* **1970**, *34*, 729-738.
5. Lalasidis, G.; Sjöberg, L. B. *J. Agric. Food Chem.* **1978**, *26*, 742-749.
6. Roland, J. F.; Mattis, D. L.; Kaing, S.; Alm, W. L. *J. Food Sci.* **1978**, *43*, 1491-1493.
7. Clegg, K. M.; McMillan, A. D. *J. Food Technol.* **1974**, *9*, 21-29.
8. Umetsu, H.; Matsuoka, H.; Ichishima, E. *J. Agric. Food Chem.* **1983**, *31*, 50-53.

Chapter 16

Microbial Transformation of Monoterpenes: Flavor and Biological Activity

Hiroyuki Nishimura[1] and Yoshiaki Noma[2]

[1]Department of Bioscience and Technology, School of Engineering, Hokkaido Tokai University, Sapporo 005, Japan
[2]Faculty of Domestic Sciences, Tokushima Bunri University, Yamashiro-cho, Tokushima 770, Japan

Microbial transformation of (-)-carvone in *Mentha spicata* and 1,8-cineole in *Eucalyptus* species was studied from the viewpoint of the flavor and biological activities. (-)-*cis*-Carveol, one of products biotransformed from (-)-carvone, was transformed by *Streptomyces bottropensis* to produce a novel compound, (+)-bottrospicatol which exhibits the germination inhibitory activity against plant seeds. Furthermore, 1,8-cineole was transformed by a strain of *Aspergillus niger* to produce 3-hydroxycineoles. Hydrogenolysis of 3-hydroxycineoles afforded *p*-menthane-3,8-diols (*cis* and *trans*) which exhibit repellent activity against mosquitoes. The flavor of microbial transformation products was evaluated. There is a growing interest among biochemists in the microbial transformation of natural products in terms of the production of economically useful chemicals. In connection with effective utilization of terpenoids which are major constituents in higher plants, the microbial transformation of (-)-carvone (ca. 70% content) in *Mentha spicata* oil and 1,8-cineole (40-70% content) in *Eucalyptus* oils has been investigated from viewpoint of the flavor and biological activities.

Growth Conditions of Microorganisms. Microorganisms and growth conditions were as follows. A liquid medium having the following composition (%, w/w): glucose 1%; meat extract 0.5%; polypeptone 0.5%; NaCl 0.3% was prepared in distilled water. Strains of *Streptomyces* A-5-1, *Streptomyces bottropensis* and *Aspergillus niger* isolated from soil were inoculated and cultured under static conditions for 3 days at 30°C. After full growth of the microorganism, (-)-carvone, (-)-*cis*-carveol and 1,8-cineole (0.4 - 1.5 g/l medium) were individually added into the medium. The microorganism were further cultivated for 7 to 15 days under the same conditions.

Chromatography and Identification of Metabolic Products. Each culture broth (12 liters) was extracted with ether (3 liters x 4) and the extract was dried over anhydrous sodium sulfate. The ether extract was analyzed by gas liquid chromatography (GLC): Shimadzu GC-4C equipped with a stainless steel column (3 m x 3 mm i.d.) packed with 10% PEG-20M on 80 - 100 mesh of Celite 545.

0097–6156/96/0637–0173$15.00/0

Metabolic products were identified by the interpretation of spectral data which were obtained by using the following instruments. Infrared (IR) and mass spectra were taken on a Hitachi-285 grating spectrometer and a JEOL JMS-D300 spectrometer, respectively. NMR spectra were measured on a JEOL JNM-FX200FT spectrometer (200 MHz) in $CDCl_3$ with $(CH_3)_4Si$ as an internal standard. Optical rotations were measured on a JASCO DIP-4 spectrometer.

Microbial Transformation of (-)-Carvone (1). Microbial transformation of (-)-carvone (**1**) by bacteria (*1 - 3*), yeast (*4*), fungi (*5*) and photosynthetic microorganism, *Euglena* species (*6*) has been investigated. It was shown that (-)-carvone (**1**) was mainly converted via (+)-dihydrocarvone (**2**) to either (+)-neodihydrocarveol (**4**) or (-)-dihydrocarveol (**5**), or via (+)-isodihydrocarvone (**3**) to either (+)-isodihydrocarveol (**6**) or (+)-neoisodihydrocarveol (**7**) (Figure 1). However, in some actinomycetes including a strain of *Streptomyces*, A-5-1 and a strain of *Nocardia*, 1-3-11, the major metabolic products of (-)-carvone were (-)-*trans*-carveol (**8**) and (-)-*cis*-carveol (**9**) by 1,2-reduction (*7*) (Figure 2).

Microbial Transformation of (-)-*cis*-Carveol (9). Carveol is known to be converted to carvone (*7*) or 1-*p*-menthane-6,9-diol (*8,9*) by microorganisms. However, (-)-*cis*-carveol (**9**) was transformed to produce novel bicyclic monoterpenes (**10** in Figure 2) by a strain SY-2-1, which was isolated from soil and identified to be *Streptomyces bottropensis* (*10*). The time course for the conversion of (-)-*cis*-carveol (**9**) is shown in Figure 3. After ten days, more than 95% of **9** was converted to products **a-d**. The major product (**a** in Figure 3, 85% yield) was separated and purified by column chromatography on SiO_2 and by preparative GLC. The IR spectrum of the unknown compound showed absorptions due to a hydroxyl group at 3420 cm^{-1} and olefin at 1640 cm^{-1}. The IR spectrum differed from that of (-)-*cis*-carveol, the starting material, in terms of the lack of absorption due to the isopropenyl group at 880 cm^{-1}. The molecula formula was determined to be $C_{10}H_{16}O_2$ from the high resolution FI-MS spectrum, 168.1139 (M^+). The ^{1}H-NMR spectrum (JEOL JNM-FX 200FT, 200 MHz) indicated the presence of a methyl group on a tertiary carbon (δ1.24, 3H, s, H-10), a methyl group on a trisubstituted double bond (δ1.69 - 1.72, 3H, H-7), a trisubstituted double bond adjacent to methylene (δ5.23, 1H, broad s, H-2), a methylene carrying a hydroxyl group (δ3.57 and 3.71, 1H each, d, *J*=11.4 Hz, H-9), and a methine between a double bond and an ethereal oxygen (δ4.05, 1H, broad d, band width=12 Hz, H-6 cf. in (-)-*cis*-carveol, band width=24 Hz; in (-)-*trans*-carveol, band width=6 Hz). The ^{1}H-NMR spectrum differed from that of (-)-*cis*-carveol in terms of the lack of any signal due to the *exo*-methylene protons at about δ4.65, the appearance of a pair of 1H-doublets (H-9) at δ3.5 - 3.8, and the high field shift of methyl proton signal to δ1.24. In addition, the change of band width of the peak of H-6 suggested some conformational changes.

In the ^{1}H-NMR spectrum of the monoacetate, which was obtained by acetylation of the unknown compound with acetic anhydride in pyridine, the carbinyl proton signals shifted to lower fields (δ4.08 and 4.17). This result confirmed the presence of a hydroxymethyl group.

Based on the evidence mentioned above, it was concluded that product (**a**) in Figure 3 was (*4R, 6R*)-(+)-6,8-oxidomenth-1-en-9-ol (**10** in Figure 2), which is a novel

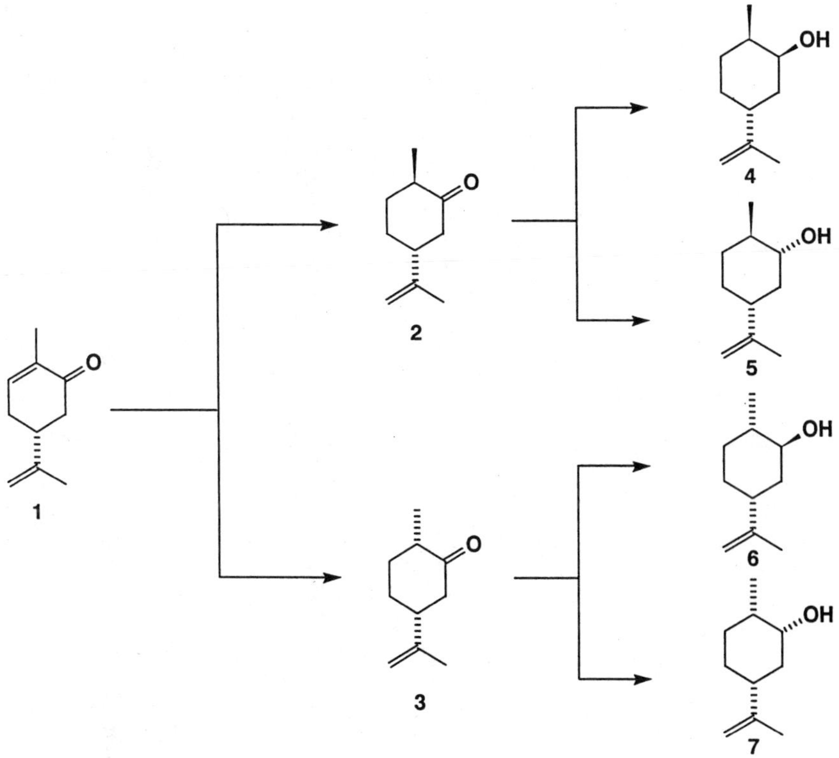

Figure 1. Reductive Transformation of (-)-Carvone (**1**) by Microorganisms.

compound, named (+)-bottrospicatol. However, the relative configuration at the C-8 substituents of (+)-bottrospicatol has not been determined by IR, mass and NMR spectra (*10*).

Absolute Configuration of (+)-Bottrospicatol (10). To determine the stereochemistry, synthesis of (+)-bottrospicatol and isobottrospicatol (C_8-epimer) was carried out according to Figure 4. (-)-Carvone (**1**) was oxidized by *m*-chloroperbenzoic acid in dry ether to give the diastereomixture of 8,9-epoxycarvone (85%), colorless oil; $[\alpha]_D^{23}$ -26.6° (*c*=0.094, $CHCl_3$); EI-MS *m/z* (70eV) 166 (M^+, 1%), 151(3), 109(100), 108(75), 82(31), 54(14), 43(18); IR ν_{max}^{neat} cm^{-1} 1660 (C=O), 1240, 890, 820 (epoxide); NMR (in $CDCl_3$) δ1.33 (3H, s, C_{10}-methyl), 1.83 (3H, s, C_7-methyl), 2.60 (2H, s, C_9-methylene), 6.70 (1H, m, C=C-H). The resulting epoxyketones were stereoselectively reduced by sodium borohydride to give 8,9-epoxy-*cis*-carveols and the successive reaction of epoxycarveols with 0.1N sulfuric acid led to the diastereomixture of bottrospicatol (68%). Two isomers (**10a** and **10b**) were separated by SiO_2 column chromatography using ether-hexane as the eluting solvent and obtained in the ratio of 6 to 4. As a result, **10a** was indicated to be identical with natural (+)-bottrospicatol

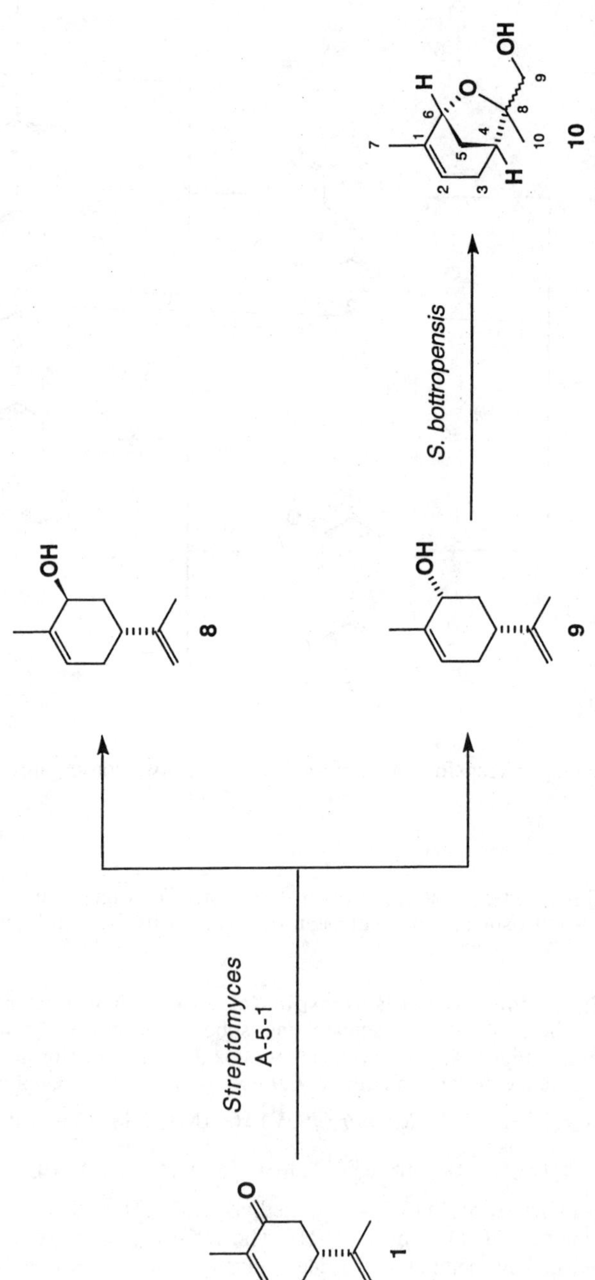

Figure 2. Major Metabolic Products of (-)-Carvone (**1**) by *Streptomyces* Species.

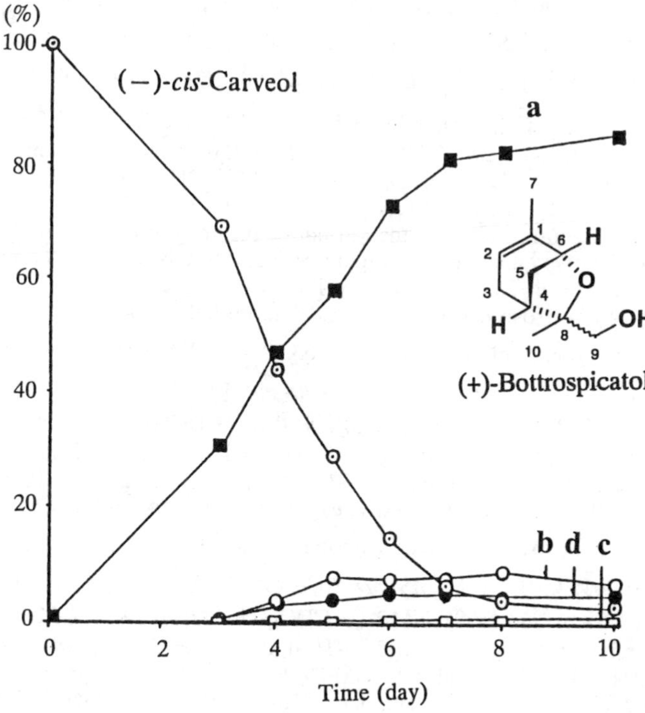

Figure 3. Time Course for the Conversion of (-)-Carveol by *S. bottropensis*, SY-2-1. Reproduced with permission from ref. 10. Copyright 1982 Japan Society for Bioscience, Biotechnology, and Agrochemistry.

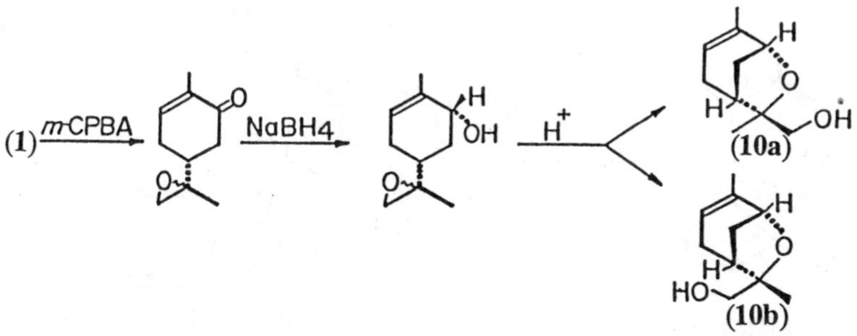

Figure 4. Synthesis of (+)-Bottrospicatol (**10a**) and Its Isomer (**10b**). Reproduced with permission from ref. 11. Copyright 1983 Japan Society for Bioscience, Biotechnology, and Agrochemistry.

concerning its *Rf* value of TLC and spectral data. Subsequently, the *p*-bromobenzoyl ester of **10a** was prepared and crystallized (Figure 5), mp 75.8 - 76.5°C; $[\alpha]_D^{16}$ +16.1° (*c*=0.050, $CHCl_3$); FI-MS *m/z* 352 (M^++2, 100%), 350 (M^+, 99); EI-MS *m/z* 137 (100%), 93 (90), 43 (50); IR ν_{max}^{KBr} cm^{-1} 1720 (C=O), 1600 (benzene C=C); NMR (in $CDCl_3$) δ1.32 (3H, s, C_{10}-Me), 1.71 (3H, dd, *J*=3.67 and 2.20 Hz, C_7-Me), 1.90 (1H, d, *J*=10.21 Hz, C_5-H*exo*), 2.22 - 2.39 (4H, m, C_3-methylene, C_4-methyne, C_5-H*endo*), 4.11 (1H, d, *J*=4.88 Hz, C_6-methyne), 4.35 (2H, s, C_9-methylene), 5.25 (1H, m, C=C-H), 7.58 (4H, d, benzene).

The absolute configuration of the (+)-bottrospicatol *p*-bromobenzoyl ester was confirmed by X-ray crystallography (Figure 6). The crystals are orthorhombic, space group $p2_12_12_1$ with *a*=7.274(1), *b*=33.531(5), *c*=6.670(2)Å, *Z*=4. The chemical structure of (+)-bottrospicatol (**10a**) was determined as (*4R, 6R, 8R*)-(+)-6,8-oxidomenth-1-en-9-ol, a colorless oil; $[\alpha]_D^{16}$ +63.7° (*c*=0.056, $CHCl_3$); FI-MS *m/z* 169 (M^++1, 42%), 168 (M^+, 100), 133 (33); EI-MS *m/z* 150 (M^+-H_2O, 0.1%), 137 (10), 94 (69), 93 (60), 79 (15), 77 (13), 43 (100); NMR (in $CDCl_3$) δ1.24 (3H, s, C_{10}-Me), 1.71 (3H, d, C_7-Me), 1.85 (1H, d, *J*=10.45 Hz, C_5-H*exo*), 2.20 - 2.35 (4H, m, C_3-methylene, C_4-methyne,C_5-H*endo*), 3.57 and 3.71 (1H each, d, *J*=10.74 Hz, C_9-methylene), 4.07 (1H, d, C_6-methyne), 5.24 (1H, broad s, C=C-H). (+)-Isobottrospicatol (**10b**) was formed as colorless needles, mp 24.0 - 24.5°C; $[\alpha]_D^{16}$ +108.6° (*c*=0.061, $CHCl_3$); EI-MS *m/z* 168 (M^+, 0.2%), 150 (M^+-H_2O, 0.1), 137 (29), 93 (60), 45 (16), 43 (100); NMR (in $CDCl_3$) δ1.32 (3H, s, C_{10}-Me), 1.73 (3H, dd, *J*=3.90 and 1.95 Hz, C_7-Me), 1.87 (1H, d, *J*=9.76 Hz, C_5-H*exo*), 1.98 (1H, s, OH), 2.13 (1H, dd, *J*=9.53 and 4.45 Hz, C_5-H*endo*), 2.18 (1H, m, C_4-methyne), 2.27 (2H, m, C_3-methylene), 3.32 and 3.41 (1H each, d, C_9-methylene), 4.01 (1H, d, C_6-methyne), 5.26 (1H, m, C=C-H).

The antipodes of (+)-bottrospicatol and (+)-isobottrospicatol were also prepared from (+)-carvone.

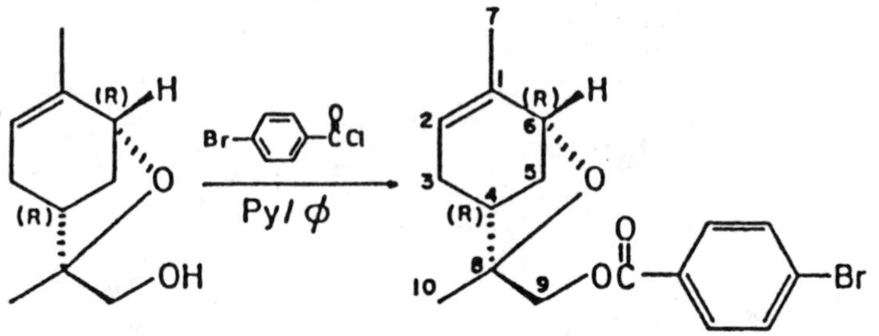

(10a)

(4R,6R,8R)-(+)-6,8-Oxidomenth-1-en-9-ol *p*-Bromobenzoyl Ester

Figure 5. Preparation of (+)-Bottrospicatol *p*-Bromobenzoyl Ester.

Biological Activity of (+)-Bottrospicatol and Related Compounds. Isomers and derivatives of bottrospicatol were prepared by the procedure shown in Figure 7. The chemical structure of each compounds was confirmed by the interpretation of spectral data. The effects of all isomers and derivatives on the germination of lettuce seeds were compared as shown in Figure 8. The germination inhibitory activity of (+)-bottrospicatol (**10a**) was the highest of isomers (upper in Figure 8) (*11*). Interestingly, (-)-isobottrospicatol (**11b**) was not effective even in a concentration of 500 ppm. (+)-Bottrospicatol methyl ether (**12**) and esters (**13**) exhibited weak inhibitory activities (lower in Figure 8). The inhibitory activity of dihydrobottrospicatol (**14**) which is prepared by hydrogenation of (+)-bottrospicatol was as high as that of (+)-bottrospicatol (**10a**).

Furthermore, an oxidized compound, (+)-bottrospicatal (**15**) exhibited much higher activity than (+)-bottrospicatol (**10a**). So, the germination inhibitory activity of (+)-bottrospicatal against several plant seeds was examined (Figure 9). The result indicates that (+)-bottrospicatal is a selective germination inhibitor.

Microbial Transformation of 1,8-Cineole (16). The monoterpene cyclic ether, 1,8-cineole (eucalyptol) is well known as a significant constituent in *Eucalyptus* and some other essential oils. It has a characteristic camphoraceous odor and pungent taste. It is extensively used in pharmeceutical preparations for external application and also as nasal spray (*12 - 15*). *Eucalyptus radiata* var. *Australiana* leaves produce the largest amount of essential oil among six hundred *Eucalyptus* species. Results of the analysis showed a 1,8-cineole (**16** in Figure 10) content in the oil of *ca.* 75% corresponding to 31 mg/g fr.wt. leaves (*16*). Studies on microbial transformation of **16** are very important in terms of the exploration of economically useful chemicals. However, the research reports are very few (*17 -20*). MacRae et al. (*17*) have reported that the metabolism of 1,8-cineole (**16**) by *Pseudomonas flava* results in oxidation at the C-2 position of **16** to give 2-*endo*-hydroxycineole (**17**), 2-*exo*-hydroxycineole (**18**) and 2-oxocineole (**19**) as shown in Figure 10.

In the course of our studies on the microbial transformation of terpenes, this paper deals with the identification of oxidation products of **16** by *Aspergillus niger* and chemical conversion to biologically active compounds.

Microbial Transformation Products from 1,8-Cineole. The culture broth of *A. niger* cultured in the presence of 1,8-cineole (**16**, 1.5 g/l) for 7 days was extracted with ether. Gas chromatographic (GC) analysis of the extract revealed the formation of five products from **16**. After the isolation of each products by column chromatography, 2-*endo*-hydroxycineole (**17**), 3-*endo*-hydroxycineole (**20**), 3-*exo*-hydroxycineole (**21**), 2-oxocineole (**19**) and 3-oxocineole (**22**) were identified from the interpretation of physico-chemical data (Figures 10 and 11) (*18*). All compounds were isolated as racemates.

(±)-2-*endo*-Hydroxycineole (**17**). The crystalline 2-*endo*-alcohol had mp 66.0 - 66.5°C and $[\alpha]_D^{22}$ ±0° (*c*=0.2, EtOH). IR ν_{max}^{KBr} cm^{-1}: 3385 (OH), 1130 (C-O-C), 1065 (alcoholic C-O). ^{1}H-NMR $\delta_{TMS}^{CDCl_3}$: 1.10 (3H, s, CH_3 at C-7), 1.20 (3H, s, CH_3 at C-9), 1.28 (3H, s, CH_3 at C-10), 2.52 (1H, m, $J_{3exo,2}$=10 Hz, J_{gem}=13 Hz, $J_{3exo,4}$=3 Hz, $J_{3exo,5exo}$=3 Hz attributed to W-conformation, C_3-H_{exo}), 3.72 (1H, ddd, $J_{2,3exo}$=10 Hz, $J_{2,3endo}$=4 Hz, $J_{2,6exo}$=2 Hz, $\underline{H}$COH). ^{13}C-NMR $\delta_{TMS}^{CDCl_3}$: 22.2 (t, C-5), 24.1 (q, C-7), 25.0 (t, C-6), 28.6 (q, C-10), 29.1 (q, C-9), 34.4 (d, C-4), 34.7 (t, C-3), 71.1 (d, C-2), 72.6 (s, C-1), 73.5 (s, C-8), MS *m/z* (%): 170 (M^+, 7.0), 155

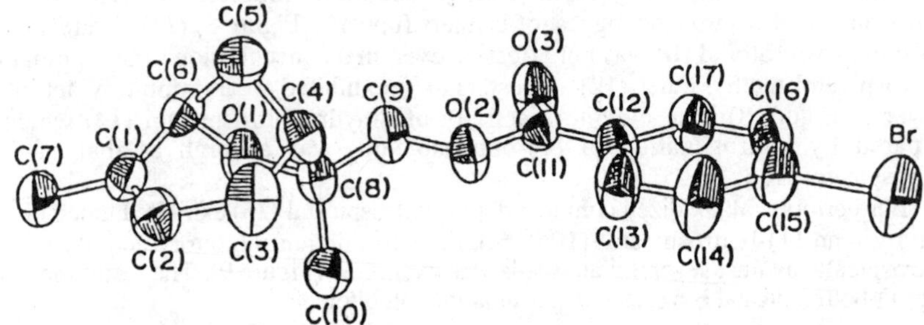

Figure 6. X-Ray Crystallography of (+)-Bottrospicatol *p*-Bromobenzoyl Ester. Reproduced with permission from ref. 11. Copyright 1983 Japan Society for Bioscience, Biotechnology, and Agrochemistry.

12

MeI, NaH

1atm H_2, PtO_2 / AcOH

14

(RCO)$_2$O or RCOCl/Py

CrO_3/Py

(+)-Bottrospicatol

13

R= CH_3, C_2H_5, n-C_3H_7

15

Figure 7. Preparation of (+)-Bottrospicatol Derivatives.

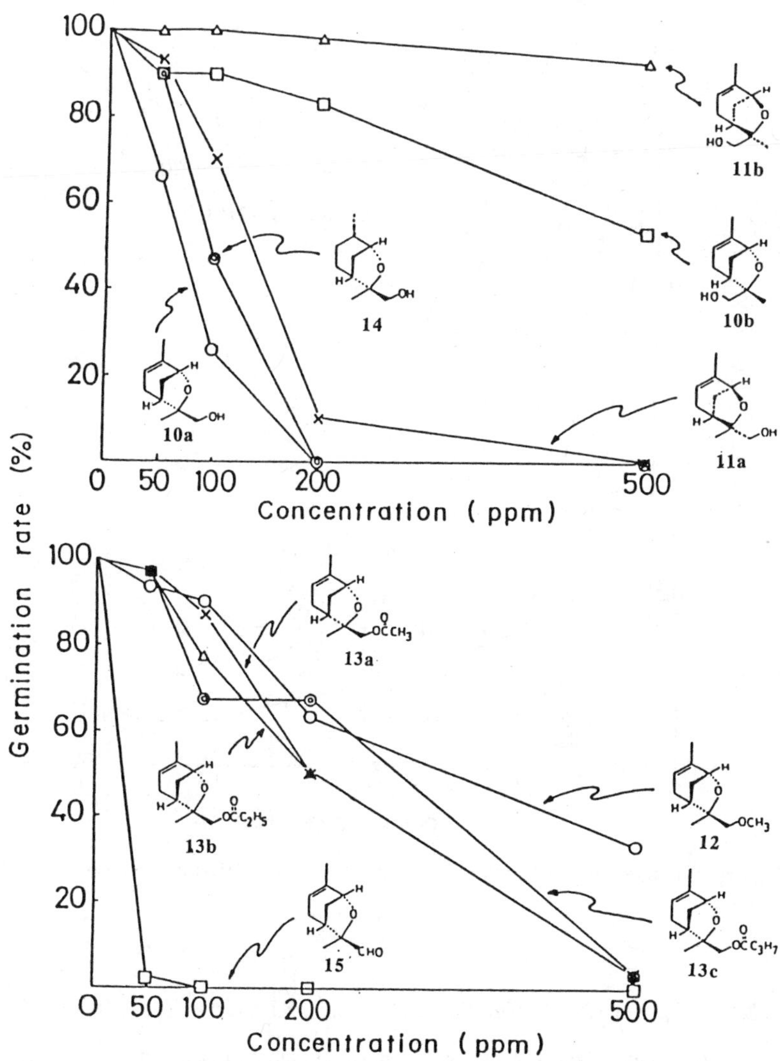

Figure 8. Effects of (+)-Bottrospicatol (**10a**) and Its Related Compounds on the Germination of Lettuce Seeds.

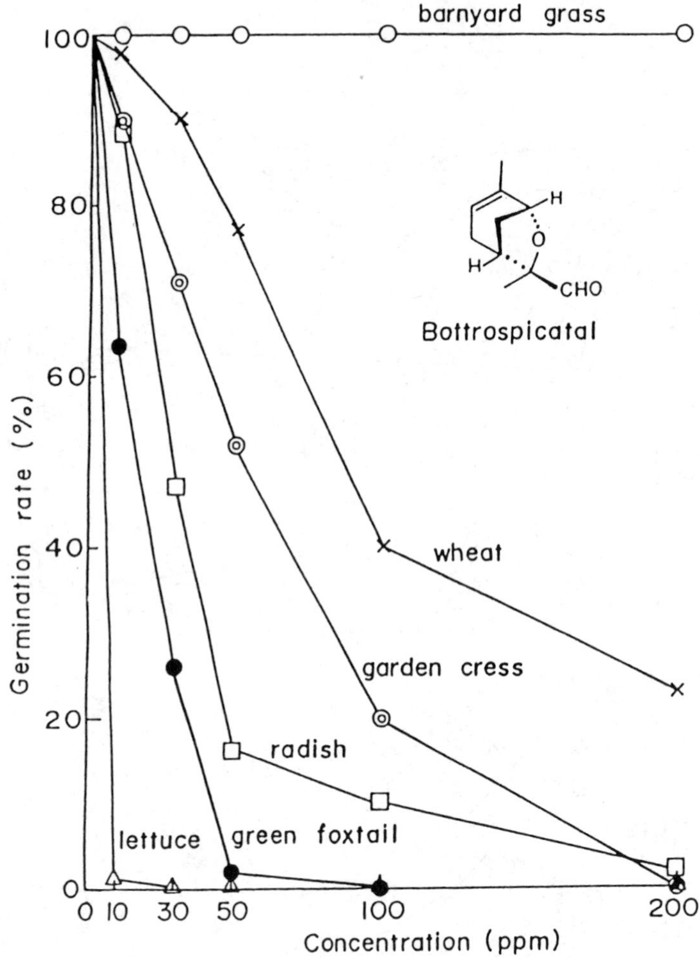

Figure 9. Effects of (+)-Bottrospicatal on the Germination of Several Plant Seeds.

(0.6), 126 (83.7), 111 (29.0), 108 (98.9), 93 (26.6), 83 (22.5), 71 (86.0), 69 (40.3), 43 (100).

(±)-3-*endo*-Hydroxycineole (**20**). A novel 3-*endo*-alcohol was isolated as colorless crystals, mp 55.0 - 55.5°C; $[\alpha]_D^{26}$ ±0° (*c*=0.1, EtOH). IR ν_{max}^{KBr} cm^{-1}: 3420 (OH), 1130 (C-O-C), 1060 (alcoholic C-O). ^{1}H-NMR $\delta_{TMS}^{CDCl_3}$: 1.07 (3H, s, CH_3 at C-7), 1.22 (3H, s, CH_3 at C-9), 1.30 (3H, s, CH_3 at C-10), 1.98 - 2.20 (2H, m, *exo*-proton signal at *ca.* 2.17 collapsed to d with J_{gem}=13 Hz on irradiation at δ4.46, C$\underline{H_2}$-COH),

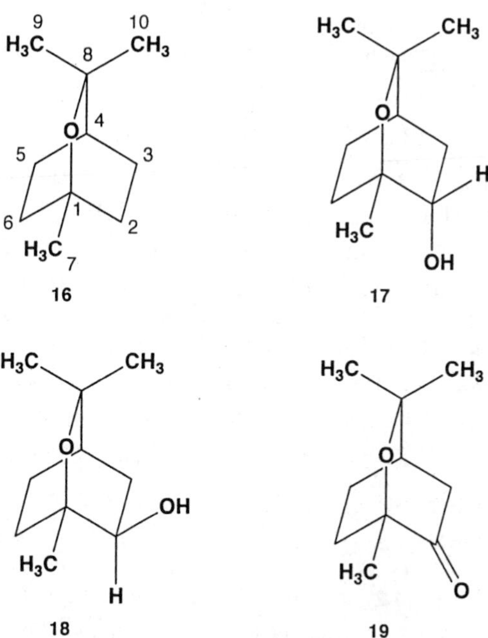

Figure 10. Microbial Transformation Products of 1,8-Cineole (**16**) by *Pseudomonus flava* (*17*).

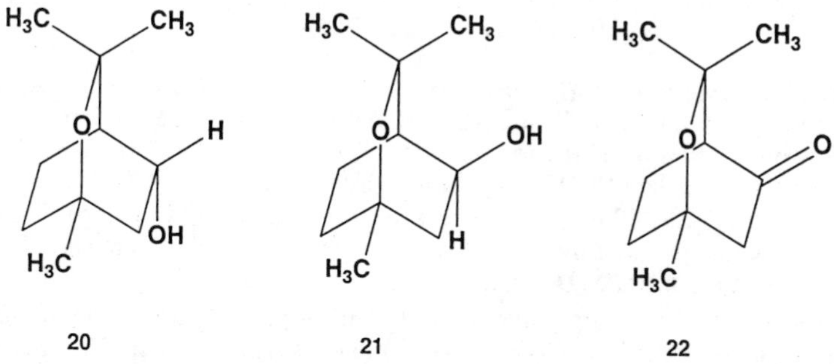

Figure 11. New Products Transformed from 1,8-Cineole by *Aspergillus niger*.

4.46 (1H, dd, $J_{3,2exo}$=10 Hz, $J_{3,2endo}$=4 Hz, H̲COH). ^{13}C-NMR $\delta_{TMS}^{CDCl_3}$: 13.9 (t, C-5), 27.1 (q, C-7), 28.3 (q, C-9), 28.9 (q, C-10), 31.0 (t, C-6), 40.3 (d, C-4), 43.0 (t, C-2), 65.2 (d, C-3), 70.9 (s, C-1), 73.3 (s, C-8). MS *m/z* (%): 170 (M^+, 13.5), 155 (4.6), 137 (4.4), 127 (4.9), 126 (2.9), 108 (15.5), 93 (24.5), 87 (28.6), 85 (21.5), 84 (22.8), 71 (13.6), 69 (25.5), 59 (14.5), 43 (100).

(±)-3-*exo*-Hydroxycineole (**21**). A novel 3-*exo*-alcohol was isolated as colorless oil: $[\alpha]_D^{24}$ ±0° (*c*=0.1, EtOH). IR ν_{max}^{neat} cm^{-1}: 3420 (OH), 1100 (C-O-C), 1058 (alcoholic C-O). ^{1}H-NMR $\delta_{TMS}^{CDCl_3}$: 1.11 (3H, s, CH_3 at C-10), 1.98 - 2.15 (2H, m, *endo*-proton signal at *ca*. 2.07 collapsed to d with J_{gem}=13 Hz on irradiation at δ4.15, C$\underline{H}_2$-COH), 4.15 (1H, ddd, $J_{3,2endo}$=10 Hz, $J_{3,2exo}$=6 Hz, $J_{3,4}$=2 Hz, H̲COH). ^{13}C-NMR $\delta_{TMS}^{CDCl_3}$: 21.5 (t, C-5), 26.9 (q, C-7), 30.2 (q, C-9), 30.5 (q, C-10), 30.9 (t, C-6), 40.8 (d, C-4), 43.3 (t, C-2), 70.2 (d, C-3), 70.9 (s, C-1), 73.4 (s, C-8). MS *m/z* (%): 170 (M^+, 2.1), 155 (46.1), 137 (6.3), 127 (6.1), 126 (2.8), 108 (8.7), 93 (30.0), 87 (8.3), 85 (12.3), 84 (6.6), 71 (7.3), 69 (9.4), 59 (14.2), 43 (100).

(±)-2-Oxocineole (**19**). The crystalline ketone had a mp of 47.0 - 48.0°C and $[\alpha]_D^{22}$ ±0° (*c*=0.09, EtOH). The ketone was identical with an oxidation product of 2-*endo*-hydroxycineole (**17**) by CrO_3/pyridine (*21*). IR ν_{max}^{KBr} cm^{-1}: 1730 (C=O), 1150 (C-O-C). ^{1}H-NMR $\delta_{TMS}^{CDCl_3}$: 1.15 (3H, s, CH_3 at C-7), 1.24 (3H, s, CH_3 at C-9), 1.39 (3H, s, CH_3 at C-10), 2.21 (1H, dd, J_{gem}=20 Hz, $J_{3endo,4}$=2 Hz, C_3-H*endo*), 2.79 (1H, dt, J_{gem}=20 Hz, $J_{3exo,4}$=3 Hz, $J_{3exo,5exo}$=3 Hz, C_3-H*exo*). MS *m/z* (%): 168 (M^+, 2.6), 140 (9.6), 111 (2.5), 83 (4.2), 82 (100), 71 (2.5), 69 (8.9), 67 (9.9), 43 (30.8).

(±)-3-Oxocineole (**22**). A novel ketone was isolated as a colorless oil; $[\alpha]_D^{23}$ ±0° (*c*=0.15, EtOH). IR ν_{max}^{neat} cm^{-1}: 1735 (C=O), 1150 (C-O-C). ^{1}H-NMR $\delta_{TMS}^{CDCl_3}$: 1.16 (3H, s, CH_3 at C-7), 1.24 (3H, s, CH_3 at C-9), 1.32 (3H, s, CH_3 at C-10), 2.24 (1H, d, J_{gem}=20 Hz, C_2-H*endo*), 2.40 (1H, dd, J_{gem}=20 Hz, $J_{2exo,6exo}$=3 Hz, C_2-H*exo*). MS *m/z* (%): 168 (M^+, 8.6), 153 (64.5), 140 (2.6), 125 (15.3), 111 (57.9), 83 (54.1), 82 (62.8), 71 (6.4), 69 (10.6), 67 (12.8), 55 (26.5), 43 (100).

Chemical Conversion of Hydroxycineoles to Economically Useful Substances. Temperature-programmed GC analysis (10% PEG-20M on Celite 545, 3 m x 3 mm i.d.) showed in order of increasing retention time: 1,8-cineole (**16**), 3-oxocineole (**22**, 80mg), 2-oxocineole (**19**, 17mg), 2-*endo*-hydroxycineole (**17**, 877mg), 3-*endo*-hydroxycineole (**20**, 809mg) and 3-*exo*-hydroxycineole (**21**, 15mg). Time course changes in the products of metabolic oxidation of **16** by *A. niger* are shown in Figure 12. A peak of the substrate (**16**) on a GC trace disappeared in 7 days.

Furthermore, 3-*exo*-hydroxycineole (**21**) and 3-*endo*-hydroxycineole (**20**) were converted to *p*-menthane-3,8-*cis*-diol (**23**) and its *trans* isomer (**24**), respectively, which had previously been isolated as allelochemicals from *Eucalyptus citriodora* (*22, 23*) by catalytic hydrogenolysis (H_2/PtO_2) in a similar manner to the reaction of 2-hydroxycineole to *p*-menthane-2,8-diol (*24*). Column chromatography (SiO_2) of reaction mixtures gave *cis* and *trans* diols (Figure 13) in ca. 15% and 12% yields, respectively. These two compounds exhibited repellent activities against mosquitoes, *Aedes albopictus* and *Culex pipiens* (*25*). The repellent effects of ester derivatives from

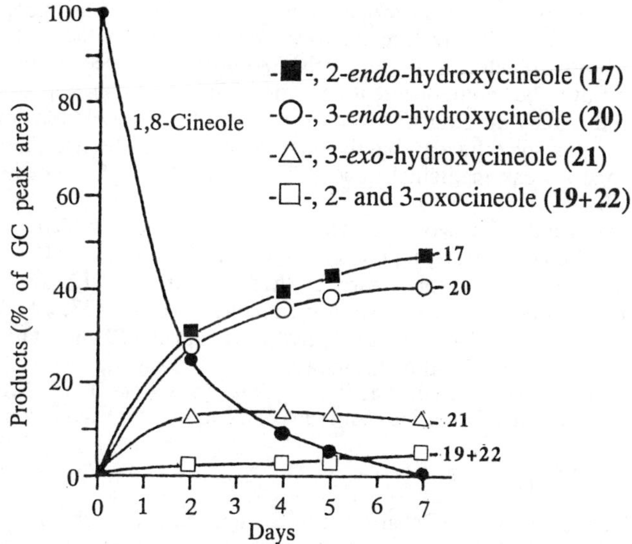

Figure 12. Time Course of Major Metabolic Products of 1,8-Cineole by a Strain of *Aspergillus niger*.

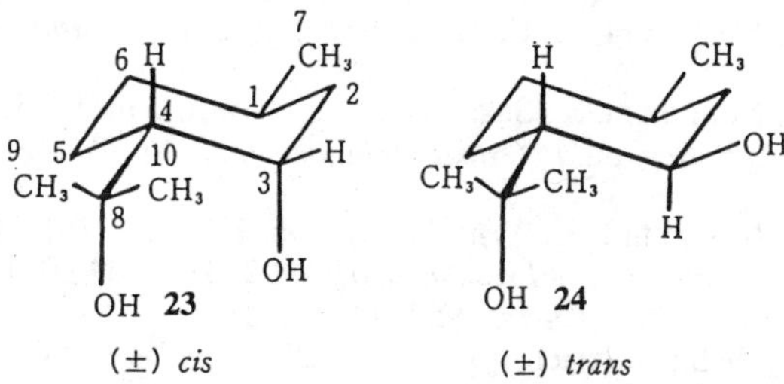

Figure 13. (±)- *p*-Menthane-3,8-diols (*cis* and *trans*).

p-menthane-3,8-diols were compared with commercially available repellent, DEET (N,N-diethyl-*m*-toluamide). Especially, the caproyl ester (C_6) of *cis*-diol (**23**) exhibited much higher activity than DEET in terms of repellency and repellent durability against mosquitoes.

Flavor of Microbial Transformation Products. The flavor of terpenoids produced by microbial transformation was evaluated. (-)-Carvone (**1**) is well known as a spearmint flavor component. Transformation products (**4** - **7**) of **1** had peppermint-like flavor although these are slightly different to each other in terms of the odor quality. The metabolites, bottrospicatols (**10a** and **10b**) *Streptomyces* species did not have a characteristic flavor but quite different activities (Figure 8). (+)-Bottrospicatal (**15**) which was produced by the oxidative reaction of (+)-bottrospicatol (**10a**) with CrO_3 in pyridine had a weak spice-flavor (slightly black-pepper like). The ester derivatives (**13** in Figure 7) had a weak medicinal flavor. The flavor of acetyl ester (**13a**) was the strongest of all.

On the other hand, 1,8-cineole (**16**) which is a significant constituent of *Eucalyptus* species has a characteristic camphoraceous odor. Oxocineoles (**19** and **22**) still possessed camphoraceous odor. However, hydroxycineoles (**17**, **18**, **20**, **21**) had slightly medicinal odor. Interestingly, *p*-menthane-3,8-diols (**23**, **24**) which were produced by catalytic hydrogenolysis of 3-hydroxycineoles (**20** and **21**) exhibited no odor to humans. *p*-Menthane-3,8-diols are fascinating to cosmetic companies since the compounds have very high repellent activity against mosquitoes. Further research will be necessary to elucidate relationships between chemical structure and flavor.

Acknowledgments

We thank Mr. Atsushi Satoh, Department of Bioscience and Technology, School of Engineering, Hokkaido Tokai University for preparation of the manuscript. This work was partly supported by the Ministry of Education of Japan (Grant 6056-0124).

Literature Cited

1. Noma, Y.; Tatsumi, C. *Nippon Nogeikagaku Kaishi.* **1973**, 47, 705-711.
2. Noma, Y.; Nonomura, S.; Ueda, H.; Tatsumi, C. *Agric. Biol. Chem.* **1974**, 38, 735-740.
3. Noma, Y.; Nonomura, S.; Sakai, H. *Agric. Biol. Chem.* **1975**, 39, 437-441.
4. Noma, Y. *Ann. Rep. Stud., Osaka Joshigakuen Junior College* **1976**, 20, 33-47.
5. Noma, Y.; Nonomura, S. *Agric. Biol. Chem.* **1974**, 38, 741-744.
6. Noma, Y.; Asakawa, Y. *Phytochemistry* **1992**, 31, 2009-2011.
7. Noma, Y. *Agric. Biol. Chem.* **1980**, 44, 807-812.
8. Dhavalikar, R.S.; Bhattacharyya, P.K. *Ind. J. Biochem.* **1966**, 3, 144-157.
9. Dhavalikar, R.S.; Rangachari, P.N.; Bhattacharyya, P.K. *Ind. J. Biochem.* **1966**, 3, 158-164.
10. Noma, Y.; Nishimura, H.; Hiramoto, S.; Iwami, M.; Tatsumi, C. *Agric. Biol. Chem.* **1982**, 46, 2871-2872.
11. Nishimura, H.; Hiramoto, S.; Mizutani, J.; Noma, Y.; Furusaki, A.; Matsumoto, T. *Agric. Biol. Chem.* **1983**, 47, 2697-2699.
12. Jori, A.; Bianchetti, A.; Prestini, P.E. *Biochem. Pharmacol.* **1969**, 18, 2081-2086.

13. Jori, A.; Bianchetti, A.; Prestini, P.E.; Garattini, S. *Eur. J. Pharmacol.* **1970**, 9, 362-368.
14. Jori, A.; Salle, E.D.; Pescador, R. *J. Pharm. Pharmacol.* **1972**, 24, 464-468.
15. Jori, A.; Briatico, G. *Biochem. Pharmacol.* **1973**, 23, 543-545.
16. Nishimura, H.; Paton, D.M.; Calvin, M. *Agric. Biol. Chem.* **1980**, 44, 2495-2496.
17. MacRae, I.C.; Alberts, V.; Carman, R.M.; Shaw, I.M. *Aust. J. Chem.* **1979**, 32, 917-922.
18. Nishimura, H.; Noma, Y.; Mizutani, J. *Agric. Biol. Chem.* **1982**, 46, 2601-2604.
19. Carman, R.M.; MacRae, I.C.; Perkins, M.V. *Aust. J. Chem.* **1986**, 39, 1739-1746.
20. Miyazawa, M.; Nakaoka, H.; Hyakumachi, M.; Kamioka, H. *Chemistry Express* **1991**, 6, 667-670.
21. Ratcliffe, R.; Rodehorst, R. *J. Org. Chem.* **1970**, 35, 4000-4002.
22. Nishimura, H.; Kaku, K.; Nakamura, T.; Fukuzawa Y.; Mizutani, J. *Agric. Biol. Chem.* **1982**, 46, 319-320.
23. Nishimura, H.; Nakamura, T.; Mizutani, J. *Phytochemistry* **1984**, 23, 2777-2779.
24. Gandini, A.; Bondavalli, R.; Schenone, P.; Bignardi, G. *Ann. Chim.* (Rome) **1972**, 62, 188-191.
25. Nishimura, H.; Mizutani, J. *Proceeding of Hokkaido Tokai University Science and Engineering* **1989**, 2, 57-65.

Chapter 17

Microbial Oxidation of Alcohols by *Candida boidinii*: Selective Oxidation

M. Nozaki, N. Suzuki, and Y. Washizu

Central Research Laboratory, Takasago International Corporation, 1–4–11 Nishi-Yawata, Hiratsuka, Kanagawa 254, Japan

Chemistry of the microbial oxidation by *Candida boidinii* SA051 was investigated. Chemo-, regio- and stereo- selectivity and substrate specificity of this microbial oxidation will be discussed.

It is known that there are some types of microbial oxidation. The alcohol dehydrogenase type of oxidation requires cofactors such as NAD, etc. For efficiency and economy reasons, it is preferable that there is a regeneration system of cofactors in the alcohol dehydrogenase type of oxidation. However, alcohol oxidase is an unidirectional (mostly flavin dependant) redox enzyme. If a cell suspension of the microorganisms having alcohol oxidase activity is used, regeneration by endogenous production relieves one of exogenous addition. This type of oxidation seems more practical. It has been reported that methylotrophic yeasts such as *Candida boidinii* (*1*), *Pichia pastoris* (*2*) and *Candida maltosa* (*3*) show alcohol oxidase activity and can oxidize alcohols effectively to the corresponding aldehydes. *Candida boidinii* SA051 (*4*) used in this study was derived from *Candida boidinii* AOU-1 by UV irradiation and had the highest alcohol oxidase activity producing formaldehyde among the mutants. In our previous work, this oxidation showed feasibility for the preparative scale production of flavor aldehydes (*5*). In this paper, substrate specificity of this system and chemistry of this microbial oxidation was examined.

Methods and Materials

Organisms. *Candida boidinii* SA051 was generously donated by Prof. Y. Tani and Prof. Y. Sakai (Kyoto University).

Biomass production. The cultures of the yeasts were carried out in a 70 L jar fermentor with a working volume of 40 L. Seed cultures for the fermentor were started from slant cultures. The slant cultures were inoculated directly into 100 mL of the basal medium containing 3% of glucose in a 500 mL Sakaguchi flask. In 1000mL of the basal medium, there were NH_4Cl 7.63g, KH_2PO_4 2.81g,$MgSO_4/7H_2O$ 0.59g, EDTA/2Na 0.45g, $CaCl_2/2H_2O$ 55.0mg, $FeCl_3/6H_2O$

0097–6156/96/0637–0188$15.00/0

37.5mg, $ZnSO_4/7H_20$ 22mg, $MnSO_4$ 17mg, Thiamine HCl 17mg, H_3BO_4 4mg, $CoCl_2/2H_2O$ 2.8mg, Na_2MoO_4 2.6mg, KI 0.6mg and Biotin 0.05mg. The cultures were incubated at 28°C with shaking (150 rpm). After 2 days growth, the cultures were used to start the 40 L scale fermentation. The volume of inoculum used was 800 mL. It was cultivated on the basal medium containing 0.3% (W/V) glucose as the carbon source to accelerate the initial growth. Methanol is required to induce the alcohol oxidase; no enzyme is formed during growth on glucose. However, if methanol is added at the beginning of fermentation, it inhibits the initial growth of the microorganisms. The impeller speed was 200 rpm and the aeration rate was 0.5 vvm. The temperature was kept at 28°C. After 16 hours, the pH of the cultured broth was pH statically adjusted to 5.0 by addition of 5N-NaOH solution. At this point the first aliquot of methanol (1% of the broth) was added. After the consumption of the initially added methanol, a second amount of methanol was added. This addition-consumption technique was repeated and showed a typical fed-batch fermentation profile. This gave a high biomass concentration (60 g dry cell weight base/L) after six feeding cycles (Figure 1).

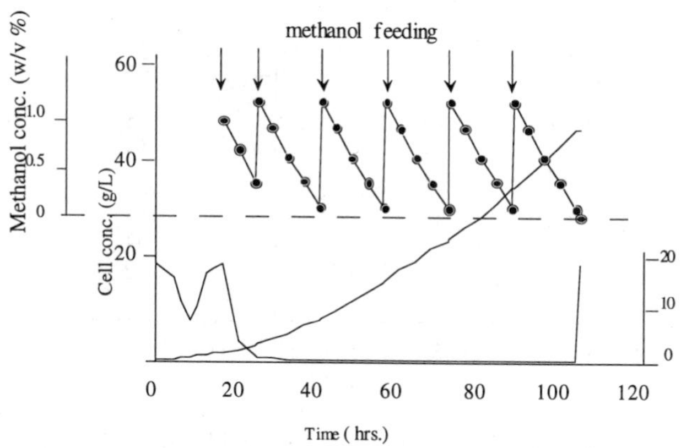

Figure 1. Biomass production.

The grown cells were harvested by centrifugation. They were washed with 0.1M potassium phosphate buffer (pH 7.5), resuspended in the buffer and stored at -20°C until use.

Conversion of alcohols. Grown cells having high alcohol oxidase activity were harvested. The cells and substrate alcohols were added to potassium phosphate buffer under pure oxygen atmosphere. The biomass concentration used was 30 g dry cell weight base /L and that of substrates was from 30 g/L. The reaction mixtures were kept at 25°C. The reaction time varied depending on substrates and conversion ratio. The reaction was then terminated by removing the cells from the reaction mixture by centrifugation. The supernatant was distilled under atmospheric pressure or extracted with ethyl acetate and concentrated at 40°C/20mmHg. This gave the crude products. The crude products were subjected to further purification (distillation or silica gel chromatography) to afford the products. The products by this microbial oxidation were analyzed by TLC, GLC and GC/MS. Some were analyzed by NMR. Conversion was calculated from GLC area ratio.

Results

Chemo-selectivity

Oxidation of Primary Alcohols. *Candida boidinii* SA051 oxidized primary alcohols to the corresponding aldehydes effectively. In the case of 3-methyl-1-butanol oxidation, generated isovaleraldehyde was up to 50 grams per liter medium. Thus high yield of the aldehyde was obtained but was the equimolar amount of H_2O_2 produced? The strain had strong catalase activity and therefore H_2O_2 accumulation was not observed. The oxidation proceeded smoothly even at high substrate concentration. Primary alcohols used and the relative conversion ratio to the corresponding aldehydes are shown in Table I. In this system, the corresponding acids were not detected.

Table I. Oxidation of primary alcohols.

Substrate	Relative activity (%)		
methanol	80	2-methyl-1-propanol	68
ethanol	100	3-methyl-1-butanol	33
1-propanol	60	2-methyl-1-butanol	15
1-butanol	80	allyl alcohol	65
1-pentanol	64	(E)-2-pentenol	61
1-hexanol	42	(E)-2-hexenol	59
1-heptanol	23	(E)-2-heptenol	52
1-octanol	19	(E)-2-octenol	37
1-nonanol	5	benzyl alcohol	11
		cinnamyl alcohol	4

substrate: 3 % w/v SA051: 30 g/L time: 2 hrs

Table I suggests that shorter chain alcohols were oxidized faster than longer chain ones. In the case of alcohols with the same carbon number, the straight chain alcohols were oxidized faster than branched chain ones and the α,β unsaturated alcohols were oxidized faster than straight chain ones. These observations were applied to the selective oxidation of C6 alcohols. The oxidation profile of C6 alcohols by *Candida boidinii* SA051 whole cells is shown in Figure 2.

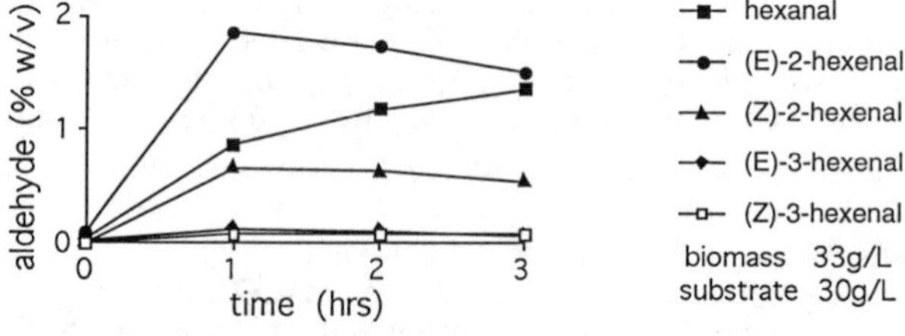

Figure 2. Oxidation profile of C6 alcohols by *Candida boidinii* SA051.

This figure demonstrates that among the C6 alcohols, (E)-2 hexenol was oxidized fastest, followed by hexanol, (Z)-2 hexenol, (E)-3 hexenol and (Z)-3 hexenol. Figure 3 shows that the oxidation profile of the C6 alcohols mixture composed of equal volumes of C6 alcohols.

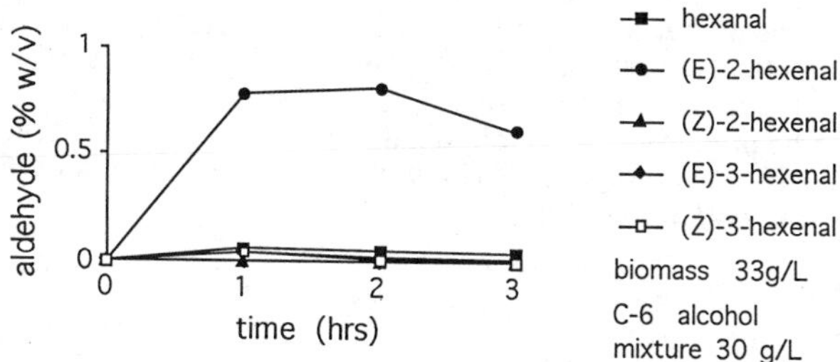

Figure 3. Oxidation profile of C-6 alcohols mixture by *Candida boidinii* SA051.

In this system, surprisingly only (E)-2 hexenol was oxidized. Thus this observation was applied to developing natural green notes. The natural C6 alcohols mixture used was composed of (E)-2-hexenol, (Z)-3-hexenol and hexanol and was isolated as the top fraction of mint oils from distillation. In terms of green note, (E)-2-hexenol is less valuable. On other hand, its aldehyde form (E)-2-hexenal is quite an important component of green notes. Figure 4 shows *Candida boidinii* SA051 could selectively oxidize (E)-2-hexenol in the mixture to desirable (E)-2-hexenal.

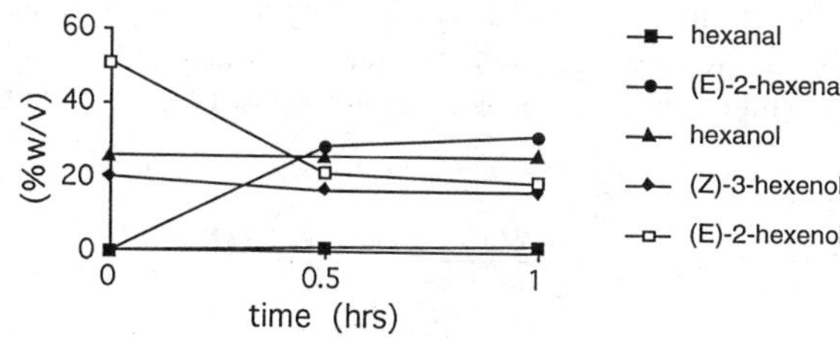

Figure 4. Oxidation profile of natural C-6 alcohols fraction.

Regio-selective Oxidation of Alcohols

Oxidation of α,ω-diols. In the field of organic synthesis, it is known that it is difficult to oxidize α,ω-diols to the corresponding ω-hydroxy aldehydes since oxidizing agents used in organic synthesis can not differentiate both terminal hydroxy groups. However, it was expected that biocatalysts could distinguish the α hydroxy group from the ω-one. Therefore, selective oxidation of α,ω-diols by *Candida boidinii* SA051 was investigated. Concerning the oxidation of 1,4-butanediol and 1,5-pentanediol, this microbial oxidation gave the corresponding lactones in the broth at yields of 24% and 34%, respectively . It is known that 4-hydroxybutanal and 5-hydroxypentanal transform themselves into the corresponding lactols easily. However, in the course of this oxidation, the lactols were not detected. This means that the lactols were rapidly oxidized to the corresponding lactones or that this oxidation afforded directly the lactones from the diols. Considering that *Candida boidinii* SA051 did not give acids from alcohols, these lactones should be produced by the microbial oxidation of the lactols. In the case of 1,6-hexanediol, this gave no product (Figure 5). This was caused by the substrate specificity of *Candida boidinii* SA051.

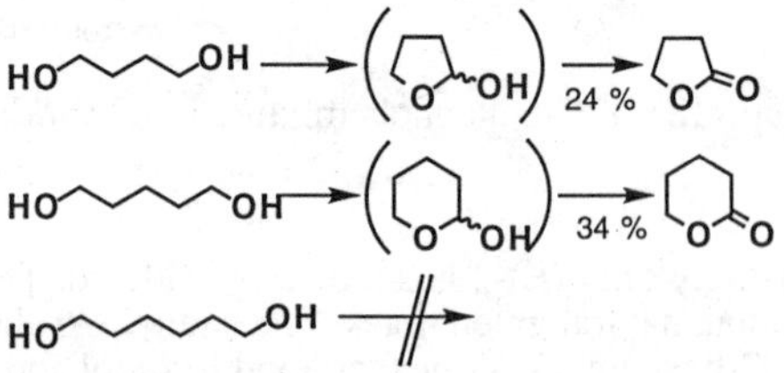

Figure 5. Oxidation of α,ω-diols 1.

In the case of 1,7-heptanediol, 1,8-octanediol and 1,9-nonanediol, these only gave the corresponding ω–hydroxy aldehydes in the broth at yields of 52%, 75% and 60%, respectively (Figure 6). The corresponding lactols were not found.

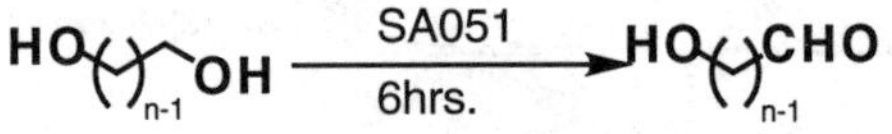

	yield*1	yield*2	max. production
n=7	52 %	14%	4.12 g/L
8	75	18	5.50
9	60	13	3.96

substrate: *1 0.1 w/v, *2 3%
SA051 : *1 30 g/L, *2 90g/L

Figure 6. Oxidation of α,ω-diols 2.

Neither dialdehydes nor dicarboxylic acids were detected and it was shown that this microbial oxidation had excellent regioselectivity . These generated ω–hydroxy aldehydes could be used as starting materials for macrocyclic musk related compounds. The yields varied depending on the substrate concentration. Lower concentrations gave better yields. However, even at 3% concentration, the conversion ratio reached about 15% and ω–hydroxy aldehydes were produced at about 5g/L medium. This figure was sufficient for production.

Stereoselective Oxidation of Alcohols

Alkane Diols Having Primary and Secondary Alcohol Groups. This microbial oxidation was also applied to alkanediols having one terminal hydroxy group. With 1,2 alkanediols, this oxidation did not proceed. With 1,3 alkanediols, the oxidation proceeded and gave the corresponding keto alcohols. This showed that the yeasts preferred the secondary hydroxy group to the primary one. The oxidation of 1,3-butanediol gave 4-hydroxy-2-butanone at 75% yield. The oxidation of 1,3-pentanediol gave 5-hydroxy-3-pentanone at 80% yield (Figure 7).

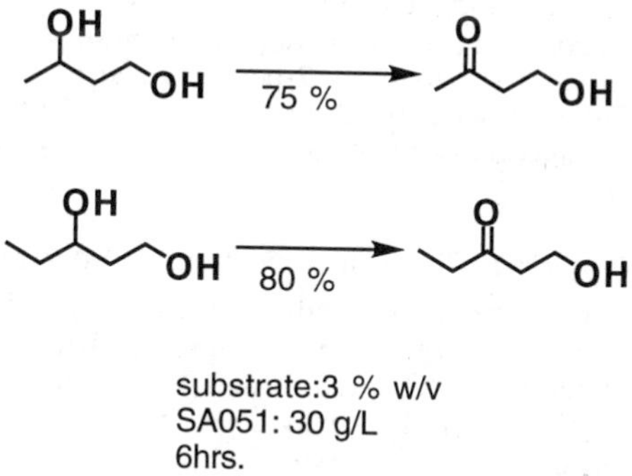

OH OH 75 % O OH

OH OH 80 % O OH

substrate:3 % w/v
SA051: 30 g/L
6hrs.

Figure 7. Oxidation of alkane1,3-diols.

In the case of 1,3-butanediol, the remaining diol had slight optical activity. This showed (S)-configuration at 27% optical purity. This was determined by comparison of optical rotation with a reference (*6*). The yeasts could differentiate the (R)- and (S)-alcohol and oxidize the (R)-alcohol predominantly. With 1,4-pentanediol, the yeasts preferred the primary hydroxy group to the secondary one and gave pentane-1,4-olide and its lactol at 6.3% yield. The afforded lactone showed slight optical activity and was the (R)-form at 5% optical purity. This was determined by comparison of optical rotation with a reference (*7*). This showed that the yeasts preferred (R)-1,4-pentanediol as the substrate.

Oxidation of *Meso* Diols. Asymmetric induction of meso and prochiral diols by lipases is very successful in the field of organic synthesis. Also it is well known that selective oxidation of prochiral or meso diols by HLADH provides oxidized products with a significant degree of enantioselectivity. However, it has not been reported that alcohol oxidases were applied to such types of oxidation. The microbial oxidation of meso diols by *Candida boidinii* SA051 was carried out and gave optically active hydroxy ketones (Figure 8).

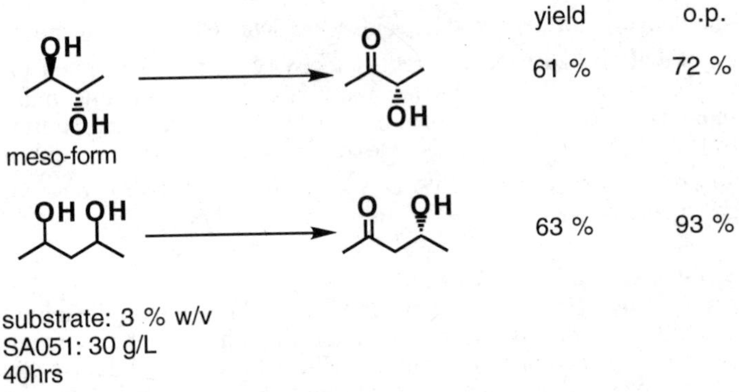

	yield	o.p.
meso-form	61 %	72 %
	63 %	93 %

substrate: 3 % w/v
SA051: 30 g/L
40hrs

Figure 8. Oxidation of meso-diols.

Oxidation of *meso* 2,3-butanediol gave (S)-acetoin at 61% yield and its optical purity was 72%. This was determined by comparison of optical rotation with a reference (*8*). This is the antipode of the product into which *meso* 2,3-butanediol was transformed by lactic acid bacteria. Oxidation of *meso* and a racemic mixture of 2,4-pentanediol afforded 4-(R)-hydroxy-2-pentanone at 63% yield and its optical purity was 93%. This was determined by comparison of optical rotation with a reference (*9*). These results proved the feasibility of the stereoselective oxidation of alcohols by the yeasts.

Oxidation of *Prochiral* Diol. The oxidation of prochiral diol 3-methyl-1,5-pentanediol did not give the corresponding ω–hydroxy aldehyde but gave the corresponding lactol at 22% yield. To determine stereochemistry of the lactol, this was chemically oxidized by Ag_2O and afforded 3-(R)-methyl-pentan-1,5-olide at 38% optical purity. (Figure 9). This was determined by comparison of optical rotation with a reference (*10*).

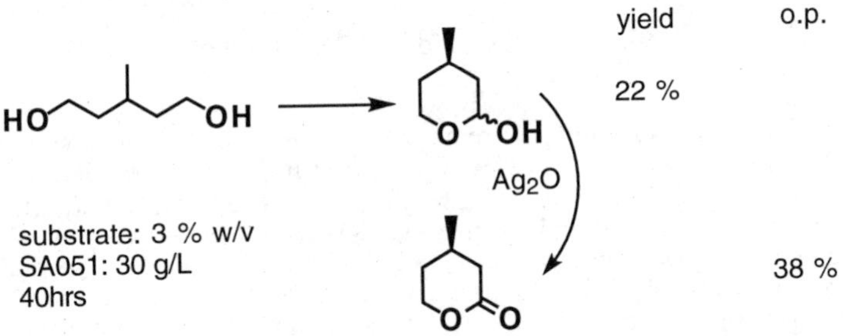

	yield	o.p.
HO–OH → lactol (OH)	22 %	
Ag_2O → lactone		38 %

substrate: 3 % w/v
SA051: 30 g/L
40hrs

Figure 9. Oxidation of 3-methyl-1,5-pentanediol.

Conclusion

The methylotropic yeast *Candida boidinii* SA051 showed excellent ability for oxidation of alcohols to aldehydes or ketones. In the production of isovaleraldehyde, the generated aldehyde was up to 50 grams/L. Also this oxidation showed reaction selectivities. It was an example of chemoselectivity that the yeasts preferred (E)-2-hexenol among various C6 alcohols and oxidized it selectively to the desired (E)-2-

hexenal. It was an example of regioselectivity that α,ω-alkanediols were oxidized to the corresponding ω-hydroxy aldehydes. It was an example of stereoselectivity that some diols were stereochemically distinguished as substrates and that some meso or prochiral diols were oxidized and gave one enantiomer predominantly. These results mentioned above showed that this microbial oxidation is useful for production of flavor chemicals and related compounds.

Acknowledgment

We thank Prof. Y. Tani and Prof. Y. Sakai for generous gift of *Candida boidinii* SA051.

Literature Cited

1. (a)Sakai, Y.; Tani ,Y. *Agric. Biol. Chem.* **1987**, *51*(9), 2617-2620. (b)Shachar-Nishri, Y.; Freeman, A. *Appl. Biochem. Biotechnol.* **1993**, *39/40*, 387-399 (c) Clark, D.S., et al. *Bioorganic & Medicinal Chem. Lett.*, **1994**, *4*, 1745-1748.
2. Murray, W.D.; Duff, S.J.B.; Lanthler, P.H. Production of natural flavor aldehydes from natural source primary alcohols C2-C7, USP, 4871669, **1989.**
3. Mauersberger, S. et al. *Appl. Microbiol. Biotechnol.* **1992**, *37,* 66-73.
4. Sakai, Y.; Tani, Y. *Agric. Biol. Chem.* **1987**, *51*(8), 2177-2184.
5 Nozaki, M.; Suzuki, N.; Washizu, Y. In *Bioflavour95*; Étiévant, P.; Schreier, P.,Eds.; INRA editions: France, 1995; pp 255-260.
6 Dictionary of Organic Compounds, 5th ed., Chapman and Hall.
7. Mori, K. *Tetrahedron* **1975**, *31,* 3011.
8. Taylor, M.B.; Juni, E. *Biochim. Biophys. Acta.* **1960,** *39*, 448.
9. Ohta, H., et al. *Agric. Biol. Chem.* **1986**, *50,* 2499.
10. Jones, J. B. *J. Am.Chem. Soc.* **1977**, *99*, 556.

Chapter 18

Alcohol Acetyl Transferase Genes and Ester Formation in Brewer's Yeast

Yukio Tamai

Central Laboratories for Key Technology, Kirin Brewery Co., Ltd., 1–13–5 Fukuura, Kanazawa-ku, Yokohama, Kanagawa 236, Japan

Alcohol acetyl transferase (AATase), which catalyses the synthesis of acetate esters from alcohols and acetyl-CoA, was purified from *Saccharomyces cerevisiae* (*sake* yeast) and its partial amino acid sequence was determined. The AATase-encoding *ATF1* genes were cloned from *sake* yeast and brewer's yeast (bottom fermenting yeast). The bottom fermenting yeast was found to have two types of *ATF1* gene; *ATF1* and a homologous gene Lg-*ATF1*, while *sake* yeast had one *ATF1* gene. Strains carrying either the *ATF1* genes carried on multiple copy plasmid or a single *ATF1* gene with the native promoter replaced by the strong constitutive *GPD* (glyceraldehyde-3-phosphate dehydrogenase) promoter produced higher concentrations of acetate esters. On the other hand, disruption of the *ATF1* gene resulted in a decrease in AATase activity and reduced acetate ester formation. Control of the *ATF1* gene expression is a novel approach for the enrichment or reduction of acetate esters in yeast strains used in the brewing industry.

The volatile esters formed during fermentation are thought to be important flavor components in beer and other alcoholic beverages. Beer contains ethyl acetate (light fruity, solvent-like flavor) and isoamyl acetate (banana flavor) at concentrations near or just below threshold values and they are considered to determine the characteristics of a beer. Isoamyl acetate is also responsible for the excellent aroma of Japanese *sake*, especially of high quality *sakes*. Therefore, a number of studies have been carried out to determine the mechanism of formation of acetate esters in the various brewing processes. Nordström carried out extensive studies of acetate ester formation in yeast and proposed that the acetate esters were synthesized by the enzymatic condensation of acetyl CoA and the alcohols formed during fermentation (*1-3*). The higher alcohols, which are the substrates for AATase, are formed by both a catabolic route (Ehrlich pathway) and a anabolic route (biosynthetic pathway of amino acids).

Yoshioka and Hashimoto were successful in partially purifying alcohol acetyl transferase (AATase), the enzyme responsible, and proposed a mechanism for acetate ester formation (*4*). As the AATase has a wide range of higher alcohol substrates, low or medium molecular weight esters are thought to be synthesized by AATase. Ester formation during fermentation is reduced by aeration of the medium or the addition of

0097–6156/96/0637–0196$15.00/0

unsaturated fatty acids. As unsaturated fatty acids are synthesized from oxygen and saturated fatty acid by Δ9-desaturase in *S. cerevisiae* (*5*), aerobic conditions increase the concentration of unsaturated fatty acids in the cell membrane. As activity of the partially purified enzyme was inhibited by unsaturated fatty acids, Yoshioka and Hashimoto proposed that the unsaturated fatty acids synthesized in aerobic conditions reduce the ester formation by the inhibition of AATase activity (*6-7*).

AATase was previously reported to be a membrane bound enzyme which was labile during purification. However, recently Minetoki et al. have purified AATase from *S. cerevisiae* and the partial amino acid sequence has been determined (*8*). The *ATF1* genes encoding AATase was cloned from the *sake* yeast *S. cerevisiae* and the bottom fermenting yeast *S. pastorianus* (*9*). *ATF1* is thought to be a useful tool for the study of *ATF1* gene expression and for the regulation of ester formation in yeast. In this report, the structure of genes cloned from the bottom fermenting yeast *S. pastorianus* is summarized and the gene expression and the regulation of ester formation during brewing is described.

Cloning of *ATF1* Genes Encoding AATase

Cloning of the K7-*ATF1* Gene from *Sake* Yeast. Two mixed oligonucleotide probes were designed and synthesized on the basis of the amino acid sequences obtained from the purified AATase (*8*). Clones carrying sequences homologous to the probes were isolated from a genomic gene library constructed from the *sake* yeast Kyokai No.7 by the plaque hybridization method. Positive clones which hybridized to both probes were chosen for further analysis. Figure 1 shows a restriction map of the DNA fragment carrying K7-*ATF1*, which encoded AATase. The two synthetic oligonucleotides hybridized to a 1.0 kb *Eco*RI-*Bam*HI fragment.

Nucleotide Sequence of the K7-*ATF1* Gene. An approximately 2.0-kb nucleotide sequence containing a *Eco*RI-*Bam*HI fragment was determined. This fragment contained an open reading frame encoding a protein consisting of 525 amino acids. All amino acid sequences obtained from the purified AATase were found within this open reading frame. The molecular weight of the protein deduced from the nucleotide sequence was calculated as being 61,059 which was consistent with the value estimated by SDS-PAGE of the purified AATase. This gene was designated K7-*ATF1*: *ATF1* derived from Kyokai No.7. Although AATase was reported to be a membrane bound enzyme (*4,8*), the hydrophobicity profile predicted by computer analysis showed that AATase has no obvious hydrophobic region which could act as a membrane anchor (*8*). Disruption of the *ATF1* gene resulted in a drastic reduction in isoamylacetate synthesis. Thus, it was concluded that *ATF1* encoded AATase in *S. cerevisiae* (Fujii, T., submitted). The *ATF1* gene was mapped to the right arm of the XV chromosome of *S. cerevisiae* (Yoshimoto, H., in preparation).

Cloning of the *ATF1* and Lg-*ATF1* Genes from the Bottom Fermenting Yeast *S.pastorianus*. To analyze the *ATF1* gene in the bottom fermenting yeast, Southern blotting was carried out using K7-*ATF1* as a probe. The results showed that the bottom fermenting yeast contained two types of gene homologous with K7-*ATF1* based upon the intensity of hybridization signal with the probe. One gene was highly homologous with K7-*ATF1*, but the other showed less homology with K7-*ATF1*. The two homologous *ATF1* genes were cloned from a gene library constructed from the bottom fermenting yeast KBY004 using a 0.4-kb *Cla*I-*Eco*RI fragment of K7-*ATF1* as a probe. Figure 1 shows the restriction maps of the strongly hybridizing gene and the weakly hybridizing gene. As the restriction map of the strongly hybridizing DNA

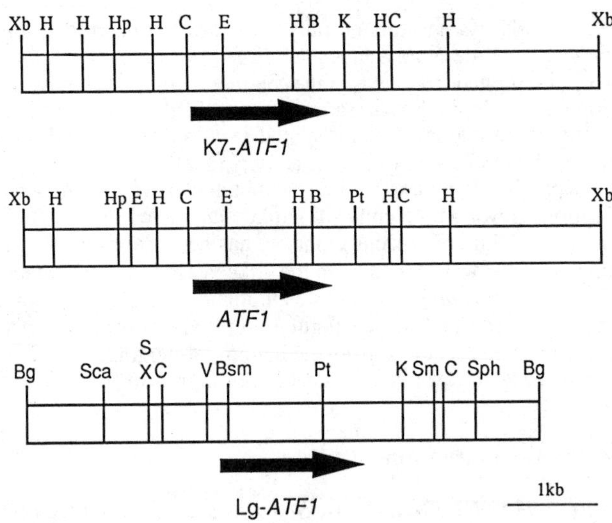

Figure 1. Restriction maps of the *ATF1* genes. The restriction enzymes used were *Xba*I (Xb), *Hin*dIII (H), *Hpa*I (Hp), *Cla*I (C), *Eco*RI (E), *Bam*HI (B), *Kpn*I (K), *Pst*I (P), *Bgl*II(Bg), *Sca*I (Sca), *Sal*I (S), *Xho*I (X), *Eco*RV (V), *Bsm*I (Bsm), *Sma*I (Sm), *Sph*I (Sph). The position and direction of the open reading frame are indicated by arrows.

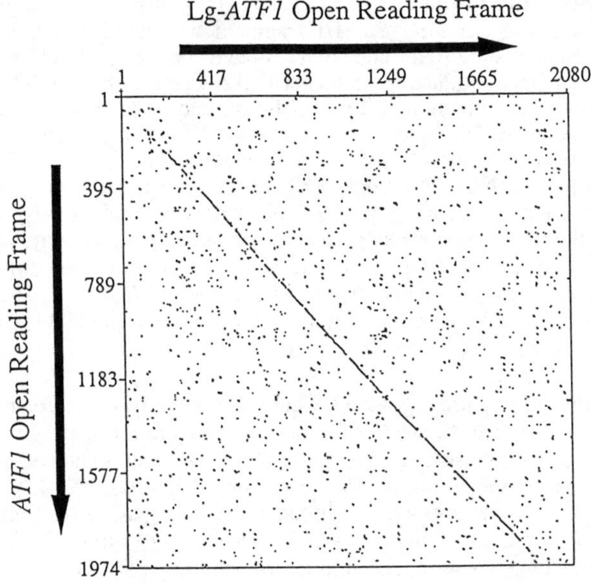

Figure 2. Harr plot comparison of nucleotide sequences of *ATF1* and Lg-*ATF1* gene. Arrows indicate coding regions of the two genes.

fragment was quite similar to that of K7-*ATF1*, it is considered that the cloned DNA fragment encodes the *ATF1* gene from the bottom fermenting yeast. The weakly hybridizing DNA differs in structure from K7-*ATF1*.

Each of these two homologous genes was cloned into a multicopy yeast-*E.coli* shuttle vector and introduced into a laboratory yeast strain. As transformants carrying these genes homologous with *ATF1* produce higher levels of AATase activity than the parental strain, it was clear that these cloned DNA fragments encoded AATase. The weakly hybridizing gene was thought to be specific to the bottom fermenting, brewery Lager yeast, and this was designated the Lg-*ATF1* gene.

Comparison of *ATF1* and K7-*ATF1* showed that these two genes share an almost identical nucleotide sequence in the region which was sequenced. Thirteen nucleotide changes were observed in the open reading frame and three amino acid changes were found in these genes. However, Lg-*ATF1* had less homology with *ATF1*. Figure 2 shows a Harr plot comparing the Lg-*ATF1* and *ATF1* genes from the bottom fermenting yeast, indicating a relatively high homology (76%) for the open reading frames. However, the N-terminal region of the open reading frame and its 5' upstream region showed reduced homology. The amino acid sequence encoded by *ATF1* shared 80% homology with that of Lg-*ATF1*. The AATase encoded by *ATF1* contained fourteen cysteines and the AATase encoded by Lg-*ATF1* had eleven cysteines. Ten of these cysteine residues were found in the same position within their amino acid sequences. The hydrophobicity of the AATases encoded by *ATF1* and Lg-*ATF1* also showed a similar profile. These results suggested that the protein structures of both AATases were analogous.

ATF1 was mapped to a region close to the right arm telomere of XV chromosome in *S. cerevisiae* using an *S. cerevisiae* prime clone grid filter (ATCC 77284). The labelled *ATF1* hybridized strongly to the 1050-kb chromosomes of *S. pastorianus*, which has the similar molecular weight as the XV chromosome of *S. cerevisiae,* and the 1000-kb chromosome of *S. pastorianus*. The labelled Lg-*ATF1* probe hybridized strongly with the 850-kb chromosome of both *S. pastorianus* and *S. bayanus*. The bottom fermenting yeast *S. pastorianus* is thought to be a hybrid between *S. cerevisiae* and *S. bayanus* (*10-11*). Therefore, *ATF1* might be derived from *S. cerevisiae* and Lg-*ATF1* might be derived from *S. bayanus* (Yoshimoto, H., in preparation).

Gene Dosage Effect of *ATF1* and Lg-*ATF1* in Brewer's Yeast

Construction of Vectors Carrying the *ATF1* Genes for Brewer's Yeasts. Laboratory strains are usually heterothallic haploid and have several genetic markers which can be used to isolate trasformants for the study of yeast molecular genetics. However, brewer's yeasts, which include bottom fermenting yeast (Lager Yeast) and top fermenting yeast (ale yeast), are diploid or polyploid. Therefore, nutritional requirement mutations used for the yeast transformation as selectable markers are very difficult to isolate. A dominant marker gene is essential to the transformation of the brewer's yeast. As shown in Figure 3, two types of vector, pYT71 and pYT77, were constructed in this study: a single copy YCp vector and a multicopy YEp vector respectively, both carrying the G-418 resistance gene as a dominant selectable marker.

To insert the genes to these vectors, the 5' upstream *Hpa*I site and the 3' downstream *Kpn*I site of the *ATFI* gene were changed to *Sal*1 by blunting and linker ligation. This 2.8-kb *Sal*1 DNA fragment was ligated to both vectors. *Sal*1 linker was ligated to the blunted *Pst*1 site in 3' downstream region of Lg-*ATF1*. The resulting 3.0-kb of *Sal*I fragment was then ligated to both vectors.

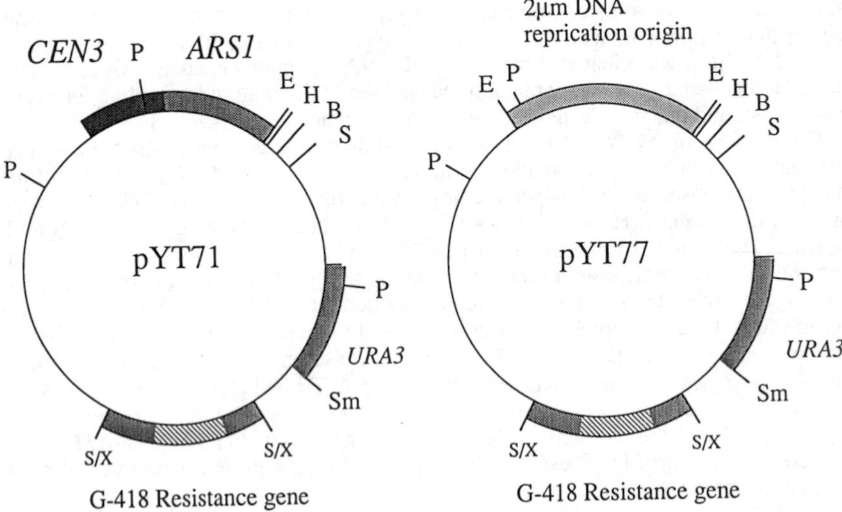

Figure 3. Structure of pYT71 and pYT77, vectors for brewer's yeast transformation. pBR322 sequence is presented by the thin line.

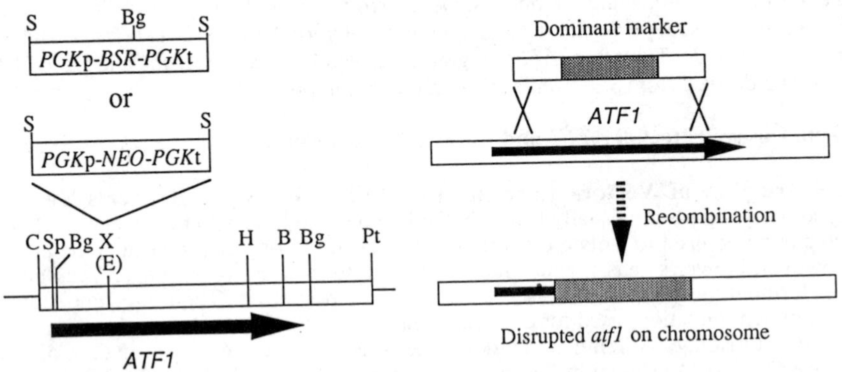

Figure 4. Construction of DNA fragments for *atf1* gene disruption and one step gene disruption. The restriction enzymes used were *Hin*dIII (H), *Cla*I (C), *Eco*RI (E), *Bam*HI (B), *Pst*I (P), *Bgl*II(Bg), *Sal*I (S), *Sph*I (Sph). The position and direction of the open reading frame are indicated by an arrow.

Laboratory Scale Fermentation Using Ale-Yeast Carrying *ATF1* and Lg-*ATF1* Genes. Transformants carrying the *ATF1* genes were pre-cultured in brewery wort, to which 50 μg/mL of G-418 had been added. Pre-culture was carried out anaerobically with agitation at 20°C for 3 days and the transformant cells were collected by centrifugation. The wort was pitched with 3.0g-wet weight cells/L without the addition of antibiotics and transferred to a 500 mL laboratory fermentation vessel (120-cm long, 2.5-cm diameter). Fermentation was carried out at 18°C for 4 days. There was no substantial difference in the fermentation performance of the transformants and the parental strain (data not shown). As shown in Table I, young beer produced by the transformants carrying the *ATF1* gene on a multicopy vector had a 13 fold greater concentration of isoamyl acetate and a 9 fold greater concentration of ethyl acetate than the parental strain. However, transformants carrying *ATF1* on single copy vector produced almost same concentration of both esters as parental strain.

Transformants carrying Lg-*ATF1* on the multicopy vector produced less acetate esters than transformants carrying *ATF1* on the multicopy vector. The enzyme activity of the transformants were compared on the third day of fermentation. All transformants carrying the *ATF1* or Lg-*ATF1* on the multicopy vector exhibited high levels of AATase activity, but the AATase activity produced by *ATF1* strains was higher than that of Lg-*ATF1* strains. Although the copy number of *ATF1* and Lg-*ATF1* on the multicopy vector was determined by Southern blot analysis using a *URA3* DNA fragment as a probe, no difference was observed (data not shown). Therefore, the expression level of the gene or the specific activity of the AATase encoded by the Lg-*ATF1* might be lower than that of the AATase encoded by *ATF1*.

Gene Disruption of Ale-Yeast *ATF1*. To reduce ester formation, the *ATF1* genes were disrupted in the top fermenting yeast (Ale-yeast). As shown in Figure 4, the *Eco*RI site within the open reading frame of *ATF1* was filled using Klenow enzyme and a *Xho*I linker was ligated. DNA fragments for the *ATF1* gene disruption were constructed by ligating the *Sal*I DNA fragment containing the G-418 resistance gene (*NEO*) or blasticidine S resistance gene (*BSR*) to the *Xho*I site. The *Bgl*II fragment was used to transform the top fermenting yeast KTY001. The homologous ends pairs with the *ATF1* locus on the chromosome, and recombination results in a chromosomal gene disruption, as shown in Figure 4. DNA was isolated from the transformants and digested with *Cla*I for Southern blot analysis. Transformants in which at least one *ATF1* gene was disrupted were used for re-transformation with the blasticidine resistance marker to isolate clones in which two *ATF1* genes were disrupted. Southern blot analysis showed that transformants carrying the two disrupted *ATF1* genes still maintained at least one *ATF1* gene (data not shown). These results suggested that Ale-yeast used in this study maintained at least three *ATF1* genes. A yeast strain yYT113 contains one disrupted *atf* gene designated *atf1::NEO*, and yYT127 contains two disrupted *atf* genes designated *atf1::NEO* and *atf1::BSR*.

Laboratory Scale Fermentation Using *ATF1* Disruptants. Transformants carrying the disrupted *atf1* genes were pre-cultured in YM10 medium (Yeast extract 1.25%, Malt Extract 1.25%, Glucose 10%), to which 50 μg/mL of G-418 had been added. Pre-culture was carried out anaerobically with agitation at 20°C for 3 days, and cells were collected by centrifugation. Brewery wort was pitched with 3.0g-wet weight cells/L without the addition of antibiotics and transferred to a 500 mL laboratory fermentation vessel (120-cm long, 2.5-cm diameter). Fermentation was carried out at 18°C for 4 days.

There was no substantial difference in the fermentation performance of the transformants and the parental strain (data not shown). As shown in Table II, young

Table I. Gene dosage effects of *ATF1* on the volatile esters and alcohols concentrations in the laboratory scale fermentation

Plasmids	Isoamyl acetate (ppm)	Ethyl acetate (ppm)	n-propyl alcohol (ppm)	isobutyl alcohol (ppm)	Isoamyl alcohol (ppm)
Parental strain	0.67	7.4	14.4	21.9	77.8
ATF1 on pYT71	0.78	8.9	16.9	18.5	69.4
ATF1 on pYT77	10.10	62.7	17.8	22.0	74.1
Lg-ATF1 on pYT71	0.69	7.8	17.1	20.3	76.8
Lg-ATF1 on pYT77	2.20	16.6	16.7	22.1	79.3

Table II. Effects of *ATF1* gene disruption on the volatile esters and alcohols concentrations in the laboratory scale fermentation

Strains	Isoamyl acetate (ppm)	Ethyl acetate (ppm)	n-propyl alcohol (ppm)	isobutyl alcohol (ppm)	isoamyl alcohol (ppm)
Parental strain	0.83	9.6	14.9	13.9	73.8
yYT113 (*atf1::NEO*)	0.72	9.1	16.9	18.5	69.4
yYT127 (*atf1::NEO,atf1::BSR*)	0.53	8.3	17.8	22.0	74.1

beer produced by the transformant carrying a single disrupted *atf1* gene contained approximately 15% less isoamyl acetate than the parental strain and there was a slight decrease in ethyl acetate production. Young beer fermented by the transformant carrying two disrupted *atf1* genes contained about 35% less isoamyl acetate than the parental strain and there is a small but significant decrease in ethyl acetate production.

Although the haploid laboratory strain TD4 has one *ATF1* gene, disruption of *ATF1* did not affect cell growth or fermentation ability (Fujii, T., submitted). This result indicated that *ATF1* was not an essential gene in *S. cerevisiae*. Compared to the parental strain, laboratory strain carrying a disrupted *atf1* gene, TD4-Δ*atf1,* only produced 30% isoamyl acetate in static culture. However, 70% production of ethyl acetate was maintained compared to the parental strain. These results showed that the AATase encoded by *ATF1* mainly participates in isoamyl acetate formation, but play a lesser role in ethyl acetate formation. These results suggested that AATases encoded by genes other than *ATF1* might be involved in the remaining ester formation in the *atf1* disruption strain. Minetoki et al showed that two separate AATase activities could be eluted from an ion exchange column (*8*). The major peak of AATase activity was purified and its amino acid sequences were used to clone the *ATF1* gene. It will be interesting to elucidate enzyme characteristics of the minor peak of AATase activity and the gene expression.

ATF1 Gene Regulation and Promoter Replacement

Transcription Regulation of *ATF1* Gene. The yeast cell adopts to circumstances by regulating protein function and gene expression. Many genes are controlled by transcription initiation. AATase enzyme activity is thought to be reduced by aeration or the addition of unsaturated fatty acids to the medium. To elucidate whether the *ATF1* gene was regulated at the transcriptional level, RNA was extracted from cells cultured in repressed or derepressed conditions. The amount of mRNA transcribed from *ATF1* was analyzed by Northern blotting. Transcription of *ATF1* was reduced to basal levels in cultures containing unsaturated fatty acids or by aeration (Kobayashi, O., in preparation).

It was also confirmed that expression of *ATF1* and Lg-*ATF1* were co-regulated in the bottom fermenting yeast. A 1-kb DNA fragment containing the *ATF1* or Lg-*ATF1* promoter region was obtained by PCR (polymerase chain reaction). Amplified *ATF1* promoter or Lg-*ATF1* promoter was ligated to the *E.coli* ß-galactosidase gene *lac*Z on the plasmid pHY428 which was derived from pYT77. The resultant plasmid, pHY429 carrying the *ATF1*::*lac*Z fusion gene or pHY430 carrying the Lg-*ATF1*::*lac*Z, was introduced into the bottom fermenting yeast. The expression of *ATF1* or Lg-*ATF1* was monitored by determining the level of ß-galactosidase activity under a variety of culture conditions. It was found that unsaturated fatty acids reduce *lacZ* expression to basal level. This result clearly indicated that *ATF1* and Lg-*ATF1* were co-regulated at the transcriptional level in *S. pastorianus* (Yoshimoto, H., in preparation).

Replacement of the *ATF1* Promoter with a Constitutive *GDP* Promotor. AATase activity is known to be reduced by the addition of unsaturated fatty acids to the medium or shaking the culture for aeration (*7*). To enhance the expression of the *ATF1* gene, a constitutive strong promoter was placed in front of the *ATF1* coding region. As genes encoding glycolysis enzymes are reported to be strongly expressed in *S. cerevisiae*, the promoter of alcohol dehydrogenase (*ADH*), phospho-glycerol kinase (*PGK*) or glyceraldehyde-3-phosphate dehydrogenase (*GPD*) are usually chosen for heterologous gene expression. In this study, the *GPD* promoter was used for *ATF1* gene expression. The 5' untranslated region of the *ATF1* gene was deleted by *Exo*III endonuclease, blunted and the *Pst*I site in 3' untranslated region was also blunted by T4 DNA polymerase. This DNA fragment carrying the *ATF1* open reading frame was ligated to the blunted *Hin*dIII site of the *GPD* promoter vector. The *Sal*1 fragment containing *GPD*p-*ATF1* was inserted into *Sal*I site of pYT71. Laboratory scale fermentation experiments were carried out by the procedure described above. As shown in Table III, the transformant carrying *GAP*p-*ATF1* on a single copy plasmid produced much greater levels of ethyl acetate and isoamyl acetate than *ATF1* on the multicopy plasmid.

Conclusion and Discussion

The *S. cerevisiae* AATase was purified and its internal peptide sequences were determined. *ATF1* genes encoding the AATase were cloned from the bottom fermenting yeast *S. pastorianus*. The two genes isolated have different structures. One is very similar to the K7-*ATF1* gene cloned from *S. cerevisiae* and the other, Lg-*ATF1*, has less homology with *ATF1*. Transformants carrying the *ATF1* or the Lg-*ATF1* on a multicopy plasmid produced greater levels of acetate esters than the parental strain in laboratory scale fermentations.

ATF1 and Lg-*ATF1* are present on two chromosomes of different molecular weight. *ATF1* strongly hybridized to the XV chromosome of *S. cerevisiae* and two chromosomes in *S. pastorianus*, but not to *S. bayanus*. Lg-*ATF1* strongly hybridized to the 850-kb chromosome of *S. bayanus* and *S. pastorianus*, but not to *S. cerevisiae*. The bottom fermenting yeast *S. pastorianus* is thought to be a natural hybrid between *S. cerevisiae* and *S. bayanus* (*10-11*). Therefore, our results suggest that *ATF1* might be derived from *S. cerevisiae* and Lg-*ATF1* might be derived from *S. bayanus*.

Transformants carrying the disrupted *atf1* gene were constructed by a one step gene disruption method. As brewer's yeasts are not haploid, dominant marker genes are essential to select transformants. In this study, the G-418 resistance gene and blasticidine S resistance gene were used for gene disruption. G-418 (geneticine) is a standard reagent in yeast genetics. However, there are no reports describing the use of Blasticidine S in yeast transformation. The blasticidine resistance gene was placed downstream of the phospho-glycerol kinase promoter (*PGK*p) and this resistance gene allowed yeast cell growth on the medium containing 100 μg/mL of blasticidine S (Tamai, Y., in preparation). Double disruption of *atf1* in brewing yeast could be obtained by the use of these two resistance genes.

ATF1 gene expression was monitored by measuring the ß-galactosidase activity produced from the *ATF1*::*lac*Z fusion gene. Interestingly, *ATF1* was repressed by aeration or the addition of unsaturated fatty acids to the medium. *ATF1* expression became constitutive by the replacement of the native promoter with the constitutive *GPD* promoter. These results suggested that *ATF1* was regulated at the transcriptional level. Unsaturated fatty acids has been reported as repressing the *OLE1* gene encoding Δ9-desaturase at the transcriptional level (*12-13*). However, the mechanism of this regulation has not been elucidated. Oxygen regulates the expression of many genes (*14*). It induces genes involved in respiration and represses genes involved in anaerobic fermentation. It will be interesting to know whether *ATF1* is the oxygen regulated genes or whether it is regulated by the higher concentration of unsaturated fatty acids synthesized in aerobic culture.

High temperature fermentation or high gravity fermentation produce greater levels of esters and high pressure in tall cylinder conical tanks reduces the ester formation in beer fermentation. Control of the *ATF1* gene expression will provide a means of maintaining beer flavor characteristics in the different brewing processes.

Table III. Volatile esters and alcohols concentrations produced by the yeast carrying *GPD*p-*ATF1* in the laboratory scale fermentation

Plasmids	Culture condition	Isoamyl acetate (ppm)	Ethyl acetate (ppm)	n-propyl alcohol (ppm)	isobutyl alcohol (ppm)	isoamyl alcohol (ppm)
pYT71	w/o Oleic acid	0.67	7.4	14.4	21.9	77.8
pYT71	+Oleic acid	0.18	5.0	17.8	15.1	77.8
GPDp-ATF1 on pYT71	w/o Oleic acid	9.24	68.4	15.3	12.8	69.4
GPDp-ATF1 on pYT71	+Oleic acid	2.16	14.5	10.7	10.0	47.6

Acknowledgement

Research and cloning of *ATF1* was carried out in collaboration with Ozeki Co., Ltd. I wish to express sincere thanks to Mr. Masaaki Hamachi and his colleagues at Ozeki Research Laboratories for their cooperation in this research and I would like to thank Dr. Reisuke Takahashi of KIRIN Brewery Co., Ltd. for permission to publish this work. I would also like to express my appreciation to Dr. Hiroyuki Yoshimoto for valuable discussion, and Miss Keiko Kanai and Mrs. Ritsuko Katoh for technical assistance.

Literature Cited

1. Nodström, K. *J. Inst. Brewing* **1961,** *67*, 173-181.
2. Nodström, K. *J. Inst. Brewing* **1962,** *68*, 398-407.
3. Nodström, K. *J. Inst. Brewing* **1963,** *69*, 142-153.
4. Yoshioka, K.; Hashimoto, N. *Agric. Biol. Chem.* **1981,** *45*, 2183-2190.
5. Stukey, J. E.; McDonough, V. M.; Martin, C. E. *J. Biol. Chem.* **1989,** *264*, 16537-16544.
6. Yoshioka, K.; Hashimoto, N. *Agric. Biol. Chem.* **1983,** *47*, 2287-2294.
7. Inoue, T.; Tanaka, J.; Mitsui, S. *Recent Advances in Japanese Brewing Technology*; Japanese Technology Reviews; Gordon and Breach Science Publisher: Reading, U.K., 1992; vol. 2, 38-39.
8. Minetoki, T.; Bogaki, T.; Iwamatsu, A.; Fujii, T.; Hamachi, M. *Biosci. Biotech. Biochem.* **1993,** *57*, 2094-2098.
9. Fujii, T.; Nagasawa, N.; Iwamatsu, A.; Bogaki, T.; Tamai, Y.; Hamachi, M. *Appl. Envir. Microbiol.* **1994,** *60*, 2786-2792.
10. Vaughan-Martini, A.; Martini, A. *Antonie van Leeuwenhoek* **1987,** *53*, 77-84.
11. Vaughan-Martini, A. *Syntem. Appl. Microbiol.* **1989,** *12*, 179-182.
12. Bossie, M. A.; Martin, C. E. *J. Bact.* **1989,** *171*, 6409-6413.
13. McDonough, V. M.; Stukey, J. E.; Martin, C. E. *J. Biol. Chem.* **1992,** *267*, 5931-5936.
14. Zitomer, R. S.; Lowry, C. V. *Microbiol. Rev.* **1992,** *56*, 1-11.

Chapter 19

Plant Biochemical Regulators and Agricultural Crops

H. Yokoyama[1,3] and H. Gausman[2]

[1]Fruit and Vegetable Chemistry Laboratory, Agricultural Research Service, U.S. Department of Agriculture, 263 South Chester Avenue, Pasadena, CA 91106

[2]Agricultural Research Service, U.S. Department of Agriculture, Amarillo, TX 79109

New plant biochemical regulators (PBRs) were observed to have profound positive effects on crop performance. For the crops that were tested, the PBRs improved both crop yield and yield quality. Negative correlations between crop yield and crop quality were not observed.

Early studies on bioregulation of plant responses by bioregulatory compounds have demonstrated the regulation of isoprenoid biosynthesis in fruits of *Citrus* spp. (*1,2*) and tomato (*Lycopersicum esculatum*) (*3*), in carotenogenic fungi (*1,4*), in cotton (*Gossypium hirsutum*) cotyledon (*5*), in rubber-producing desert shrub guayule (*Parthenum argentenum*) (*6,7*), and photosynthetic bacteria (*8,9*). Bioinduction of lycopene synthesis in lemon fruits by MPTA (N,N-diethylaminoethyl-4-methylphenylether) was shown to require nuclear gene transcription and translation of the poly A+ gene transcripts on 80S ribosomes (*2*). The mode of action of DCPTA (N,N-diethylaminoethyl-3,4-dichlorophenylether)- and MPTA-induced carotenoid accumulation in cotton cotyledons (*5*) involved the selective inhibition of *zeta*-carotene dehydrogenation and the inhibition of lycopene cyclase by MPTA. These observations indicate that the activities of bioregulatory agents on carotenoid biosynthesis involve (1) indirect general biosynthetic pathway induction effects that are mediated through nuclear gene expression, and (2) direct inhibitory effects on the cyclase enzymes involved in tetraterpenoid (carotenoid) biosynthesis. The inhibitory effects of MPTA and DCPTA on tetraterpenoid biosynthesis indicate that the individual bioregulatory agents may induce specific biological responses in crop plants. DCPTA treatments have induced significant enhancement of the biomass and phenology of cotton (*10*), increased root-shoot ratio of cotton seedlings (*11*) and radish (*Raphanus sativus*) (*12*), increased the seed yield and quality of soybean (*Glycine max*) (*13*). In the above studies, DCPTA was applied as either a seed treatment or as a foliar spray to early seedling plants just after they emerged from the seed.

Studies (*14*) have shown that a ^{14}C-labeled N,N-diethylaminoethyl analog of DCPTA, when applied to guayule leaves and stems, was completely catabolized within four days of application. No specific ^{14}C-labeled catabolites were detected. However,

[3]Current address: 975 Ellington Lane, Pasadena, CA 91105

the stimulatory effect of DCPTA on rubber accumulation in guayule stem tissue is generally not observed until two to three months after treatment (*7*). These observations are indicative of the indirect role of DCPTA in rubber accumulation. Fungal studies (*15*), using the carotenogenic mold *Phycomyces blakesleeanus,* have indicated that DCPTA has an indirect role in the bioinduction of tetraterpenoids and a direct role in the inhibition of transformation of acyclic lycopene to bicyclic *beta*-carotene. In the mycelia cultured on GAY (glycine-asparagine-yeast) media containing DCPTA, the acyclic lycopene is bioinduced and is seen as the main pigment. However, when small pieces (1 mm^2) of mycelia cultured on GAY media containing DCPTA were transferred to and cultured on GAY media without the bioregulatory compound DCPTA, enhancement of bicyclic *beta*-carotene was observed, and it accumulated as the main pigment in the mycelia instead of the acyclic lycopene. These results suggest that the enhancement of crop performance by bioregulatory compounds may involve secondary effector (promoter) compounds and that the putative effector(s) control long term crop performance.

Studies (*16*) have shown that, as compared with controls, application of DCPTA to seedling spinach plants increased the total chloroplast volume per cell, or the chloroplast compartment size (CCS) of mature leaves. In DCPTA-treated leaves, coordinated increases in thylakoid development and stromal area per chloroplast were observed as compared with those of untreated leaves. In addition, starch grains of DCPTA-treated leaves were reduced in size as compared with controls. Extractable chlorophyll (Chl) per gram of fresh weight of leaf tissue and per unit leaf area were increased significantly in DCPTA-treated plants as compared to controls. Coordinated increases in Chl a and Chl b were observed in DCPTA-treated plants as compared with controls. However, statistically similar Chl a/b ratios for all DCPTA treatments suggested that Chl accumulation in DCPTA-treated spinach is regulated by the CCS and not by specific increases in either Chl a or Chl b synthesis. Electrophorectic analysis of thylakoid membrane proteins indicated that the light-harvesting chloroplast protein II (LHCP II) concentration per mg of total Chl was increased in thylakoids isolated from DCPTA-treated plants as compared with control preparations from untreated plants (*16*). Ribulose 1,5-bisphosphate carboxylase/oxygenase (Rubisco) activity *in vitro* in mature leaves of DCPTA-treated spinach plants was increased significantly. The enhanced biomass gain of mature DCPTA-treated spinach was due to a significant acceleration of relative growth rate and not to an extension (days) of exponential growth. In addition, the enhanced biomass gain observed in DCPTA-treated plants was manifested in all plant parts. Compared with controls, the accelerated leaf area development of DCPTA-treated plants would significantly increase light interception per plant during exponential plant growth, which in turn would help to increase photosynthetic productivity and subsequent crop growth rate (*17*). Promotive effects of DCPTA have been observed on the vegetative groweth rate of sugar beets (*Beta vulgaris*) (*18*), spinach (*16*), and *Phalaenopsis* (*19*).

In this chapter we will present results of three new and improved bioregulatory compounds, or plant biochemical regulators (PBRs), on several agricultural crops.

Bioregulation of Agricultural Crops

Crop yield and crop quality are often determined by the same regulatory mechanisms that control crop growth rate and vegetative plant development (*17,20,21*). In PBR-treated tomato plants as shown in Table 1, harvestable crop yield is increased as much as 2.25 times that of the control (0.8 kg $plant^{-1}$).

Table 1. Effect of PBRs on Fruit Yield and Quality of Tomato (*Lycopersicum esculatum* cv. Pixie)

Treatment		Brix	Carotenoids	Fruit				
PBR*	μM		mg (g fresh wt)$^{-1}$	total (plant)$^{-1}$	size g(fruit)$^{-1}$	%ripe	yield kg(plant) total	ripe
1	30	8.2±.04	186±7	40	41	90	1.6	1.26
1	3	7.9±.04	158±4	35	39	88	1.4	1.23
2	30	7.0±.03	160±8	46	37	74	1.7	1.26
2	3	7.2±.04	156±9	46	37	72	1.7	1.22
3	3	8.4±.04	191±8	47	39	92	1.8	1.66
3	0.3	7.3±.04	172±7	38	40	89	1.5	1.34
4	0	5.1±.07	78±8	27	24	31	0.8	0.25

*1. FVCL-1 2. FVCL-8 3. FVCL-9 4. Control

Foliar application of solutions containing 0.1% (w/v) Kinetic nonionic wetting agent at early seedling stage (3-4 leaf). Fruit harvested 3 months after planting. Results represent means of 6 replicate plants. Determination of Brix and carotenoid content made on fully mature fruits. Greenhouse grown in 2 gal pots.

In each crop that was tested, in addition to tomato, PBR enhanced yield resulted from improved seedling vigor and leaf canopy development during exponential crop growth of treated as compared with untreated plants. In DCPTA-treated bush bean (*Phaseolus vulgaris*), eggplant (*Solarium melongena*), paprika pepper (*Capsicum annum*), and tomato enhanced yield was associated with acceleratyed development of secondary branches during exponential growth and with improved fruit set at crop maturity (*3,18,22*). Moreover, the harvestable yield of PBR-treated plants was associated with increased vegetative growth when compared to controls. Negative correlations between crop yield and crop quality were not observed. When compared with controls, the soluble solids content were increased at crop harvest. In addition, the carotenoid contents of treated tomato fruits were increased as compared with controls. FVCL-9 appears to be very effective at the lower concentration levels of 1 (3 μM) and 0.1 (0.3 μM) ppm, indicative of its high biological activity. The sweetness and flavor intensity of ripe fruits were positively correlated with the total soluble solids content (*23,24*). However, negative correlations of total fruit yield with total soluble solids content of ripe fruits are often observed (*24*). That is, agricultural treatments (chemical, cultural, environmental) that tend to increase tomato fruit yields often produce mature fruits with a reduced total soluble solids content. This study shows the overall promotive effects of the new PBRs upon fruit yield and fruit quality of fresh market tomato. For example, both total soluble solids (degrees Brix) and ripe fruits harvested 60 days after seed planting were increased significantly for PBR-treated plants, including foliar applications of FVCL-1, FVCL-8 and FVCL-9, over those of their respective controls. The largest numerical increases in fruit ripening relative to the controls were observed in 3 μM FVCL-9 treated plants. Increases, though at slightly lower levels, occurred also with FVCL-1 and FVCL8 treated plants. The red-ripe fruit yields of 3 µM FVCL-9 treated plants were six times those of the controls. Foliar application of the PBRs increased the total soluble solids and carotenoid contents of ripe tomato fruits.

PBR treatment increased the biomass of spinach, Table 2. The largest increases (about two-fold) were observed with 3 µM FVCL-9 treated plants as compared with controls. The spinach leaves of the treated plants had a much deeper green coloration than leaves of the untreated plants. These observations were confirmed by the increased chlorophyll (Chl) content of PBR-treated plants over that of controls as shown in Table 2. There were parallel enhancement effects of FVCL-9 on total Chl accumulation and on biomass gains. Similar efects have been observed in DCPTA-treated blue spruce (*Pices pungens*) (*25*) and sugar beet (*18*). These results indicated that PBRs have an influence on the photosynthetic capacity of crop plants.

Table 2. Effect of PBR on Spinach (*Spinacea oleracea* cv. New Zealand)

Treatment		Biomass	Total Chlorophyll
PBR*	µM	dry wt gm	mg (g fresh wt)$^{-1}$
1	30	10.07±.31	1.98±.11
2	30	10.21±.39	1.82±.12
3	3	10.59±.43	1.69±.15
4	0	5.76±.65	1.28±.17

*1. FVCL-1 2. FVCL-8 3. FVCL-9 4. Control

Single foliar application of PBR solutions with 0.1% (w/v) Kinetic wetting agent at early seedling stage (2-3 true leaf). Harvested 60 days after planting. Greenhouse grown in 2 gal pots; 8 replicate plants.

Table 3 shows that the new PBRs have profound effects on biomass and root formation in radish. Foliar treatment of radish resulted in greatly enhanced root and biomass development at crop harvest as compared with controls.

FVCL-9 at both 3 µM and 0.3 µM levels was the most effect in increasing biomass about two-fold and promoting root formation almost three-fold, as compared with respective controls.

The PBR effect on the biomass of the monocotyledenous corn plant (*Zea mays*) are presented in Table 4. Foliar applications of the three PBRs, FVCL-1, FVCL-8, and FVCL-9, were made on corn seedlings, grown in pots under greenhouse conditions, at the early postemergence-growth stage. Thirty-five days later the plants were sacrificed for biomass determinations. As compared with the control (0 ppm treatment) 3 µM FVCL-1 gave the largest increase in biomass production, ranging in magnitude from 1.7 to 2.4-fold increases, the leaf blades of treated plants were larger with thiker culans than untreated plants, accounting for much of the biomass increases. These greenhouse results are important because corn as a monocotyledenous plant was affected by PBRs as has been previously noted for dicotyledenous plants.

Table 3. Effect of PBR on Radish (*Raphanus sativus* cv. Red Devil)

Treatment		Biomass	Root
PBR*	μM	wet wt (g)	wet wt (g)
1	30	55.4±3.3	25.3±1.8
1	3	57.7±4.7	25.2±2.3
2	30	59.0±4.6	23.9±1.5
2	3	54.0±3.5	16.5±1.3
3	3	72.9±2.9	41.3±4.0
3	0.3	70.2±2.6	40.4±2.8
4	0	37.7±2.8	14.6±1.2

*1. FVCL-1 2. FVCL-8 3. FVCL-9 4. Control

Foliar application at early seedling stage (2-3 true leaf). Nonionic wetting agent used was Kinetic at 0.1% (w/v). All radishes harvested 32 days after planting; 6 replicate plants. Greenhouse grown in 2 gal pots.

Table 4. Effect of PBR on Biomass of Corn (*Zea mays* cv. Early Xtra Sweet)

Treatment		Biomass
PBR*	μM	dry wt g
1	30	57.9±3.4
1	3	72.5±2.3
2	30	52.6±2.5
2	3	53.4±1.5
3	3	52.6±2.5
3	0.3	53.5±1.5
4	0	30.1±3.4

*1. FVCL-1 2. FVCL-8 3. FVCL-9 4. Control

Foliar application of PBR and control solutions with 0.1% (w/v) nonionic wetting agent Kinetic at early seedling stage (6-8 cm tall). Harvested 35 days after planting; 8 replicate plants. Greenhouse grown in 2 gal pots.

Treatments with PBRs FVCL-1 and FVCL-8 increased Valencia orange (*Citrus sinensis*) fruit diameter significantly whereas DCPTA treatment effects were comparable to those of the controls (Table 5).

Table 5. Comparison of PBR Effects on Orange (*Citrus sinensis* cv. Valencia) Fruit

Treatment		Fruit Diam.	Juice	Vitamin C	Serum	Peel	Percent of Fruit	
PBR*	µM	mm	ml g^{-1}	mg(100ml)$^{-1}$ fresh fruit wt	Brix	mm	Juice	Peel+Pulp
1	75	65.4 a	0.53 a	56.8 a	14.8 a	4.1 b	51.9 a	46.9 a
2	75	65.2 a	0.53 a	56.2 a	14.6 a	4.2 b	51.8 a	46.7 a
3	75	64.1 b	0.49 b	46.4 b	12.3 ab	4.7 ba	51.4 a	47.1 b
4	0	63.9 b	0.49 b	42.6 c	10.6 b	4.9 a	50.7 b	47.4 b

*1. FVCL-1 2. FVCL-8 3. DCPTA (N,N-diethylaminoethyl-3,4-dichlorophenylether) 4. Control

Foliar spray applied to entire tree shortly after fruit set (0.5-1.5 cm diameter fruit size). Higher concentrations (75 µM) were applied to compensate for the thicker layer of wax coating the leaves and young fruits of citrus trees. PBR and control solutions contained 0.1% (w/v) Kinetic nonionic wetting agent and were applied as a single foliar treatment during early fruit development. Fruits harvested at full maturity. Means associated with the same letter are not significantly different, according to Duncan's multiple range test ($p<5\%$).

All treatments reduced peel thickness as compared to that of the controls; largest reductions were induced by FVCL-1 and FVCL-8. Improved juice recovery was generally related to a reduction in peel thickness, although with both FVCL-1 and FVCL-8 treatments increased fruit sizes were observed in relation to the respective controls. DCPTA-treated fruits were the same size as those of the controls. Both FVCL-1 and FVCL-8 treatments increased juice recovery per fresh fruit weight in Valencia oranges and significantly improved degrees of Brix and vitamin C content when compared with control fruits. These results indicate that fruits treated with FVCL-1 and FVCL-8 were sweeter with an improved nutritional quality. Moreover, the new PBRs, FVCL-1 and FVCL-8, have improved bioregulatory activity over that of DCPTA.

Visual observations indicated that both endocarp and flavedo of FVCL-1 and FVCL-8 treated fruit exhibited enhanced pigmentation compared with untreated fruit, suggesting a deeper coloration caused by an increased accumulation of carotenoid pigments, Table 6. these results also indicated that the new PBRs enhanced fruit color more than did DCPTA.

Valencia orange trees that were treated with FVCL-1 abd FVCL-8 generally had the most improvement in flavedo and endocarp carotenoid development when compared with control fruits. Most importantly, all orange fruits of trees which were treated with FVCL-1 and FVCL-8 matured approximately one month earlier than did fruits of untreated trees.

Table 6. Comparison of Effect of PBR on Carotenoid Pigment (Color) of Orange (*Citrus sinensis* cv. Valencia) Fruit

Treatment		Carotenoid Content (μg/g dry wt)	
PBR*	μM	Endocarp	Peel
1	75	266	302
2	75	224	263
3	75	174	205
4	0	156	186

*1. FVCL-1 2. FVCL-8 3. DCPTA 4. Control

Results presented in this table are from the fruits described in Table 5.

Conclusion

In PBR-treated crops that were examined herein, crop yield and yield quality were positively correlated, suggesting influences of enhanced photosynthate supply and of photosyntate utilization similar to those observed with DCPTA (*16*). Negative correlations of crop yield and crop quality were not observed. These results suggest that application of bioregulatory compounds to developing green plants caused increased carbon dioxide assimilation within the photosynthetic pathway, thereby increasing the amount of carbon available for synthesis of total biomass and of individual plant constituents. Furthermore, application of bioregulatory compounds at an early stage of plant or fruit development during the period of active division and before the completion of cell differantiation enhances the plant's genetic expression so as to tap the unused biological potential of the plant. Thus, as new cells develop under the influence of the bioregulatory compounds, they will have an increased capacity to form and store valuable materials and to form increased plant tissues. The effects of the new PBRs FVCL-1, FVCL-8 and FVCL-9 on tomato and citrus fruit development and quality performance illustrate the basic regulatory function(s) that are modulated by their application to various agricultural crops.

Literature Cited

1. Yokoyama, H.; Hsu, W.J.; Poling, S.M.; Hayman, E.P. *Carotenoid Chemistry and Biochemistry;* Pergamon Press: Oxford, **1982**, pp 371-385.
2. Benedict, C.R.; Rosenfeld, C.L.; Mahan, J.R.; Madhavan, S.; Yokoyama, H. *Plant Sci.* **1985,** *41*, 169.
3. Keithly, J.H.; Yokoyama, H.; Gausman, H. W. *Plant Growth Regul.* **1990,** *9*, 127.
4. Hsu, W.J.; Yokoyama, H.; Coggins, C. *Phytochemistry* **1972,** *11*, 2985.

5. Greenblatt, G.A.; Benedict, C.R.; Madhavan, S.; Yokoyama, H. *Plant Physiol.* **1987,** *83,* S-805.
6. Yokoyama, H.; Hayman, E.P.; Hsu, W.J.; Poling, S.M. *Science* **1977,** *197,* 1076.
7. Benedict, C.R.; Reibach, P.H.; Madhavan, S.; Stipanovic, R.V.; Keithly, J.H.; Yokoyama, H. *Plant Physiol.* **1983,** *72,* 897.
8. Hayman, E.P.; Yokoyama, H.; Chichester, C.O.; Simpson, K.L. *J. Bacteriol.* **1974,** *20,* 1339.
9. Hayman, E.P.; Yokoyama, H. *J. Bacteriol.* **1976,** *127,* 1030.
10. Gausman, H.W.; Burd, J.D.; Quesenberry, J.; Yokoyama, H.; Dilbeck, R.; Benedict, C.R. *Bio/Technology* **1985,** *3,* 255.
11. Gausman, H.W.; Yokoyama, H.; Quisenberry, J.; Keithly, J.H.; Burd, J.D. *Plant Growth Regul. Soc. Am.* **1988,** *16,* 6.
12. Keithly, J.H.; Yokoyama, H.; Gausman, H.W. *Amer. Hort. Sci.* **1992,** *117,* 294.
13. Yokoyama, H.; DeBenedict, C.; Hsu, W.J.; Hayman, E.P. *Bio/Technology* **1984,** *2,* 712.
14. Kelly, K.M.; Gilliland, M.G.; Van Staden, E.R.; Peterson-Jones, J.C. *Plant Growth Regul.* **1987,** *5,* 25.
15. Yokoyama, H.; Keithly. J.H. *Quality Factors of Fruits and Vegetables: Chemistry and Technology;* American Chemical Society: Washington, D.C., **1989,** pp 65-70.
16. Keithly, J.H.; Yokoyama, H.; Gausman, H.W. *Plant Biochemical Regulators* Marcel Dekker, Inc.: New York, **1991,** pp 223-245.
17. Gifford, R.M.; Jenkins, C.L.D. *Photosynthesis, Vol. II Development of Carbon Metabolism and Plant Productivity* Academic Press: New York, **1982,** pp 419-457.
18. Keithly, J.H.; Yokoyama, H.; Gausman, H.W. *Plant Sci.* **1990**, *68,* 57.
19. Keithly, J.H.; Yokoyama, H. *Plant Growth Regul.* **1990,** *9,* 16.
20. Patrick, J.W. *HortScience* **1988,** *23,* 33.
21. Daie, J. *Hort. Rev.* **1985,** *7,* 69.
22. Kobayashi, H.; Keithly, J.H.; Yokoyama, H. *J. Jpn. Soc. Hortic. Sci.* **1990,** *59,* 115.
23. Stevens, M.A.; Kader, A.A.; Albright-Horton, M.A.; Algazi, M. *J. Amer. Soc. Hort. Sci.* **1977,** *102,* 680.
24. Stevens, M.A. *Plant Breeding Rev.* **1986,** *4,* 273.
25. Keithly, J.H.; Kobayashi, H.; Yokoyama, H. *Plant Growth Regul. Soc. Am. Q.* **1990,** *18,* 55.

METHODOLOGY

Chapter 20

Analytical Methodology in Biotechnology: An Overview

Gary R. Takeoka[1], Akio Kobayashi[2], and Roy Teranishi[1]

[1]Western Regional Research Center, Agricultural Research Service, U.S. Department of Agriculture, 800 Buchanan Street, Albany, CA 94710
[2]Laboratory of Food Chemistry, Ochanomizu University, 2–1–1 Ohtsuka, Bunkyo-ku, Tokyo 112, Japan

For every advance in scientific knowledge, there must a development of a method or technique to uncover new information. Biotechnology is the use of biological agents to provide us with foods and flavors with better quality and/or higher yields. In order to follow changes brought about by biotechnology, we must have better methods of evaluating these changes for food quality or for toxicity aspects. In this section, various methods for isolation, identification and characterization of food constituents altered by biotechnology are discussed.

Michael Faraday in 1830 (*1*) said, "Chemistry is necessarily an experimental science; its conclusions are drawn from data, and its principles supported by evidence derived from facts. A constant appeal to facts, therefore, is necessary; and yet so small, comparatively, is the number of these presented to us spontaneously by Nature, that were we to bound our knowledge by them, it would extend but to a very small distance, and in that limited state be exceedingly uncertain in its nature. To supply the deficiency, new facts have been created by *experiment*, the contrivance and hand of the philosopher having been employed in their production and variance."

Chemistry is still an experimental science. New reactions must be devised and experimental methods extended and improved, or new ones invented, to provide new data. Modern biotechnology is a much more recent area of investigation and therefore demands even more innovation to accumulate new facts by experimentation. If modern technology is to help produce more and better foods, chemistry must be utilized to increase acceptability of such new foods.

Recent developments in analytical methodology have enabled flavor chemists to make considerable advances in recent years (*2*). Methods developments have progressed to the point that ripening of fruit can be followed quantitatively as the fruit progresses from green to ripe. Compounds contributing the green aroma decrease in concentration while compounds contributing to the ripe fruit aroma increase as nectarines go from green, green-red, shipping ripe, to full ripe (*3*). Maximum aroma

0097–6156/96/0637–0216$15.00/0

of apples is reached approximately two weeks after harvest, and the drop in aroma has been followed as apples are stored (*4*). One of the first sensory analysis of apple aroma was done with Red Delicious apple essence (*5, 6*), but now sensory evaluation utilizing aroma values of different varieties of apples has been studied (*7*).

Currently, there is much interest in flavor research in determining what compounds are actually responsible for the characteristic flavor of a food. Researchers have utilized odor unit values (compound concentration/odor threshold) to determine the contribution of individual constituents to the overall flavor of a food. For example, citrus peel oil aroma quality has been characterized using logarithmic odor unit values (*8*). Tamura et al. (this volume) discuss the suitability of the detection threshold and the recognition threshold in determining the limited odor unit for characterizing citrus aroma quality.

Advantages and disadvantages of simultaneous distillation and extraction and dynamic headspace sampling methods for isolating volatiles from foods are compared by Buttery and Ling (this volume). A high flow dynamic headspace sampling technique with "closed loop stripping" was found to be most practical, rapid, and comprehensive for studying seasonal development of new cultivars of corn and other food crops. This technique combined with the addition of excess sodium sulfate permitted the quantitaive analysis of very water soluble and polar flavor compounds such as 2,5-dimethyl-4-hydroxy-3(2H)-furanone (Furaneol®) and 3-hydroxy-2-methyl-4-pyrone (maltol). This method will lead to the identification of additional highly water soluble odorants which are not effectively isolated by conventional sample preparation methods.

Accurate quantitative data are a prerequisite for determining the contribution of individual constituents to the overall food flavor. A stable isotope dilution assay is an accurate method for quantitation of flavor constituents which are unstable, inefficiently extracted, and/or which occur at trace concentrations. Grosch and co-workers (*9*, *10*, *11*, *12*) have effectively used this method to quantitate key flavor compounds in a variety of foods. In this method deuterated compounds with only a slight mass difference to the constituents of interest are synthesized and used as internal standards. The food sample is spiked with the deuterated internal standards and the volatiles are isolated. Since the chemical and physical properties such as volatility, reactivity and chromatographic behavior of the analyte and its deuterated analogue are nearly identical, these internal standards are ideal for correcting for any losses during sample preparation. Stable isotope-dilution mass spectrometry has been used by Allen (this volume) to follow the concentration of methoxypyrazines in berry maturity and vine growing conditions of Sauvignon blanc and Cabernet Sauvignon grapes. Biochemical pathways of the origin of pyrazines has been suggested from the observations of the abundance of the pyrazines determined by this stable isotope-dilution mass spectrometry method.

Study of precursors (*13*, *14*, *15*) and how enzymatic action releases characteristic volatiles from precursor compounds has progressed only because of the advance in methods of handling glycosides. Various modern countercurrent chromatographic techniques (*16*) have been successfully employed in the isolation of flavor precursors such as glycosides (*17*, *18*, *19*). The biotechnological possibilities, perspectives and limitations, from the knowledge of the pathways of the biogeneration of important norisoprenoid compounds, such as β-damascenone and the theaspiranes, are discussed by Winterhalter (this volume). Tremendous progress in the chirospecific analysis of these desirable flavor compounds has been achieved, primarily due to the development of modified cyclodextrin capillary columns (*20*) as well as advances in multidimensional gas chromatography (MDGC) coupled with mass spectrometry (*21*). Liquid chromatographic-thermospray mass spectrometric analysis of crude plant extracts containing phenolic and terpene glycosides has been described by Hostettman and co-workers (*22*). The development of atmospheric pressure chemical ionization and electrospray ionization, liquid chromatography coupled with standard mass spectrometry and with tandem mass spectrometry, has opened further dimensions in the field of bio-organic analysis. These tools now provide elucidation of structures of a variety of flavor precursors and will influence flavor biotechnology in many ways. Importantly, these analyses can be performed on very small samples such as a single strawberry.

Analytical methods for non-volatile flavor compounds have been developed. To extend understanding of flavor development during fruit ripening, these methods have been used to study changes in total soluble solids during the ripening of melons grown under hydroponic conditions (Wyllie et al., this volume). One of the first steps in the safety assessment of genetically modified foods is compositional analysis. A major interest in genetically modified potatoes is the concentration of the naturally occurring alkaloids, which are toxic if concentrations exceed certain limits. A broad spectrum of genetically modified potatoes, from greenhouse experiments to field trials, has been screened for steroidal glycoalkaloids for safety considerations (Engel et al., this volume). Methods developed for studying the safety aspects of genetically modified foods have been discussed (*23*).

Processes of interest to food and flavor industries are: sterilization of juices without adverse effects, such as those caused by hydrolyses of various food constituents, esterification and transesterification, rearrangements, isomerizations, condensations, etc. The microwave oven is found in many homes because of the convenience of rapid heating by this method. Advantages of microwave technology are: rapid heating and quenching, minimal temperature gradients, elimination of wall effects, etc. Continuous and batchwise laboratory-scale microwave reactors developed for controlled heating under pressure are described by Strauss and Trainor (this volume).

Literature Cited

1. Faraday, M. *Chemical Manipulation;* 2nd edition; John Murray: London, 1830; p 1.
2. Teranishi, R. In *Flavor Chemistry: Trends and Developments*; Teranishi, R.; Buttery, R. G.; Shahidi, F., Eds.; American Chemical Society Symposium Series No. 388; American Chemical Society: Washington, DC, 1989; pp 1-6.
3. Engel, K.-H.; Ramming, D. W.; Flath, R. A.; Teranishi, R. *J. Agric. Food Chem.* **1988,** *36*, 1003-1006.
4. Dirinck, P.; De Pooter, H.; Schamp, N. In *Flavor Chemistry: Trends and Developments*; Teranishi, R.; Buttery, R. G.; Shahidi, F., Eds.; ACS Symposium Series No. 388; American Chemical Society: Washington, DC, 1989; pp 23-34.

5. Flath, R. A.; Black, D. R.; Guadagni, D. G.; McFadden, W. H.; Schultz, T. H. *J. Agric. Food Chem.* **1967,** *15*, 29-35.
6. Guadagni, D. G. In *Correlation of Subjective-Objective Methods in the Study of Odors and Taste;* Special Technical Publication No. 440; American Society for Testing and Materials: Philadelphia, PA, 1968; pp 36-48.
7. Petersen, M. A.; Poll, L. In *Aroma: Perception, Formation, Evaluation;* Rothe, M., Kruse, H.-P., Eds.; Deutsches Institut für Ernährungsforschung, Potsdam-Rehbrücke, 1995, pp 533-543.
8 Tamura, H.; Yang, R.-H.; Sugisawa, H. In *Bioactive Volatile Compounds from Plants*; Teranishi, R.; Buttery, R.G.; Sugisawa, H., Eds.; ACS Symposium Series No. 525; American Chemical Society: Washington, DC, 1993, pp 121-136.
9. Schieberle, P.; Grosch, W. *J. Agric. Food Chem.* **1987,** *35*, 252-257.
10. Guth, H.; Grosch, W. *Lebensm.-Wiss. u.-Technol.* **1990,** *23*, 513-522.
11. Sen, A.; Laskawy, G.; Schieberle, P.; Grosch, W. *J. Agric. Food Chem.* **1991,** *39*, 757-759.
12. Schieberle, P.; Gassenmaier, K.; Guth, H.; Sen, A.; Grosch, W. *Lebensm.-Wiss. u.-Technol.* **1993,** *26*, 347-356.
13. Williams, P. J.; Sefton, M. A.; Wilson, B. In *Flavor Chemistry: Trends and Developments*; Teranishi, R.; Buttery, R. G.; Shahidi, F., Eds.; ACS Symposium Series No. 388; American Chemical Society: Washington, DC, 1989; pp 35-48.
14. *Flavor Precursors: Thermal and Enzymatic Conversions*; Teranishi, R.; Takeoka, G. R.; Guentert, M., Eds.; ACS Symposium Series 490; American Chemical Society: Washington, DC, 1992, 258 pp.
15. *Progress in Flavour Precursor Studies*; Schreier, P.; Winterhalter, P., Eds.; Allured: Carol Stream, IL, 1993, 507 pp.
16. Conway, W.D. *Countercurrent Chromatography: Apparatus, Theory & Applications*; VCH Publishers, Inc.: New York , NY, 1990.
17. Humpf, H.-U.; Schreier, P. *J. Agric. Food Chem.* **1992,** *40*, 1898-1901.
18. Krammer, G.E.; Buttery, R.G.; Takeoka, G.R. In: *Fruit Flavors: Biogenesis, Characterization and Authentication*; Rouseff, R.L.; Leahy, M.M., Eds.; ACS Symposium Series 596; American Chemical Society: Washington, DC, 1995; pp 164-181.
19. Skouroumounis, G.; Winterhalter, P. *J. Agric. Food Chem.* **1994,** *42*, 1068-1072.
20. Koenig, W.A. *Gas Chromatographic Enantiomer Separation with Modified Cyclodextrins*; Huethig: Heidelberg, 1992, 168 pp.
21. Full, G.; Winterhalter, P.; Schmidt, G.; Herion, P.; Schreier, P. *J. High Resol. Chromatogr.* **1993**, *16*, 642-644.
22. Wolfender, J.L.; Maillard, M.; Hostettmann, K. *J. Chromatogr.* **1993,** *647*, 183-190.
23. *Genetically Modified Foods: Safety Aspects;* Engel, K.-H.; Takeoka, G.; Teranishi, R., Eds.; ACS Symposium Series 605, American Chemical Society: Washington, DC, 1995, 243 pp.

Chapter 21

Existence of Different Origins for Methoxypyrazines of Grapes and Wines

M. S. Allen[1], M. J. Lacey[2], and S. J. Boyd[3,4]

[1]Ron Potter Centre for Grape and Wine Research, Charles Sturt University, P.O. Box 588, Wagga Wagga 2678, Australia
[2]CSIRO Division of Entomology, G.P.O. Box 1700, Canberra 2601, Australia
[3]Environmental and Analytical Laboratory, Charles Sturt University, P.O. Box 588, Wagga Wagga 2678, Australia

2-Ethyl-3-methoxypyrazine has been identified, in addition to 2-isobutyl-3-methoxypyrazine and smaller quantities of 2-*sec*-butyl-3-methoxypyrazine and 2-isopropyl-3-methoxypyrazine, in grapes and wines of the grape variety Cabernet Sauvignon. Measurements by stable isotope-dilution mass spectrometry indicate a pathway to 2-ethyl-3-methoxypyrazine that is different from that of 2-isobutyl-3-methoxypyrazine. The abundance of the ethyl component does not show the dependence that the isobutyl compound has on grape variety, berry maturity, and vine growing conditions. Although 2-ethyl-3-methoxypyrazine is usually well below its sensory detection threshold, it has also been found in a wine at a high level (>1000 ng/L) that exceeds its sensory detection threshold and may contribute to wine aroma.

Amongst the winemaking varieties of the European grapevine *Vitis vinifera* are some, such as Cabernet Sauvignon and Sauvignon blanc, that have fruit with a distinct vegetative or herbaceous aroma. This aroma can add flavor complexity and varietal character to their wines, but in wines produced from the fruit of cool regions, or from fruit of vigorous vines, or in years with poor ripening, this vegetative/herbaceous aroma can be overpowering and undesirable. The aroma arises from trace quantities of 2-methoxy-3-(2-methylpropyl)pyrazine (**1**) (isobutylmethoxypyrazine). Isobutylmethoxypyrazine is a potent flavorant (sensory detection threshold: 1-2 ng/L) of bell peppers (*1, 2*) and a component that contributes to the characteristic aroma of some vegetables (*3, 4*).

[4]Current address: Faculty of Environmental Sciences, Griffith University, Nathan Campus, Kessels Road, Brisbane 4111, Australia

0097–6156/96/0637–0220$15.00/0

Isobutylmethoxypyrazine in Grapes and Wines

Tentative identification of isobutylmethoxypyrazine in grapes was first reported in 1975 (*5*). Its concentration is so low, at parts per trillion level (typically 1-35 ng/L in either grapes or wine of Cabernet Sauvignon), that rigorous quantitative analysis had to await the application of stable isotope-dilution mass spectrometry together with the development of a strategy that limits the volume of volatile solvent to 1 mL to minimize methoxypyrazine loss during solvent evaporation (*6*). This gas chromatography-mass spectrometry (GC-MS) approach uses the synthetic trideuterated isobutylmethoxypyrazine (**5**) as internal standard. Selectivity is gained by selected ion monitoring and by positive ion chemical ionization with ammonia; by these means a detection limit of < 0.1 ng/L is possible (*7*). Such quantitative analysis has shown that isobutylmethoxypyrazine always occurs in grapes and wines of the varieties Cabernet Sauvignon and Sauvignon blanc. Furthermore, sensory studies have confirmed that the level of isobutylmethoxypyrazine typically found in Sauvignon blanc or Cabernet Sauvignon wines can contribute to wine aroma (*8, 9*).

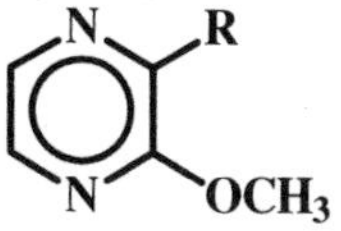

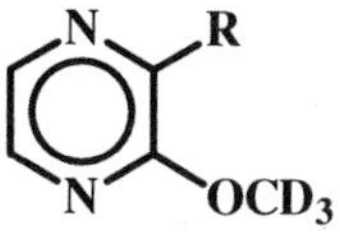

N R OCH$_3$ N R OCD$_3$

1 R = $CH_2CH(CH_3)_2$
2 R = $CH(CH_3)CH_2CH_3$
3 R = $CH(CH_3)_2$
4 R = CH_2CH_3

5 R = $CH_2CH(CH_3)_2$

6 R = $CH(CH_3)_2$

Origin of Isobutylmethoxypyrazine. There can be little doubt that isobutylmethoxypyrazine is an endogenous vine metabolite. Its occurrence displays a clear dependence upon the variety, physiology and environment of the vine. First, its occurrence in grapes is variety-specific. For example, it has always been detected in over 200 analyses of Cabernet Sauvignon and Sauvignon blanc grapes and wines, but so far it has not been found in those of the variety Pinot noir. Secondly, the abundance of isobutylmethoxypyrazine depends upon the fruit developmental stage; there is a consistent decrease of concentration during ripening (Figure 1). In warm areas, its concentration at fruit maturity and grape harvesting may be only 1% of that present in the berry at véraison (the onset of ripening, at which softening, coloring and sugar accumulation commence) (*7*). Further evidence is that climate has a consistent influence on isobutylmethoxypyrazine level. Although other influences are also important and must add to the data variability, a clear trend towards higher levels in wines of cooler regions is evident (Table I) (*10*). Finally, shading of the fruit by the leaf canopy also influences the isobutylmethoxypyrazine concentration (Table II). At different fruiting positions within the vine, increased shading (indicated by increasing leaf layer index in Table II) provided increased levels of isobutylmethoxypyrazine

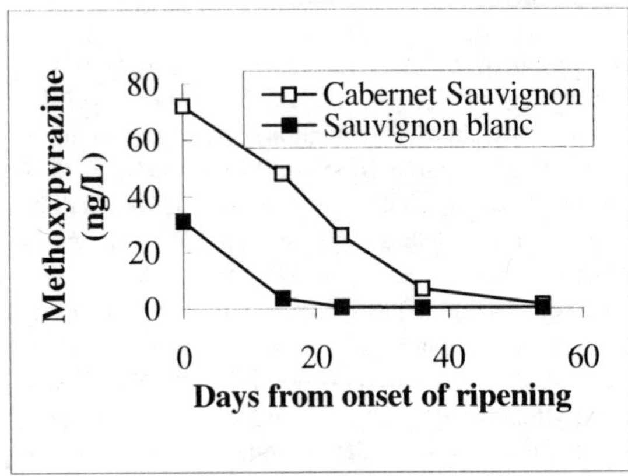

Figure 1. Berry isobutylmethoxypyrazine concentration during ripening. Veraison is day 0 (Sauvignon blanc) and day 6 (Cabernet Sauvignon). (Reproduced with permission from ref. 24. Copyright 1990 Winetitles.)

Table I. Relationship between Methoxypyrazine Level and Growing Region Mean January Temperature (MJT)[a,b]

Wine Region[c]	*Isobutylmethoxypyrazine*[d] *(ng/L)*	*M J T (°C)*
Hunter Valley, NSW	3.6	22.7
Griffith, NSW	6.2	23.6
Leeton, NSW	7.6	23.6
Seymour, Vic.	9.1	21.2
McLaren Vale, SA	11.2	19.8
Frankland, WA	12.3	20.0
Mudgee, NSW	17.1	22.7
Yarra Valley, Vic.	26.1	18.0
Hawkes Bay, NZ	27.6	18.8
Auckland, NZ	27.6	17.9
Hawkes Bay, NZ	28.6	18.8
Mornington Peninsula, Vic.	56.3	18.0

[a]Adapted from ref. 10. [b]MJT is a measure of mean growing temperature in the southern hemisphere. [c]NSW = New South Wales, Vic = Victoria, SA = South Australia, WA = Western Australia; NZ =New Zealand. [d]Values are for Australian and New Zealand Cabernet Sauvignon-based wines, and are presented in order of increasing methoxypyrazine concentration to facilitate comparison with MJT.

that could not be explained simply by delayed maturity. As vigorous vines have dense leaf canopies, such shading contributes to the association of a high level of herbaceous/vegetative wine aroma with vine vigor. This has been confirmed in a study of Cabernet Sauvignon vines growing in different soils; a strong relationship was found between high vine vigor, low light intensity in the canopy, the intensity of vegetative aroma and the concentration of isobutylmethoxypyrazine (*11*).

This response of isobutylmethoxypyrazine to vine variety and growing conditions generates an important link, evident to the wine taster, between the wine flavor and the grape variety, region of origin, and vineyard practices. It has been proposed (*11*) that photodegradation of methoxypyrazines (*12*) causes the decrease of methoxypyrazine concentration with ripening and with increased light exposure. Should this occur, an endogenous vine origin for isobutylmethoxypyrazine and tight control over its initial concentration is still implicated by the behavior outlined.

Other Methoxypyrazines in Grapes and Wines

Minor Methoxypyrazines. Stable isotope-dilution mass spectrometry has also allowed quantitative analysis of 2-methoxy-3-(1-methylpropyl)pyrazine (**2**) (*sec*-butylmethoxypyrazine) and 2-methoxy-3-(1-methylethyl)pyrazine (**3**) (isopropylmethoxypyrazine) in grapes and wines. The corresponding trideuterated isopropylmethoxypyrazine (**6**) is used in addition to the trideuterated isobutylmethoxypyrazine **5** to provide more accurate quantitative analysis of the isopropyl compound (*7, 10, 13*). The sensory detection threshold of *sec*-butylmethoxypyrazine and isopropylmethoxypyrazine are very similar to that of isobutylmethoxypyrazine, but the *sec*-butyl- and isopropyl- components are relatively minor methoxypyrazines, typically 10% or less of the isobutylmethoxypyrazine concentration. This suggests that they are less important to wine flavor than isobutylmethoxypyrazine; although isopropylmethoxypyrazine may be more important, in red wine, than its sensory detection threshold suggests (*9*), and its more earthy aroma (*3, 9, 14, 15*) may contribute to nuances of wine aroma. The origin of *sec*-butyl- and isopropylmethoxypyrazines probably reflects that of isobutylmethoxypyrazine since a high level of isobutylmethoxypyrazine usually leads also to an increased level of these more minor methoxypyrazine components.

Ethylmethoxypyrazine. Unlike the isobutyl-, isopropyl- and *sec*-butylmethoxypyrazines, the existence of 2-ethyl-3-methoxypyrazine (**4**) (ethylmethoxypyrazine) in fruits and vegetables has not been consistently confirmed in the past. While evidence for its existence was found, unambiguous identification by mass spectrometry was prevented by the trace concentration of this component. Of all the methoxypyrazines so far mentioned, it has an aroma most similar to raw potatoes (*16*) and synthetic material adds potato flavor to processed potato products (*18*). It is also described as earthy (*15*), and it has a sensory detection threshold of 425 ng/L (17). Given its aroma, it is not surprising that its possible occurrence in potatoes has been the subject of several studies.

Tentative evidence for its presence in the steam volatile oil of potatoes was first reported by Buttery *et al.* (*16*). However, later work by these authors concluded that the previously identified component was actually isopropylmethoxypyrazine, and they indicated that it was not possible to confirm the existence of 2-ethyl-3-methoxypyrazine by mass spectrometry, although an earthy aroma at the appropriate retention time was detected (*14*). They estimated that, if present, its concentration was less than 1 ng/L. Meigh *et al.* (*19*) identified ethylmethoxypyrazine in extracts of potato peel by GC retention data, but there was again insufficient material to determine a mass spectrum. Nursten and Sheen (*20*) concluded that ethylmethoxypyrazine does exist in potato sprout essence. This conclusion was based on the GC retention time similarity of a potato-like odor component to that of synthetic ethylmethoxypyrazine and on the relative intensity of three characteristic ions in the mass spectrum. However, contamination by other components prevented a full mass spectral verification. In grapes, a study of Sauvignon blanc juice (*21*) identified pyrazine-like aromas in a part of the chromatogram that correlated with elution of ethylmethoxypyrazine and isopropylmethoxypyrazine, and ethylmethoxypyrazine was tentatively identified by GC-MS with selected ion monitoring, but again full mass spectral verification was not possible.

Adding to the uncertainty over the existence of ethylmethoxypyrazine is that if the biosynthesis of methoxypyrazines in vines follows a previously postulated pathway (*3*), in which an amino acid is the source of the alkyl side chain, as already identified in bacteria by the incorporation of valine (*22*) or pyruvate (*23*) into isopropylmethoxypyrazine, then the biosynthesis of ethylmethoxypyrazine would require 2-aminobutyric acid, an amino acid not normally found in proteins, although it does exist in grapes and vegetables.

During routine analysis for the isobutyl-, *sec*-butyl-, and isopropylmethoxypyrazines in Cabernet Sauvignon grapes and wines, a further component was identified. This component has now been verified as ethylmethoxypyrazine by comparison of its full-scan mass spectrum and its retention time with those of synthetic material on GC stationary phases of widely different polarity. Support for the chromatographic purity of the analyzed peak was provided by consistency of the component concentration, determined by stable isotope-dilution MS, across different columns and to both CI and EI ionization modes (Allen, M. S., Boyd, S. J., in preparation).

Ethylmethoxypyrazine and Isobutylmethoxypyrazine Abundance

Quantitative analysis of ethylmethoxypyrazine, using the closely eluting trideuterated isopropylmethoxypyrazine **6** as internal standard, has shown that ethylmethoxypyrazine is usually present with isobutylmethoxypyrazine in Cabernet Sauvignon grapes and wines, and that the concentration of the ethyl compound is usually in the range 1 - 10 ng/L. However, whenever the influence of grape variety, berry development or shading has been examined, the behavior of ethylmethoxypyrazine has contrasted dramatically with that of isobutylmethoxypyrazine. First, ethylmethoxypyrazine has been found in wines of Pinot noir grapes, but this grape variety and its wines does not provide isobutylmethoxypyrazine. Furthermore, the concentration of ethylmethoxypyrazine in these Pinot noir wines has been comparable to the level typically found in Cabernet Sauvignon wines. Secondly, in Cabernet Sauvignon grapes, the concentration of ethylmethoxypyrazine showed no clear relationship to

Table II. Effect of Fruit Shading on Grape Methoxypyrazine Concentration at Two Ripening Stages[a]

Shading (Leaf layer index)[b]	*Isobutyl-methoxypyrazine (ng/L)*		*Ethyl-methoxypyrazine (ng/L)*	
	Jan. 25[c]	*Feb. 24*	*Jan. 25*	*Feb. 24*
0	39.3	3.1	1.8	4.6
1	52.8	4.8	2.7	3.5
2	65.5	7.0	36.8	2.6
3 or more	111.1	11.1	21.2	4.1

[a]Concentration in freshly extracted juice of Cabernet Sauvignon grapes. [b]Increased leaf layer index indicates increased fruit shading. [c]Dates are southern hemisphere growing season.

vine behavior. Although increased shading within the vine led to higher levels of isobutylmethoxypyrazine, no consistent trend was evident for ethylmethoxypyrazine (Table II). Furthermore, while isobutylmethoxypyrazine steadily decreased in concentration with increasing ripeness, ethylmethoxypyrazine showed erratic variation (Table III). Finally, there can be very wide variation of the concentration of ethylmethoxypyrazine. In one wine, it was above 1000 ng/L, a level that exceeds its sensory detection threshold of 425 ng/L (*17*), this suggests that this component may occasionally influence wine flavor.

These data clearly indicate an origin for ethylmethoxypyrazine that differs from that of isobutylmethoxypyrazine. In view of the production of some methoxypyrazines by microorganisms, sometimes in relatively high concentrations (*22*, *23*), a microbial origin may exist. Indeed, replicate determinations of ethylmethoxypyrazine in the wine that provided a level of >1000 ng/L for this component have shown a bottle variation that identifies that bottle-specific microbial contamination, possibly *via* the cork, may play a role. However, where microbial methoxypyrazine production has been reported, isopropylmethoxypyrazine has been produced rather than ethylmethoxypyrazine (*22, 23*). Moreover, freshly expressed grape juice samples from sound berries also contain ethylmethoxypyrazine, so a vine origin cannot be entirely ruled out.

Isopropylmethoxypyrazine origins. Isopropylmethoxypyrazine is a minor component when compared with isobutylmethoxypyrazine, but it responds to vine variety and growing conditions in a similar manner. However, the wine with a very high level of ethylmethoxypyrazine also contained a concentration of isopropylmethoxypyrazine that slightly exceeded that of isobutylmethoxypyrazine. This elevated level of isopropylmethoxypyrazine suggests that it may be a component that can originate by the pathway that provides ethylmethoxypyrazine as well as the pathway that leads to isobutylmethoxypyrazine.

Table III. Effect of Ripening on Grape Methoxypyrazine Concentration for Two Vine Pruning Systems[a]

Date[b]	*Isobutyl-methoxypyrazine (ng/L)*		*Ethyl-methoxypyrazine (ng/L)*	
	Minimal pruning	*Spur Pruning*	*Minimal pruning*	*Spur Pruning*
Jan. 25	111.0	188.5	21.2	13.9
Feb. 1	63.1	122.6	2.3	38.0
Feb. 8	45.6	90.7	1.4	4.9
Feb. 24	11.1	18.3	4.1	2.9
Mar. 3	10.5	16.3	3.2	3.0
Mar. 9	6.5	9.9	2.4	47.1

[a]Concentration in freshly extracted juice of Cabernet Sauvignon grapes.
[b]Dates are southern hemisphere growing season.

Conclusion

Ethylmethoxypyrazine occurs in grapes and wines. Its concentration is usually exceedingly small, of similar magnitude to that of isobutylmethoxypyrazine, but it can occasionally occur at much higher concentration. Ethylmethoxypyrazine does not show the clear relationship to vine variety, berry development and canopy light penetration that isobutylmethoxypyrazine does. The likely source of ethylmethoxypyrazine in grapes and wine is not clear, although its origin evidently differs from that of isobutylmethoxypyrazine. It may have a plant or microbial origin, or arise as an artifact of the isolation conditions. Further work is in progress to elaborate how this component is produced.

Literature Cited

(*1*) Buttery, R. G.; Seifert, R. M.; Lundin, R. E.; Guadagni, D. G.; Ling, L. C. *Chem. Ind.* (London) **1969**, 490-491.

(*2*) Buttery, R. G.; Seifert, R. M.; Guadagni, D. G.; Ling, L. C. *J. Agric. Food Chem.* **1969**, *17*, 1322-1327.

(*3*) Murray, K. E.; Shipton, J.; Whitfield, F. B. *Chem. Ind.* (London) **1970**, 897-898.

(*4*) Murray, K. E.; Whitfield, F. B. *J. Sci. Food Agric.* **1975**, *26*, 973-986.

(*5*) Bayonove, C.; Cordonnier, R.; Dubois, P. *C. R. Acad. Sci. Paris Ser. D* **1975**, *281*, 75-78.

(*6*) Harris, R. L. N.; Lacey, M. J.; Brown, W. V.; Allen, M. S. *Vitis* **1987**, *26*, 201-207.

(*7*) Lacey, M. J.; Allen, M. S.; Harris, R. L. N.; Brown, W. V. *Am. J. Enol. Vitic.* **1991**, *42*, 103-108.

(*8*) Allen, M. S.; Lacey, M. J.; Harris, R. L. N.; Brown, W. V. *Am. J. Enol. Vitic.* **1991**, *42*, 109-112.

(*9*) Maga, J. A. In *Flavors and Off-Flavors, Proceedings of the 6th International Flavor Conference;* Charalambous, G., Ed.; Elsevier: Amsterdam, 1990; pp 61-70.

(*10*) Allen, M. S.; Lacey, M. J.; Boyd, S. *J. Agric. Food Chem.* **1994**, *42*, 1734-1738.

(*11*) Noble, A. C.; Elliot-Fisk, D. L.; Allen, M. S. In *Fruit flavors: biogenesis, characterization and authentication*; R. L. Rouseff and M. M. Leahy, Eds.; ACS Symposium Series 596, American Chemical Society: Washington, D.C., 1995; pp 226-234.

(*12*) Heymann, H.; Noble, A. C.; Boulton, R. B. *J. Agric. Food Chem.* **1986**, *34*, 268-71.

(*13*) Allen, M. S.; Lacey, M. J.; Boyd, S. *J. Agric. Food Chem.* **1995**, *43*, 764-772.

(*14*) Buttery, R. G.; Ling, L. C. *J. Agric. Food Chem.* **1973**, *21*, 745-746.

(*15*) Parliment, T. H.; Epstein, M. F. *J. Agric. Food Chem.* **1973**, *21*, 714-716.

(*16*) Buttery, R. G.; Seifert, R. M.; Ling, L. C. *J. Agric. Food Chem.* **1970**, *18*, 538-539.

(*17*) Seifert, R. M.; Buttery, R. G.; Guadagni, D. G.; Black, D. R.; Harris, J. G. *J. Agric. Food Chem.* **1970**, *18*, 246-249.

(*18*) Guadagni, D. G.; Buttery, R. G.; Seifert, R. M.; Venstrom, R. M. *J. Food Sci.* **1971**, *36*, 363-366.

(*19*) Meigh, D. F.; Filmer, A. A. E.; Self, R. *Phytochemistry* **1973**, *12*, 987-993.

(*20*) Nursten, H. E.; Sheen, M. R. *J. Sci. Food Agric.* **1974**, *25*, 643-663.

(*21*) Augustyn, O. P. H.; Rapp, A.; van Wyk, C. J. *S. Afr. J. Enol. Vitic.* **1982**, *3*, 53-60.

(*22*) Gallois, A.; Kergomard, A.; Adda, J. *Food Chem.* **1988**, *28*, 299-309.

(*23*) Leete, E.; Bjorklund, J. A.; Reineccius, G. A.; Cheng, T.-B. In *Bioformation of Flavours;* Patterson, R. L. S.; Charlwood, B. V.; MacLeod, G.; Williams, A. A., Eds.; Royal Soc. Chem.: Cambridge, UK, 1992; pp 75-95.

(*24*) Allen, M. S.; Lacey, M. J.; Harris, R. L. N.; Brown, W. V. *Aust. N. Z. Wine Ind. J.* **1990**, *5*, 44-46.

Chapter 22

Development of Flavor Attributes in the Fruit of *C. melo* During Ripening and Storage

S. Grant Wyllie, David N. Leach, and Youming Wang

Center for Biostructural and Biomolecular Research, University of Western Sydney, Hawkesbury, Richmond, New South Wales 2753, Australia

To extend our understanding of flavor development during fruit ripening the changes in total soluble solids (TSS), pH, sugars, free amino acids, organic acids, total volatiles and a suite of soluble minerals including sodium, phosphorus, calcium and magnesium have been measured in fruit of cv. Makdimon melons during development, ripening and storage. Changes in flesh pH, total sugars, total volatiles, the amino acids valine, serine, threonine, proline and the soluble minerals potassium and sodium were strongly correlated (>0.87) with those of TSS. Therefore, if TSS is taken as an indicator of fruit maturity and flavor quality as is often the case, then each or all of the above parameters may also be determinants of fruit quality. The influence of fruit maturity at harvest and postharvest storage time on the changes in many of these flavor related parameters and how these may relate to the development of fruit quality is also examined.

Many of the complex biochemical changes that occur in a fruit during ripening are not yet fully understood. Outward manifestation of these changes may be observed in morphological characteristics and in many cases by the detection of a typical ripe fruit aroma. However, many of the changes which may contribute to the final flavor quality of the fruit can only be revealed by chemical analysis. The question then becomes which components should be measured and how may their impact on the flavor be assessed. Some guidance may be obtained from the sensory studies which have been carried out on melons (*1*, *2*) or on other fruit, for example, tomatoes (*3*). However, the correlations obtained are usually not convincing and may indicate that, among other things, important flavor parameters were not measured and thus not incorporated into the correlation calculations. Therefore, a study of further fruit constituents whose changes in concentration correlate closely with those of components known to be important flavor determinants (e.g. TSS in melons) may provide insight into their potential flavor contribution either in their own right or as precursors for other flavor components. Secondly and more fundamentally, an examination of the changes in this range of components may allow conclusions to be drawn about some of the chemical and/or biochemical processes taking place during growth, development and ripening of the fruit. Such information will be a key

0097–6156/96/0637–0228$15.00/0

prerequisite to the application of genetic engineering techniques for the manipulation of ripening or flavor characteristics.

Experimental.

Melons used in this study were from authenticated seed, cv Makdimon, obtained from commercial seed producers. They were grown in a weather protected environment using a hydroponic system with a controlled nutrient mix. Flowers were tagged on the day of anthesis with inspections being carried out every day during the most intense flowering periods. Fruits that set on the same day were used as sample material. All data obtained were referred to fruit age as days after anthesis (daa).

Development study. Collection of data was started at 20 days after anthesis and continued at 3~4 day intervals until abscission occurred. Three samples of each maturity rating were analyzed in duplicate.

Storage study. Melons were harvested at 35 days after anthesis (7 days before fully ripe)(E1); 40 days after anthesis (2 days before fully ripe)(E2); and 42 days after anthesis (fully ripe, full slip)(E3). Sample melons from each harvest were either immediately analyzed or stored at 12°C for 8(E1), 5(E2) and 3(E3) days respectively. Each of the latter sample groups were then stored for a further two days at room temperature (25°C) before analysis. Three samples from each stage were analyzed in duplicate.

Melon samples were taken from the whole fruit by cutting it into longitudinal sections, after which the edible portion (middle-mesocarp) removed and processed.

Amino acids, Sugars and Organic Acids were analysed as previously described (*4*).

Total Volatiles determination. Volatiles composition and concentration in the samples of melon flesh were determined using a headspace sampler (HP 19393) coupled to a gas chromatograph (HP 5890). Approximately 5g of flesh was obtained by using a cork borer to remove a number of cores from around the equator of the fruit. This was immediately sealed in a headspace vial, placed in the headspace sampler at a bath temperature of 100°C, equilibrated for 15min before analysis on an HP 5890 GC equipped with an FID detector and fitted with a 30m x 0.32mm i.d. OV1 fused silica column.

pH Measurement. The pH of the puree obtained by homogenizing mesocarp tissue (50g, 2min.) was measured with a Beckman Φ50 pH Meter.

Total Soluble Solids Determination. The purée (50 g) was centrifuged at 3,500 r.p.m for 10 min. The total soluble solids (expressed as degrees Brix) of the supernatant was measured using an ABBE Refractometer maintained at 20°C and calibrated against pure water.

Soluble Mineral Analysis. The supernatant (1.38 mL) prepared from centrifuging melon purée was diluted with 0.2 M HCl to 10 mL. The resulting solution was analyzed directly using a Labtam Plasmalab Coupled Polychromator/monochromator (Melbourne, Australia), configured to measure B, Cu, Zn, Ca, Fe, Mn, Mg, Na, K, P, S. External calibration standards were run between every 10 samples.

Statistical Analysis. Statistical analyses were performed using the Microsoft Excel package.

Results and Discussion.

In the following discussion the commonly accepted parameter total soluble solids (TSS) has been chosen as an indicator of fruit maturity even though it is not an absolutely reliable indicator of fruit quality. The changes in concentration of TSS were correlated with changes in concentrations of other components measured, to ascertain how these relate to the development and ripening process.

The components measured in this present study are summarized in Table I.

The measured components may also be grouped on the basis of the correlation of their changes during ripening with that of those in TSS. Components which had highly significant correlation coefficients (>0.87, r crit.(p<0.001, df=30)= 0.554) are shown in Table II.

Table I. Range of components measured during fruit development and ripening.

Group	Components Measured.
Carbohydrates	Total sugars, sucrose, fructose, glucose, inositol
Free Amino Acids	Val, Leu, Ile, Ser, Thr, Pro, Pip, Ala, Gly, Gaba, Met, Phe, Asp, Tyr, Asn, Gln, Cit+Orn
Soluble Minerals	K, Na, Mg, P, S, Ca, Mn, B, Fe, Zn, Cu.
Representative Aroma Volatiles	Ethanol, ethyl acetate, 2-methylbutyl acetate, ethyl propionate, butyl acetate, 2-methylthioethylacetate, methyl-3-(methylthio)propionate
Organic Acids	Citric acid, succinic acid
Chemical Parameters	Total soluble solids(TSS), pH

Table II. Components whose concentration changes show strong correlations (> 0.87) with TSS.

Variable	TSS	TS	pH	TV	Val	Ser	Thr	Pro	K	Na
TS	0.95									
pH	0.94	0.87								
TV	0.87	0.91	0.82							
Val	0.90	0.89	0.88	0.88						
Ser	0.91	0.93	0.88	0.91	0.98					
Thr	0.92	0.88	0.89	0.83	0.98	0.96				
Pro	0.88	0.87	0.83	0.86	0.96	0.97	0.96			
K	0.93	0.87	0.87	0.75	0.91	0.90	0.93	0.91		
Na	0.92	0.85	0.89	0.80	0.92	0.91	0.92	0.90	0.95	
Gaba	0.88	0.84	0.88	0.64	0.78	0.77	0.80	0.78	0.83	0.79

Table III. Components whose concentration changes show strong (>0.80) correlations with both TSS and with each other.

Variable	TSS	Val	Leu	Ile	Ser	Thr	Pip
Val	0.90						
Leu	0.84	0.95					
Ile	0.80	0.92	0.97				
Ser	0.91	0.98	0.90	0.84			
Thr	0.92	0.98	0.93	0.90	0.96		
Pip	0.79	0.90	0.86	0.79	0.92	0.87	
Pro	0.88	0.96	0.92	0.85	0.97	0.96	0.92

r crit. (P < 0.001) = 0.55.

This group contains the total sugars, pH, total volatiles and the amino acids valine, serine, threonine, proline and γ-aminobutyric acid (Gaba) together with the soluble minerals sodium and potassium. A second group shown in Table III contains many of the free amino acids; valine, leucine, isoleucine, serine, threonine, pipecolic acid and proline, whose concentration changes correlate significantly with those of TSS (>0.80) and with those of each other (>0.80).

Of the other minerals measured only magnesium, phosphorus and iron show significant positive correlations with TSS while calcium exhibited a strong negative correlation (-0.88)(data not shown).

Changes in Free Amino Acids. Of particular interest are the amino acids, alanine, valine, leucine, isoleucine and methionine which are known to be the precursors of many of the volatile compounds especially esters which are known (*4,5*) to be key contributors to the aroma profile of ripe high quality melons. These amino acids, all of whose concentration changes show strong correlation with that of TSS, exhibit significant increases in concentration during ripening provided the fruit is not harvested before it is fully ripe. Thus they may become readily available as substrates for the series of enzymatic transformations that leads to their transformation to volatile esters (*6,7*). The strong correlation between the total volatiles generated in ripening melons and the concentrations of the large majority of the amino acids measured (Table II) supports the view that the aroma profile of a melon may be dependent in terms of both amount and composition on the availability of these amino acids as substrates. The processes bringing about these increases in amino acid concentrations are not clear. One possibility is that as the requirement for fruit growth diminishes, protein synthesis also diminishes and the balance between protein degradation and synthesis shifts causing an increase in the concentration of amino acids in the amino acid pool. Another possible explanation is that some amino acid biosynthetic pathways are switched on as part of the ripening process. Available evidence (*6,7*) strongly suggests that the changes observed in amino acid concentrations during fruit ripening are subject to selective control and are not simply the result of generalized proteolysis or amino acid synthesis reactions.

Changes in the behavior of selected chemical parameters with different harvest and storage regimes. To explore the relationship between harvest time, storage time and fruit quality, the influence of harvest time and subsequent storage treatment on a selected group of the range of parameters discussed above was investigated. The results showing the changes in the behavior of these parameters under the three harvest time storage regimes employed are illustrated and summarized in Figures 1 to 6.

Figure 1 illustrates the influence of harvest time and storage on the accumulation of TSS in melon fruit and provides a reference for the behavior of other fruit components. Since melons have no reservoir of starch (*8*) they are dependent on the import of photosynthate from the leaves to increase their sugar concentration during ripening and as illustrated in the figure the accumulation ceases immediately the connection is severed. Thus fruit harvested at 35 days after anthesis and stored for a further ten days (E1) shows no significant changes in total soluble solids over this period. Fruit left on the vine for a further five (E2) and seven (E3) days, respectively show marked increases in TSS followed by some decrease during the storage period. This decrease may be caused by the utilization of the sugars as an energy supply for the greatly increased metabolic activity that accompanies ripening or by a decrease in other constituents that contribute to the TSS value. This overall pattern therefore represents the behavior expected for those components whose changes are mediated by translocation from the plant. The changes in the free amino acids on the whole do not conform to this particular pattern. They exhibit a range of responses to the particular harvest and storage protocol used, all of which indicate that translocation is

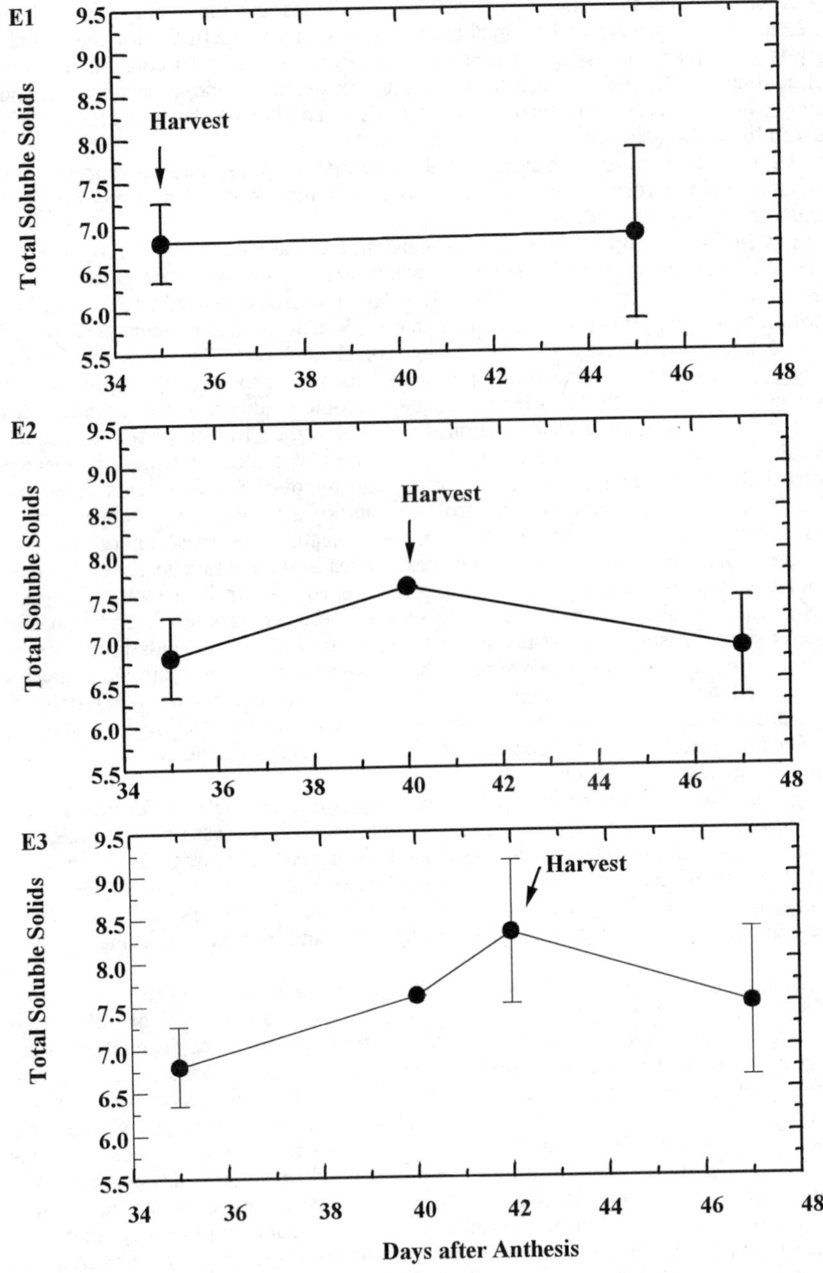

Figure 1. Changes in the total soluble solids for three different harvest and storage regimes.

not a significant factor in determining how their concentrations are controlled in the ripening fruit. Some groups having similar but not identical characteristics can be identified for the amino acids investigated. Figure 2 illustrates the behavior of one such group using leucine as an example. This pattern of behavior is also representative of that of alanine, threonine, serine, valine, isoleucine, tyrosine, proline and phenylalanine. In general the concentration of these amino acids increases both during ripening and on storage after harvest even if the fruit is picked prematurely, although in this case the final concentration values reached are somewhat lower than that achieved in fully ripe harvested samples. Therefore the particular pathways leading to the production of these acids are not particularly influenced by harvest time. Figure 3 illustrates the behavior of a second group containing methionine, glycine, pipecolic acid and glutamic acid using methionine as the example. Here premature harvest (E1) results in a very small increase or a decrease in the concentration of these particular amino acids during storage. However, harvest at a time closer to fully ripe and at fully ripe results in an increase in their concentration during the storage period with the final concentration of the amino acid at the end of the storage period being markedly higher than that achieved in the early harvest sample. This suggests that the metabolic pathways which produce this group of amino acids are not activated when melons are harvested early and are controlled by different means from those illustrated in Figure 2.

One amino acid, aspartic acid, whose concentration changes are shown in Figure 4 shows quite different responses to the other amino acids. Here the two prematurely harvested samples (E1, E2) show an increase in aspartic acid concentration on storage but that harvested fully ripe shows a decrease.

These variations in behavior indicate that harvesting melons at different stages of maturity causes subsequent biochemical events involved in amino acid accumulation to follow markedly different pathways. Recent work shows that melon fruit harvested up to ten days before commercial maturity exhibits climacteric behavior with respect to ethylene production showing that at least this aspect of ripening is not completely inhibited by premature separation from the plant(*9*). However, the amount of ethylene produced is dependent on maturity at harvest and fruit harvested five days prematurely generated only about half of the amount of ethylene produced by fruit harvested two days before maturity. Also the lag time required to initiate ethylene production after harvest depended on maturity and was longer for prematurely harvested fruit. Changes in the content of the phytohormone abscisic acid were also correlated with that of ethylene. However whether the different maturity related metabolic responses observed above result from the action of these or other plant hormones awaits further study.

Figure 5 shows the influence of harvest time and storage on total volatiles production. Since the production of many but not all of the aroma volatiles is linked to amino acid precursors it may be expected that the total volatiles behavior may reflect that of the amino acids especially those which supply many of the carbon skeletons for the esters found in melons. The pattern of behavior for the total volatiles is generally in accord with that of these amino acids and does very clearly illustrate the profound influence of harvest time on the generation of the aroma profile. Fruit harvested only two days before fully ripe develops only about one quarter of the total volatiles concentration shown a few days postharvest, by a fully ripe sample. This rather dramatic difference may reflect the inability of prematurely harvested fruit to accumulate sufficient concentrations of required volatiles substrates because certain metabolic responses have not been activated.

The pH developed in a fruit can have a number of important influences on the quality of the product. It can have a direct effect on the consumers perception of the taste of the fruit. Also it can influence the activity of enzymes involved in the ripening process and can hence exert some control on the extent of reaction in some biochemical pathways. It should be kept in mind that the pH measured in this work

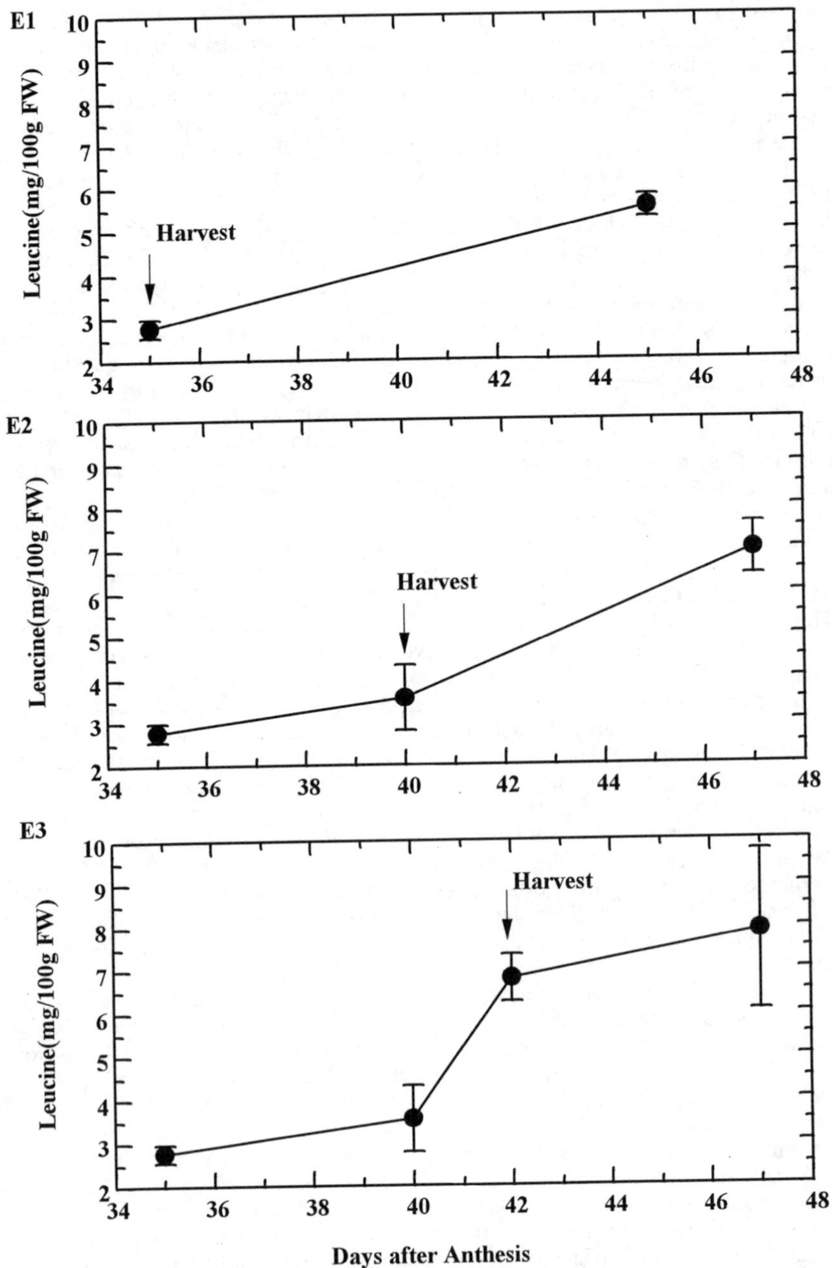

Figure 2. Changes in the concentration of leucine for three different harvest and storage regimes.

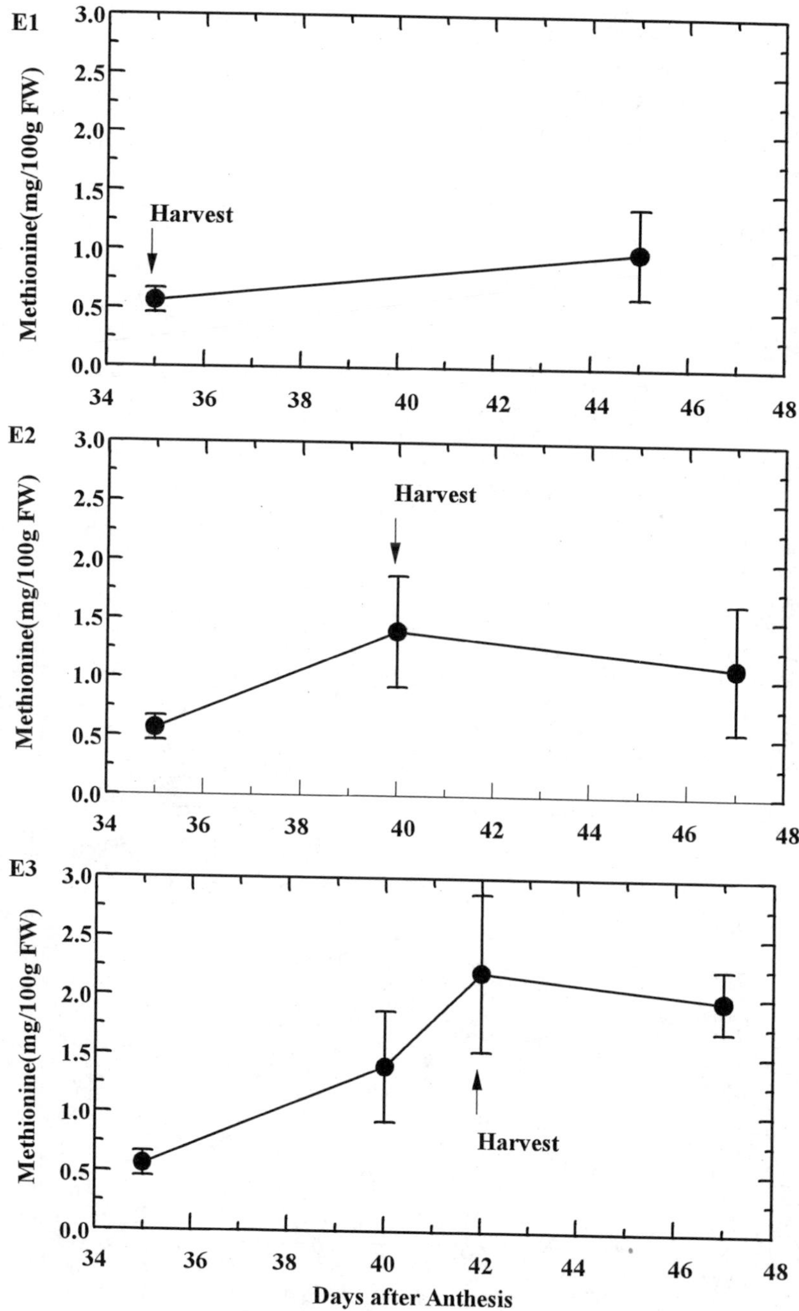

Figure 3. Changes in the concentration of methionine for three different harvest and storage regimes.

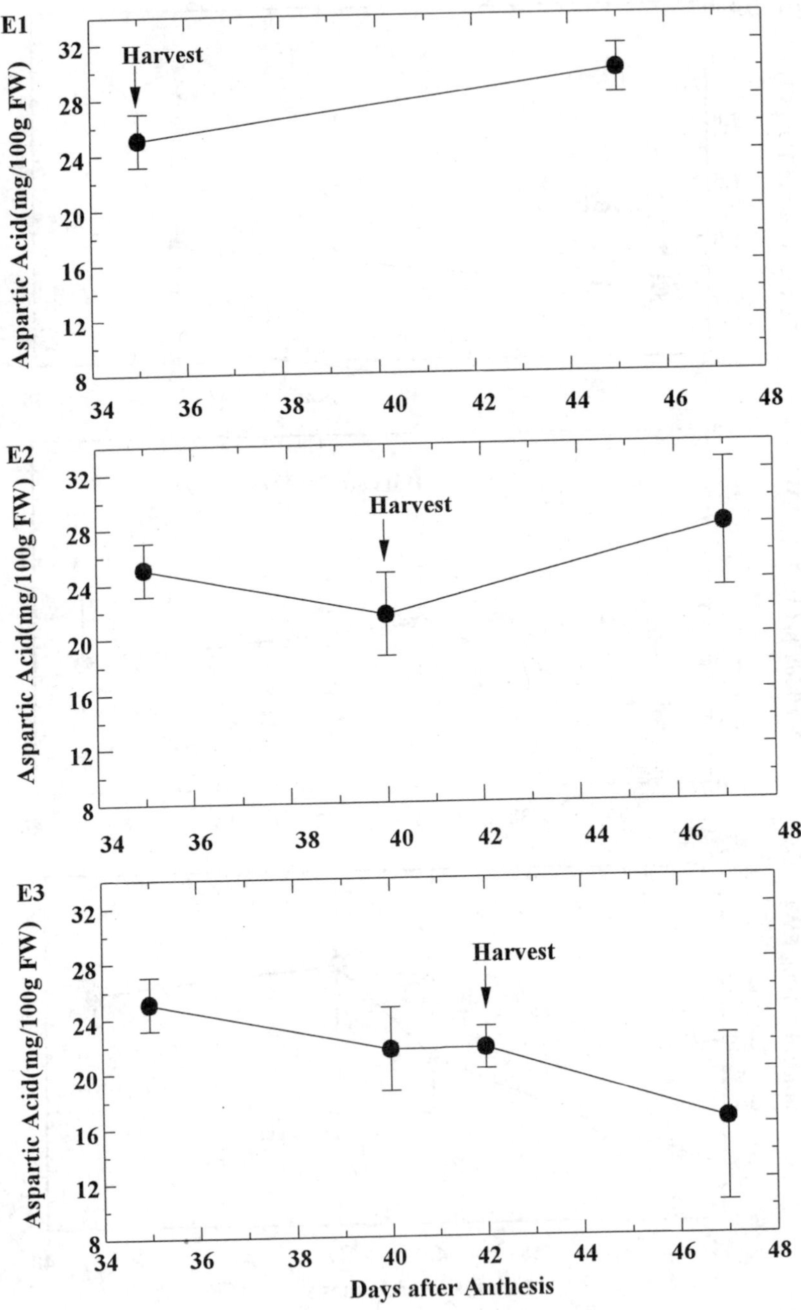

Figure 4. Changes in the concentration of aspartic acid during three different harvest storage regimes.

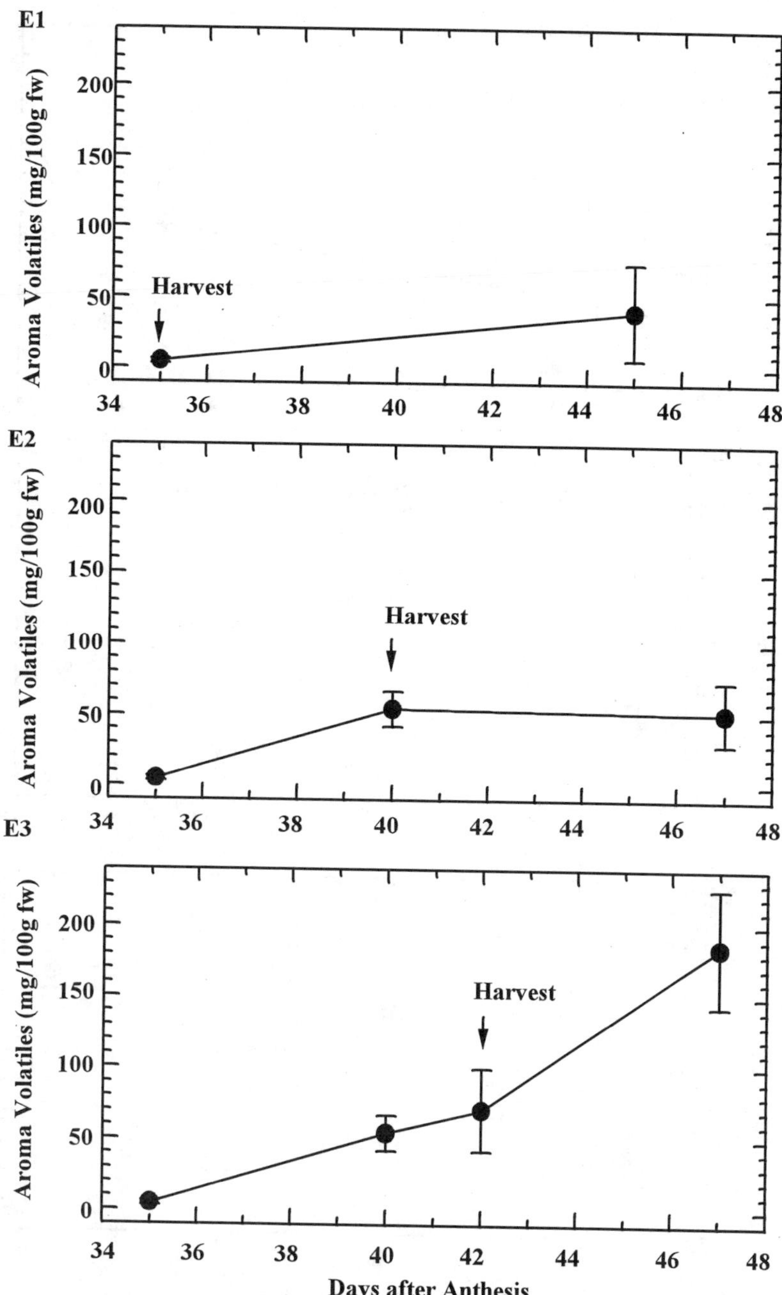

Figure 5. Changes in the concentration of selected aroma volatiles during three different harvest and storage regimes.

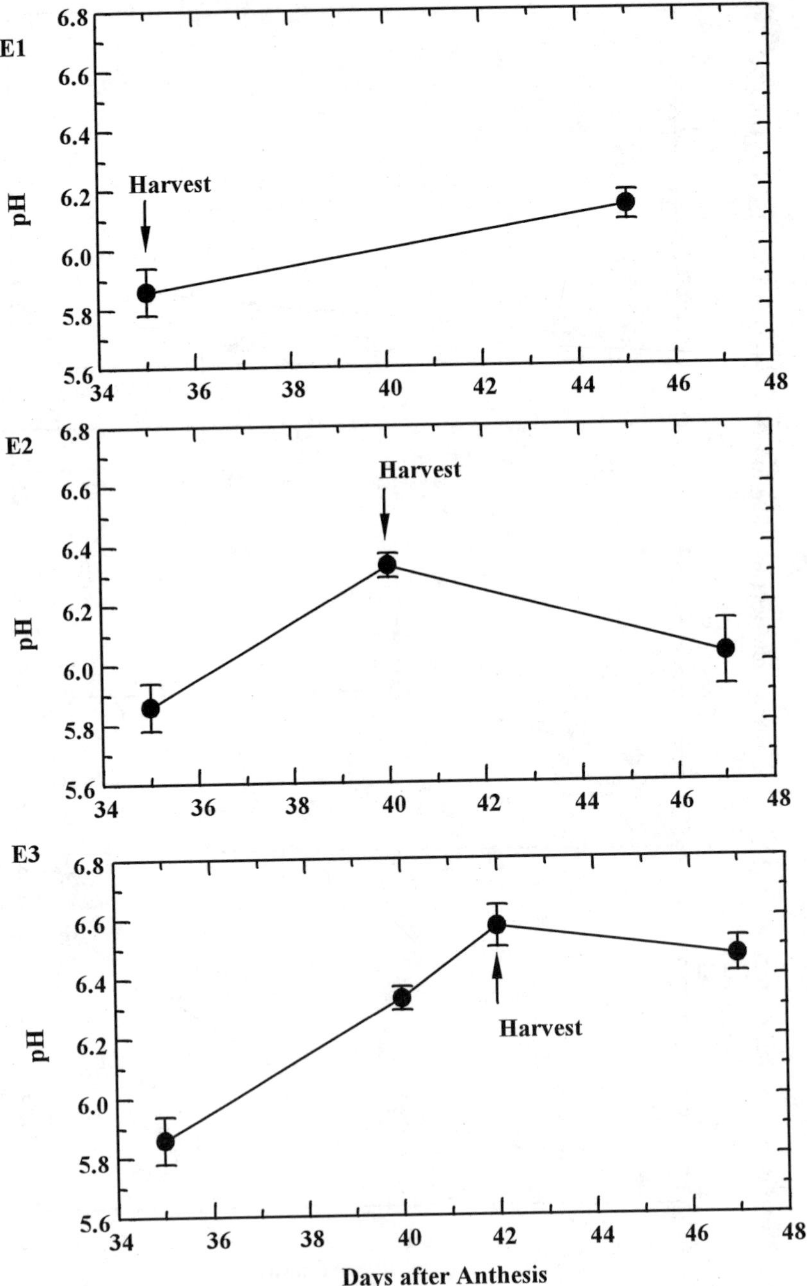

Figure 6. Changes in pH during three different harvest and storage regimes.

may not be representative of that found in all cell compartments. However, the pH may play an important role in determining the aroma profile. Figure 6 shows how the stage of maturity at harvest strongly influences the pH developed in the fruit during ripening and storage. The marked decrease in pH in fruit harvested only two days before maturity and stored is notable The much lower pH reached in fruit harvested before the fully ripe stage and stored may be an important determinant in influencing final fruit quality by either direct or indirect means.

Summary.

The anecdotal and increasingly scientifically confirmed evidence that tree/vine ripened fruit has the best flavor quality is clearly confirmed for melons. The results demonstrate that for the melons utilized in this study (cv Makdimon) the harvest time window available to harvest fruit of maximum quality is quite small, two to three days at most, and that the penalty for premature harvest is high because of the greatly diminished values for total sugars, total volatiles and pH, to name but a few parameters of known sensory importance. Unfortunately, fruit picked at full maturity especially the highly aromatic cultivars such as cv. Makdimon tends to have a short shelf life which inhibits their commercial acceptance.

At a more detailed level the results raise a number of intriguing questions about the biochemical changes occurring in ripening fruit. The data obtained suggests that the biosynthesis of specific amino acids is switched on within particular developmental time frames during ripening and that separation from the plant can impair the attainment of normal concentrations of many compounds.

Acknowledgments.

The authors gratefully acknowledge the financial assistance of the American Chemical Society, the Faculty of Science and Technology, and the Hawkesbury Foundation, University of Western Sydney, Hawkesbury. They also thank Mr. B. Corliss, Connoisseurs Choice, Sydney for the supply of authenticated melons and Mr. R. Finlayson, Department of Chemistry, University of New South Wales, Australia for ICP analysis. One of us (YW) thanks the Australian International Development Bureau for a scholarship.

Literature Cited.

1. Yamaguchi, M; Hughes, D.L.; Yabumoto, K.; Jennings, W.G. *Scientia Horticulturae*. **1977**, *6(1)*, 59-70.
2. Mutton, L.L.; Cullis, B.R. ; Blakeney, A.B. *J. Sci. Food Agric.* **1981**, *32*, 385-391.
3. Petró-Turza, M. *Food Reviews International.* **1986-87**, *2 (3),* 309-351.
4. Wyllie, S.G.; Leach, D.N.; Wang, Y.M.; Shewfelt, R.L. In *Fruit Flavors;* Rouseff, R. L. and Leahy, M. M., Eds.; ACS Symposium Series 596, American Chemical Society: Washington, DC, **1995**; pp 248-257
5. Wyllie, S. G.; Leach, D. N. *J. Agric. Food Chem.* **1990**, *38*, 2042-2044.
6. Tressl, R. ; Drawert, F. *J. Agric. Food Chem.* **1973**, *21*, 560-565.
7. Perez, A.G., Rios, J.J., Sans, C.; Olias, J.M. *J. Agric. Food Chem.* **1992**, *40*, 2232-2235.
8. Pratt, H.K. Melons. Ch. 5. In *The Biochemistry of Fruits and their Products*; Hulme, G.C., Ed.; Academic Press: London, **1970**; Vol. 2, pp 207-232.
9. Larrigaudiere, C., Guillen, P. ; Vendrell, M. *Postharvest Biology and Technology*. **1995**, *6*, 73-80.

Chapter 23

Methods for Isolating Food and Plant Volatiles

Ron G. Buttery and Louisa C. Ling

Western Regional Research Center, Agricultural Research Service, U.S. Department of Agriculture, 800 Buchanan Street, Albany, CA 94710

This chapter discusses some more recent variations of methods for isolation of volatiles from food and plant materials. For particular problems there are advantages to each of the three main types of isolation methods, direct extraction, steam distillation and dynamic headspace. Direct solvent extraction is the only method which is reasonably efficient in isolating components of both high and low water solubility. Because food and plant volatiles are usually water soluble at their ppm concentrations their isolation by steam distillation does not fit the theory's required non-miscible conditions and this may be better considered a type of dynamic headspace isolation. By adapting ideas from a recently published direct solvent extraction method, which used excess sodium sulfate to bind all water in aqueous foods, the authors discovered an effective dynamic headspace method for isolating Furaneol and other water soluble volatiles.

The first step in the analysis of food or plant volatiles is to isolate and concentrate them from these often largely aqueous materials. This is normally necessary before GC-MS or other analysis is possible. This subject has been covered in a number of previous reviews (*1,2*). The present discussion will concentrate on recent variations of the main isolation methods. There are several main method types that have been used for isolating such volatiles. These can be organized in the following way: 1. Direct Solvent Extraction: 2. Steam Distillation and Steam Distillation Continuous Extraction (SDE): 3. Dynamic and Static Headspace: 4. Other Methods. Basically all methods require transfer of the volatile components to a trapping system of some sort with rejection of water. There are several desirable qualities of isolation methods. These include that the methods be (a) comprehensive i.e. isolate (ca. quantitatively) all volatiles, (b) that they do not cause any changes in the nature of the volatiles, and (c) that they be fast enough to be practical for the analysis of large numbers of samples.

Direct Solvent Extraction

Frequently it is possible to extract the volatile components directly from the intact food or plant material. This is particularly true with thin materials such as leaves or relatively dry powders such as ground coffee, tea or wheat flour. More often it is necessary to chop or grind the material in some way (e.g. with a blender) to put it into a finely divided form. Sometimes with dry materials a Soxhlet apparatus is used (*3*). With highly aqueous materials such as fruits and vegetables, a continuous liquid-liquid extractor can be used (on the blended filtered product).

One group of researchers who have used direct solvent extraction very effectively are Grosch, Schieberle and coworkers (*4,5*). The exact extraction method used by these authors has varied but often, if the food was relatively dry (e.g., popcorn ; *3*) it was ground in a blender after being frozen in liquid nitrogen. The ground material was then frequently placed under ether or dichloromethane overnight and extracted further using a Soxhlet apparatus. A food of high water content (*5*), after being saturating with salt, was often extracted directly with solvent and the extract dried over sodium sulfate. Another interesting method, used by some of these authors (*6*), was to add the aqueous food to ca. 8 times its weight of anhydrous sodium sulfate and extract the resultant powder in a Soxhlet apparatus. The sodium sulfate binds the water making the product granular, allowing Soxhlet extraction, and facilitating extraction of even highly water soluble volatiles.

An advantage of the direct extraction method (*3-6*) is that volatiles of both low and high water solubility are isolated in one operation. Other commonly used methods such as steam distillation or dynamic headspace are not effective in isolating highly water soluble compounds such as Furaneol or maltol from mostly aqueous food and plant materials whereas these compounds can often be isolated efficiently by direct solvent extraction.

Some researchers (*7*) have used direct solvent extraction with two or more solvents used in sequence, e.g., first extraction with the non polar solvent Freon followed by extraction with the more polar ethyl acetate. This can give some pre-fractionation of the volatiles.

Most early methods developed for isolation of Furaneol and related very water soluble volatiles (*8*) involved first blending the fruit (sometimes frozen in liquid nitrogen) and filtering. The filtrate was then extracted continuously in a liquid-liquid extractor with ether. After drying over Na_2SO_4 and removal of most of the ether the concentrate was then directly analyzed by gas chromatography. No mention was made of any pre-separation of the volatile and non-volatile extracted compounds.

Separation of Volatiles from Non-Volatiles After Direct Extraction. One of the problems with direct extraction is that non-volatiles such as waxes, fats and carotenoids are also extracted. In most cases the volatiles are not present in large enough quantity to separate them from the non-volatiles by ordinary distillation. A transfer of the volatiles to another container under high vacuum is usually possible, but the presence of viscous waxy material can retard the movement of the volatile molecules to the surface and result in a non-quantitative transfer. If the extract is spread in a thin film or is vigorously stirred, the molecules do not have as far to diffuse to the surface, and the vaporization is more efficient.

Grosch, Schieberle and coworkers (*4,5*) have used a unique "high vacuum sublimation" method to separate the volatile material from the non-volatile. The exact method has varied somewhat with different products but, generally (5), the extract, after being dried over anhydrous Na_2SO_4 and concentrated to ca. 120 ml, was placed in a 1 L flask connected to a series of traps cooled to ice, dry ice and liquid nitrogen temperatures. The concentrate was frozen in liquid nitrogen for 30 minutes. A high vacuum source (ca. 10^{-4} mm Hg; 10 mPa) was then placed at the end of the series of traps. The 1 L flask was then left at room temperature for several hours while the extract sublimed to the cold traps. Eventually the flask was warmed to 50°C for 2 more hours.

The present authors (*9*) in studies of Furaneol in tomato, after direct ether extraction of the filtered tomato, transferred the extract to a 5 L flask and carefully evaporated the ether while rotating the flask in a warm water bath to spread the concentrate as a thin film around the inside walls. The volatiles were then swept to a Tenax trap by a flow of nitrogen or air using essentially the dynamic headspace approach. The method gave good recoveries of the relatively high boiling Furaneol. An obvious drawback is that some of the more volatile components such as C6 aldehydes show considerable loss from the evaporation of the solvent.

One of the disadvantages of direct solvent extraction and indeed any other method that uses solvent is that many of the very volatile compounds such as acetaldehyde and dimethyl sulfide are lost when the solvent is evaporated.

In some cases where direct extraction has been used, no mention has been made of separating the volatiles from the non-volatiles. If the concentrates were injected directly into the gas chromatograph, then there is a real question whether the volatiles analyzed are not artifacts produced by thermal decomposition of the non-volatiles. Some authors (*10*) have used liquid adsorption chromatography to separate the volatile components from the higher molecular weight compounds which are often more strongly adsorbed.

Steam Distillation and SDE

Methods using some type of atmospheric or reduced pressure steam distillation have been historically the most commonly used, and essential oils were (*11*) and are still frequently isolated this way on a commercial scale . The SDE or Likens Nickerson extraction method (*12,13*) was an extension of the Essential Oil industry's technique of cohobation where the condensed steam is separated from the volatile oil using a simple trap and returned to the still in order to minimize loss of water soluble volatiles. SDE had actually been described earlier in practical organic chemistry textbooks (*14,15*). The Likens Nickerson head though was simpler and easier to construct and control than the earlier organic chemist's version. Other versions of the head have been designed and used (*2*).

The generally accepted theoretical treatment of steam distillation has the condition that the compounds involved are immiscible with water. The equation usually given is of the type shown below. Composition of distillate:

$$\text{wt. cmpd.} / \text{wt. } H_2O = p^o \text{cmpd.} \times \text{Mol. wt. cmpd.} / \; p^o H_2O \times \text{Mol. wt. } H_2O \qquad (1)$$

Where p^o is the vapor pressure of the pure compound at the distillation temperature (ca. 100°C at atmospheric). However, with most food and plant materials the volatiles occur at such a low concentration, of the order of mgs. / Kg or less, that they are soluble in the water at such concentrations. Their vapor pressure p above the aqueous system then is as shown below (*16*):

$$p = p^o \times N \times \gamma \qquad (2)$$

Where N is the mole fraction of the compound in the aqueous solution and γ is the activity coefficient of the compound in water. For many compounds $\gamma \cong 1 / Ns$ where Ns is the mole fraction solubility of the compound in water. A fixed temperature T is assumed. For a 1 ppm solution N, the mole fraction, will be less than 10^{-6}. Therefore p, the actual vapor pressure of the compound above the aqueous system, would be very much lower than p^o and hence the composition of the distillate would be much lower than predicted by equation l. This is, however, partly compensated for by the fact that γ may be quite large.

For such dilute solutions it may be more accurate to view the process as a type of dynamic headspace isolation, where the sweep gas is steam, rather than true steam distillation. Each compound forms a concentration corresponding to p from equation (2) in the steam above the aqueous food or plant medium. This steam atmosphere is continually being displaced to the condenser and then as liquid to the extracting solvent with SDE. Calculations show (*17*) that for a reasonable rate of reflux (5 ml liquid/ minute) in an SDE apparatus the flow rate of steam to the condenser is of the order of 6 L per minute.

Although SDE is probably the most common steam distillation procedure used for food or plant volatiles, the classical method of steam distillation is also sometimes used followed by direct or continuous extraction with solvent. This often results in rather large quantities of steam distillate to be handled. The classical method does have some advantages using steam distillation under reduced pressure because, as the extraction occurs later at atmospheric pressure, a lower boiling solvent can be used. SDE can be operated also under reduced pressure (*13,18*), but a higher boiling solvent such as hexane or tert-butyl methyl ether is necessary. With SDE for stable operation it is necessary to have the reduced pressure controlled e.g. with a "Cartesian diver" mercury regulator.

It is notable that SDE is less efficient under vacuum (*13,19*). The reason is readily apparent from equation 2 because p^{o} and hence p would be much lower because of the lower temperature used in the vacuum steam distillation. Any increase in steam flow rate under vacuum would not be enough to compensate for the large decrease in p^{o}.

The addition of water to a dry food (e.g. dry beans, crackers, cereals) causes the release of more volatiles. The reason for this is not clear. It may be that volatiles encapsulated by starch molecules are released when water is added. It could also be that water displaces the volatiles from some type of hydrogen or other chemical bonding. The abundance of water with steam distillation makes it very effective for the release of such bound volatiles.

Dynamic and Static Headspace

Static headspace isolation normally involves taking a sample of the equilibrium headspace (a few ml) immediately above the food. This can be directly injected onto the GC column or more usually first concentrated on an adsorbent trap. The GC analysis of this small sample can give useful information such as the detection of rancidity in a food by measuring hexanal concentration (*20*). Static headspace can also be useful for the analysis of very volatile compounds such as acetaldehyde and dimethyl sulfide. However, in order to get enough material into the headspace, the sample frequently has to be heated to 60-100° C which, in some cases, could give an unrealistic picture of the volatiles of the food or plant material. Static headspace is a very rapid method, but it does not give a comprehensive analysis of the volatiles, and in the case of foods, may miss the most important.

Dynamic headspace isolation usually involves sweeping the food or plant material with a relatively large volume of gas and, in a separate part of the apparatus, trapping the volatiles from the gas stream. The equilibrium of the volatiles between the headspace gas and the food or plant material, is continually being renewed. Usually the sample is stirred or otherwise agitated to insure that equilibrium is brought about rapidly. Normally, as with static headspace, a solid adsorbent is used to trap the volatiles. Adsorbents which ignore the water are the most useful. General methods using solid adsorbents have been covered in a number of reviews (*21,22*). Probably the most commonly used adsorbent is Tenax. Other adsorbents which also have been popular are Porapak Q and Chromosorb 101-105. Occasionally solvents have been used as the trapping medium (*17, 23-25*).

Dynamic headspace, although not as fast as static headspace, gives an isolation which approaches being comprehensive. Although some elegant studies have been carried out with only a few mgs of activated charcoal (*26*), the dynamic headspace method, especially with commercial "purge and trap" units, typically uses ca.100-200 mg of adsorbent, ca. 50-200 mL/min sweep gas flow rates and thermal desorption. We have used rather large adsorbent traps ca. 10g of Tenax, with high sweep gas flow rates (ca. 3 L / minute) and solvent desorption. Thermal desorption, although convenient and rapid, has the disadvantage of causing molecular change in some important unstable food and plant compounds such as (Z)-3-hexenal, germacrene-D, 2-acetyl-1-pyrroline and 2-acetyltetrahydropyridines. This may be due to the metal parts frequently used in thermal desorption units. An all glass or Teflon system may overcome this.

Closed Loop Stripping. Grob and coworkers, some years ago (*26,27*), introduced a practical method for recycling sweep gas using an inert pump which has the advantage that any "breakthrough" of compounds is recycled to the head of the trap. We have adopted this method with some modern all Teflon diaphragm pumps and Teflon tubing under high flow (ca. 3-6 L / min.) conditions with N_2, argon or air (see Figure 1). In his analysis of hydrocarbons in water supplies, Grob warned against using Teflon because he found that it was often contaminated with hydrocarbons. We have not found this to be a problem with analyses of food and plant volatiles. The hydrocarbon background from well cleaned currently manufactured Teflon is well below that of the normally encountered food and plant volatiles. The inertness of Teflon is important because many unstable aroma compounds are changed by contact with metals.

Model System Theory. In contrast to static headspace, which looks only at the concentration of the volatiles in the headspace, the aim of dynamic headspace is to transfer as much as possible of the volatiles in the food to the trap. Some understanding of the transfer can be obtained by considering model systems. As fruits, vegetables and plant materials have high concentrations (70-90%) of water, it is useful to consider model systems of solutions of volatile components in water. With such systems it is possible to calculate the total volume of sweep gas (Vg) needed to transfer a certain percentage (P) of the dissolved volatile component to the trap (*22,28*). This can be done with the following relation:

$$Vg = -Vw \, . \, K^{-1} . \ln [(100-P) / 100] \tag{3}$$

where Vw is the total volume of the solution and K is the air to water partition coefficient for that compound at the temperature under consideration usually 25° C. For 67% to be transferred this simplifies to Vg = Vw / K and for a 95% transfer to Vg = 3Vw / K. Equation 3 and the simplified forms would also apply to model systems where volatiles are dissolved in vegetable oil instead of water and where K= air/vegetable oil partition coefficient. The same relations should also apply for oil water mixtures where the air to mixture partition coefficient $Kam = [Fw/Kw+Fol/Kol]^{-1}$ (*28*) where Fw and Fol are the fraction of water and oil respectively and Kw and Kol are the air to solution partition coefficients for the particular compound in water and oil, respectively.

Although not perfectly applicable to actual complex food and plant systems, such calculations can give us some idea of the sweep rates and times of trapping needed. It is interesting to compare the volume of sweep gas needed to transfer 67% of 3 common food and plant components in a water solution model system where Vw = 200 ml. At 25°C for hexanal $K = 8.7 \times 10^{-3}$; for hexanol $K = 7.0 \times 10^{-4}$; for 2-phenylethanol based on published data a calculation shows $K \cong 4 \times 10^{-5}$. To transfer 67% to the trap (at 25°C) this calculation shows hexanal would require 23 L or 8 min. at 3L/ min. sweep gas flow rate; hexanol would require 280 L or 93 min. at 3L/ min; 2-phenylethanol would require 5000 L or 28 hours at 3 L/ min. It can be seen that relatively water soluble

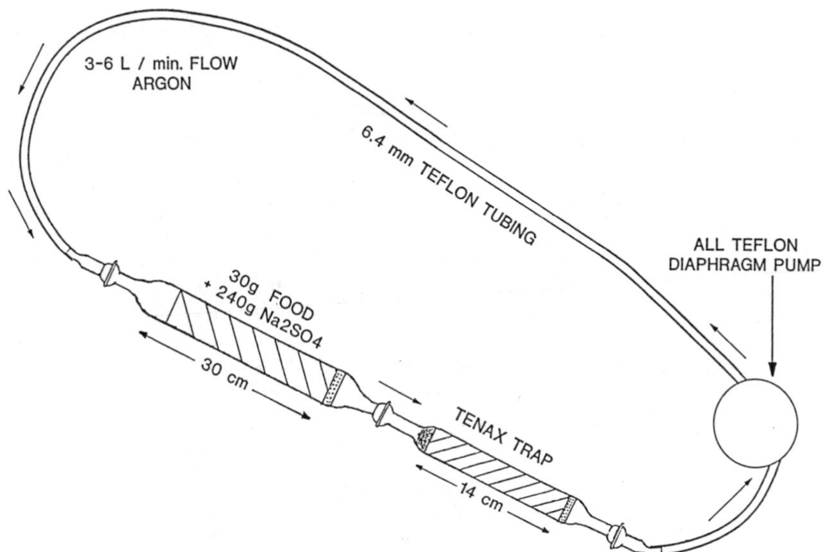

Figure 1. Apparatus for "closed loop" dynamic headspace isolation method with binding of water in food using excess anhydrous sodium sulfate.

compounds such as 2-phenylethanol require considerably more time to approach a quantitative transfer. "Salting out" by saturating a blended aqueous food with NaCl or $CaCl_2$ can increase K and so shorten the time needed. Heating would also do this, of course, but is undesirable because some unstable volatiles are lost and others are produced by the heating.

Dynamic Headspace Furaneol Isolation Using Excess Na_2SO_4. The most difficult volatiles to isolate from aqueous foods include the sugar related compounds such as Furaneol and Maltol. Until recently the accepted method to isolate these compounds for analysis was to extract the blended, filtered product directly with ether (e.g. in a continuous liquid-liquid extractor). This often required another step to separate volatiles from non-volatiles before GC injection.

By adapting the approach of Guth and Grosch (*6*), using an excess of anhydrous sodium sulfate to bind water mentioned above, we have developed a one step dynamic headspace isolation method which gives practical recoveries of Furaneol and Maltol for quantitative analysis. This new approach (see Figure 1) for the isolation of Furaneol is described below.

A weighed amount (usually 30g) of the product (e.g. tomatoes or strawberries) is placed in a blender. An internal standard (usually 1 ml of 50.0 ppm Maltol) is added and the mixture blended for 30 seconds. It is then slowly and thoroughly stirred into ca. 8 times the amount (usually 240g) of anhydrous sodium sulfate in a 1 L beaker. The resulting coarse powder is then packed into a Pyrex tube 30cm long by 3.9 cm O.D. containing a coarse fritted disk at the lower end. This tube is then connected to a large Tenax trap via ball joints and the volatiles passed through the Na_2SO_4 mixture onto the trap using a Grob type "closed loop stripping" system with argon or nitrogen sweep flow rate at ca. 6 L / minute. The process is continued for 3 hours, and the trap is then extracted with ether. The ether is concentrated and analyzed by GC in the normal way.

Using a model system with water solutions of Furaneol and adding an internal standard to the ether extract it was found that the average absolute recovery for Furaneol was 19% and for Maltol 32%. With Maltol as internal standard, the method was applied to the analysis of Furaneol in fresh tomatoes and fresh strawberries and gave quantitative data consistent with published data for these products.

The sodium sulfate, if well stirred with the product, remains granular (ca. 1 mm diam.) and allows the high flow of sweep gas. Although the method gave good recoveries of most compounds, the authors noticed that aldehydes and methyl ketones were not recovered quantitatively. Aldehydes and methylketones are known to complex with some sulfur oxygen compounds (e.g., bisulfite), but the authors do not know of any other complexing of this type with sodium sulfate. There is the possibility that the slight acidity and sugars of the tomato and fruits might modify the sodium sulfate so that complexing can occur with aldehydes and ketones.

Other Methods

There are many variations of the 3 main methods discussed above. Some other methods seem different but are really only a variation of these three.

Cryogenic Vacuum Trapping. One method that had been found useful in a number of studies of flower volatiles was called "cryogenic vacuum trapping" (*29*). Flowers were placed in a flask attached to a large cold finger type trap filled with liquid nitrogen. With a vacuum on the system, the isolation was allowed to proceed for one hour. The flower flask was then replaced with a flask containing solvent (at atmospheric pressure). Refluxing solvent extracted the volatiles from the ice on the still cooled cold finger. This method seems to be a type of vacuum steam distillation since the water vapor atmosphere around the flowers would be at a higher pressure than that around the cold finger. Volatile compounds in this atmosphere would be transported to the cold finger along with the flow of water vapor.

Freeze Concentration. A method that seems to have considerable potential but has been rarely used is freeze concentration. A blended, filtered aqueous food is frozen carefully so that ice freezes out leaving the volatiles in solution in the remaining liquid. The volatiles are then isolated by solvent extraction or dynamic headspace. As most food contains at least ca. 5-10% soluble solids, the concentration factor could only be of the order of 10 which is quite small in comparison to the concentration factors obtainable with the main methods 1-3. Freeze concentration may have more application in the concentration of essences (*30*) which are, of course, free from dissolved solids. However, a point would be reached where some of the less soluble volatiles such as terpene and sesquiterpene hydrocarbons (solubility ca. 1 ppm) come out of solution.

Freeze Drying. Freeze drying or lyophilization, with trapping of the volatiles that come over with the water vapor (*31*), would seem to have some advantages because of the mild conditions used. However, there is considerable selection of volatiles because the intact tissues act as a barrier to many volatiles whereas they are permeable to water. Freeze drying can also be considered a type of vacuum steam distillation.

Conclusions

The choice of the isolation method depends on the product being studied and the type of information needed.

The advantage of direct solvent extraction is that it is capable of isolating a wide range of compounds including very water soluble volatiles such as Furaneol. Most other methods do not give as comprehensive an isolation. On the other hand, it is time consuming and not satisfactory for compounds whose volatility is less than or close to that of the solvent.

The steam distillation and SDE methods are relatively fast and give good recoveries of many compounds. However, very water soluble volatiles are recovered only poorly or often not in detectable amounts. Atmospheric SDE, because of the relatively high temperature required, can cause large changes in the pattern of volatiles obtained.

Dynamic headspace isolation is a relatively mild method and can be carried out in a relatively short time. With thermal desorption it can be used for a wide range of compound volatilities. However, thermal desorption can also cause molecular changes to sensitive molecules. Solvent desorption is milder and applicable to large traps but has the same drawback as the other methods which use solvent in that very volatile compounds are lost or obscured. The use of closed loop stripping and binding of water with excess Na_2SO_4 markedly increases the effectiveness of high flow dynamic headspace isolation for very water soluble volatiles.

Literature Cited

1. Sugisawa, H. *Flavor Research: Recent Advances.* Eds. R. Teranishi, R. A. Flath, H. Sugisawa. Marcel Dekker Inc., New York, **1981**, pp 11-51.
2. Teranishi, R.; Kint, S. *Flavor Science: Sensible Principles and Techniques.* Eds. T. E. Acree; R. Teranishi. ACS Books, Washington, D. C., **1993,** pp 137-167.
3. Schieberle, P. *J. Agric. Food Chem.* **1995,** *43*, 2442-2448.
4. Guth, H; Grosch, W. *Fat Science and Technology.* **1989**, *91*, 225-230.
5. Sen, A; Laskawy, G.; Schieberle, P.; Grosch, W. *J. Agric. Food Chem.* **1991,** *39*, 757-759.
6. Guth, H; Grosch, W. *Lebensm. -Wiss. u.-Technol.*, **1993**, *26*, 171-177.
7. Roberts, D. D.; Acree, T. E. in *Thermally Generated Flavors: Maillard, Microwave and Extrusion Processes.* Eds. T. H. Parliment, M. J. Morello, R. J. McGorrin, ACS symposium series No. 543, Washington, D. C. 1994, pp 71-79.
8. Pickenhagen, W.; Velluz, A.; Passert, J-P.; Ohloff, G. *J. Sci. Food Agric.* **1981**, *32,* 1132-1134.
9. Buttery, R. G.; Takeoka, G. R.; Krammer, G. E.; Ling, L. C. *Lebensm.-Wiss. u. - Technol.*, **1994**, *27,* 592-594.
10. A. Yamane; A. Yamane; Shibamoto, T. *J. Agric. Food Chem.* **1994**, *42*, 1010-1012.
11. Guenther, E. *The Essential Oils.* Third Edition. John Wiley and Sons, Inc. New York, **1948**, pp 160.
12. Nickerson, G. B.; Likens, S. T. *J. Chromatog.* **1966**, *21,* 1-4.
13. Schultz, T. H.; Flath, R. A.; Eggling, S. B.; Teranishi, R. *J. Agric. Food Chem.* **1977**, *25,* 444-449.
14. Vogel, A. L.; *Practical Organic Chemistry* Third Edition. John Wiley and Sons, Inc. New York, **1962**, pp 225.
15. Wiberg, K. B. *Laboratory Techniques of Organic Chemistry.* McGraw-Hill, New York, **1960**, pp 183.
16. Buttery, R. G.; Bomben, J. L; Guadagni, D. G.; Ling, L. C. *J. Agric. Food Chem.* **1971**, *19,* 1045-1048.
17. Buttery, R. G.; Ling, L. C. *Progress in Flavor Precursor Studies.* Eds. P. Schreier; P. Winterhalter. Allured Publ. Corp. Carol Stream, IL, **1993**, pp 137-146.
18. Buttery, R. G.; Seifert, R. M.; Guadagni, D. G.; Black, D. R.; Ling, L. C. *J. Agric. Food. Chem.* **1968**, *16*, 1009-1015.

19. Buttery, R. G.; Ling, L. C.; Bean, M. M. *J. Agric. Food Chem.* **1978**, *26*, 179-180.
20. Buttery, R. G.; Teranishi, R. *J. Agric. Food Chem.* **1963**, *11*, 504-507.
21. Dressler, M. *J. Chromatog.* **1979**, *165*, 167-206.
22. Nunez, A.; Gonzalez, L. F.; Janak, J. *J. Chromatog.* **1984**, *300,* 127-162.
23. Rapp, A; Knipser, W. *Chromatographia.* **1980**, *13*, 698-702.
24. Williams, P. J.; Strauss, C. R.; Wilson, B. *J. Agric. Food Chem.* **1980**, *28*, 766-771.
25. Williams, P. J. *J. Chromatog.*, **1982**, *241*, 432-433.
26. Grob, K; Grob, G. *J. Chromatog.* **1974**, *90*, 303-313.
27. Grob, K.; Zurcher, F. *J. Chromatog.* **1976**, *117*, 285-294.
28. Buttery, R. G.; Guadagni, D. G.; Ling, L. C. *J. Agric. Food Chem.*, **1973**, *21*, 198-201.
29. Joulain, D. In *Progress in Essential Oil Research.* Ed. E. -J. Brunke. Walter de Gruyter. Berlin, **1986**, pp 57-67.
30. Kepner, R. E.; van Straten, S.; Weurman, C. *J. Agric. Food Chem.* **1969**, *17*, 1123-1125.
31. Heatherbell, D. A.; Wrolstad, R. E.; Libbey, L. M. *J. Agric. Food Chem.* **1971**, *36*, 219-221.

Chapter 24

Modern Biotechnology in Plant Breeding: Analysis of Glycoalkaloids in Transgenic Potatoes

Karl-Heinz Engel[1], Werner K. Blaas[2], Barbara Gabriel[2], and Mathias Beckman[2]

[1]Lehrstuhl für Allgemeine Lebensmitteltechnologie, Technische Universität München, D–85350 Freising-Weihenstephan, Germany
[2]Bundesinstitut für Gesundheitlichen, Verbraucherschutz und Veterinärmedizin, Postfach 330013, D–14191 Berlin, Germany

Steroidal glycoalkaloids in potatoes are among the most prominent naturally occurring food toxicants. A mathematical correlation between tuber size and alkaloid content has been established, which allows the assessment of different potato varieties in terms of glycoalkaloid content by calculation for a normalized weight, independent from the tuber size analyzed. Potatoes modified by means of recombinant DNA techniques have been investigated with this method. Inhibition of amylose biosynthesis by anti-sense RNA expression had no effect on the glycoalkaloid content. However, insertion of an invertase gene from yeast caused a reduction of the concentrations of these critical food toxicants. These studies may serve as a basis for the application of the principle of "substantial equivalence" to the safety evaluation of transgenic potatoes.

Modern biotechnology in plant breeding practices increasingly includes the application of recombinant DNA techniques (*1*). Genetically modified crops and their products have been commercialized or are about to enter the market (*2-4*). Genetic engineering in the development of foods raises a series of issues. In addition to basic ethical concerns, major emphasis has been placed on questions related to the release of genetically modified organisms into the environment (*5*). Due to the increasing transition from controlled field experiments to the placing of products on the market, food safety aspects are gaining importance (*6*). National and international organizations and authorities are in the process of developing strategies for the safety assessment of foods produced via recombinant DNA techniques (*7, 8*). The Organization for Economic Cooperation and Development (OECD) has elaborated the principle of "substantial equivalence", which is based upon comparison of the novel food to its traditional counterpart. If the new food or food component is found to be "substantially equivalent" to an existing food, it can be treated in the same manner with respect to safety (*9*).

0097–6156/96/0637–0249$15.00/0

The World Health Organization (WHO) has developed a protocol on how to apply this principle to foods or food components from plants derived by modern biotechnology (*10*). The determination of "substantial equivalence" is based on a molecular characterization on the DNA level; it further includes the evaluation of agronomic traits, and it concludes with a chemical characterization of the new food or food component. The major focus of this chemical analysis is on naturally occurring toxicants and critical nutrients.

One of the most prominent examples of naturally occurring food toxicants is steroidal glycoalkaloids in potatoes (*11*). Potatoes have been subjected to a wide spectrum of genetic modifications by means of recombinant DNA techniques (*12-14*). This paper describes studies on the influence of genetic engineering on the amounts of naturally occurring glycoalkaloids which may serve as a basis for the application of the concept of "substantial equivalence" in the safety assessment of transgenic potatoes.

Potato alkaloids. The cultivated potato (*Solanum tuberosum* L.) contains two major glycoalkaloids, α-chaconine and α-solanine. The two components both contain the C_{27} steroidal aglycone solanidine; they differ only in the sugar moieties included in the trisaccharide part. α-Solanine and α-chaconine form up to 95% of the glycoalkaloids present in potatoes. Data on occurrence, chemistry, analysis, and toxicology of the steroidal glycoalkaloids present in potatoes have been comprehensively reviewed (*11, 15*).

A modified version of the procedures described by Bushway et al. (*16*) and Carman et al. (*17*) has been applied to investigate α-chaconine and α-solanine in potatoes: the compounds were extracted from potato tubers using an acidified mixture of chloroform and methanol. After evaporation of the organic solvents, the residue was redissolved in aqueous heptanesulfonic acid. This crude extract was purified using a C_{18} reversed phase cartridge. Separate quantification of the two alkaloids was finally achieved by HPLC. With this method, the glycoalkaloid content in experimentally grown and in commercially available potatoes was determined (*18*). Differences based on genetic variability as well as on geographic origin were observed. As examples, data obtained for varieties purchased from local grocery stores are presented in Figure 1.

Dependence of potato alkaloid content on tuber size. It has long been known that the alkaloid content of potatoes is inversely related to the size of the tubers (*19, 20*). The major portion of glycoalkaloids in potato tubers is located within the first mm from the outside surface, and decreases toward the center of the tubers (*21*). Therefore, glycoalkaloid concentrations are higher in small tubers due to the surface to mass ratio (*22*). This dependence of glycoalkaloid content on the tuber size reflected in the data presented in Figure 1 makes it difficult to compare analytical data. Because of the lack of linearity, it is not possible to take the average of tuber sizes investigated and to relate it to the average of the glycoalkaloid amounts determined. Therefore, the first goal was to establish a mathematical correlation between tuber size and glycoalkaloid content, thus establishing a basis for a comparison of glycoalkaloid data independent from the tuber size analyzed.

The investigation of a wide spectrum of different potato cultivars with tuber weights ranging from less than 10 g to more than 280 g revealed that the relationship between tuber size and alkaloid concentration is based on the equation shown in Figure 2a. The fact that this function is well suited to describe the correlation between tuber size and alkaloid content in potatoes becomes even more evident after transforming the equation as shown in Figure 2b. Plotting of the natural logarithm of the alkaloid content versus the natural logarithm of the tuber weight shows a linear relationship between these parameters.

The coefficient "a" is a characteristic of each variety; it determines the position of the graphs on the y-axis and reflects the glycolkaloid concentration in the potato. Coefficient "b" determines the slope of the curves. Analyses of a broad spectrum of different potato varieties revealed that b averages -0.5 for α-solanine and -0.42 for α-chaconine, respectively (*23*). This correlation offers the possibility to calculate the alkaloid content for a certain size tuber on the basis of the results obtained from the analysis of any other tuber size. Glycoalkaloid contents (GA_1 and GA_2) of different tuber sizes (weights m_1 and m_2) are determined by the following equations:

$$\frac{GA_1}{GA_2} = \frac{e^{a + b \cdot \ln m_1}}{e^{a + b \cdot \ln m_2}} \quad [1]$$

$$GA_2 = GA_1 \cdot e^{b (\ln m_2 - \ln m_1)} \quad [2]$$

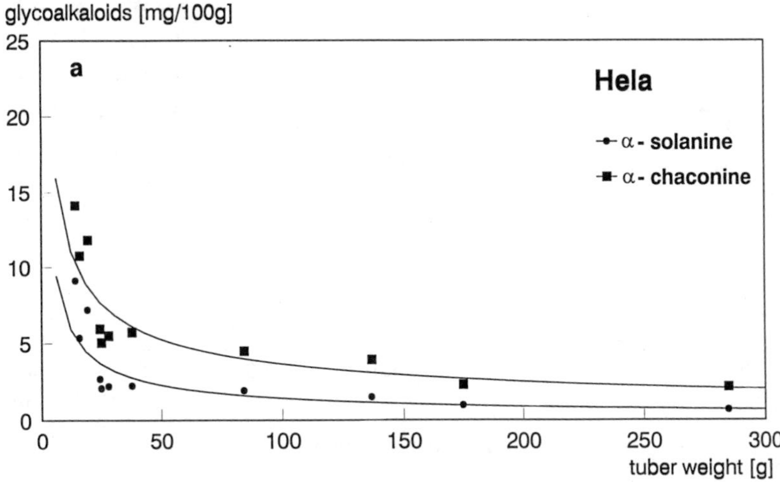

Figure 1. Glycoalkaloids determined in commercial potato varieties.

Continued on next page

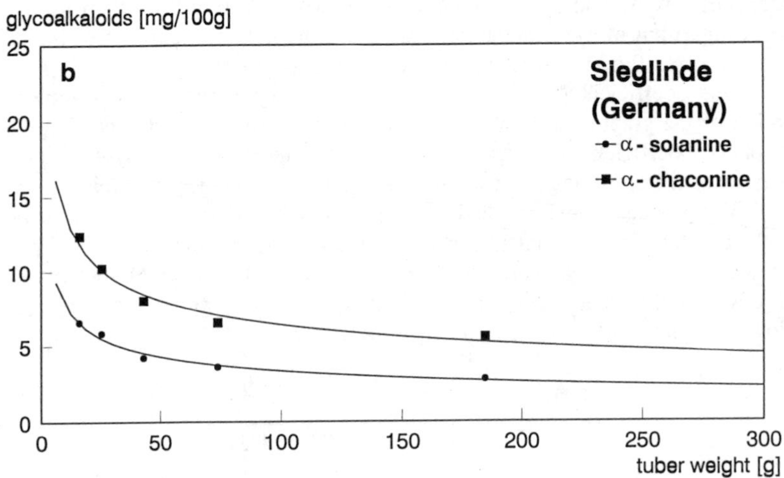

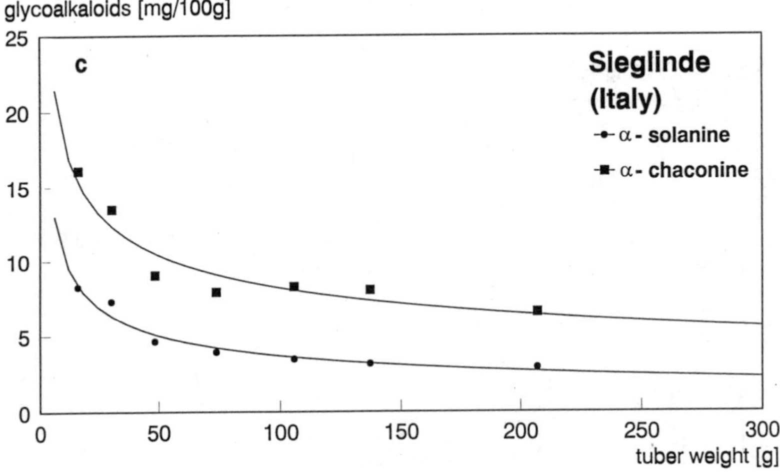

Figure 1. *Continued*

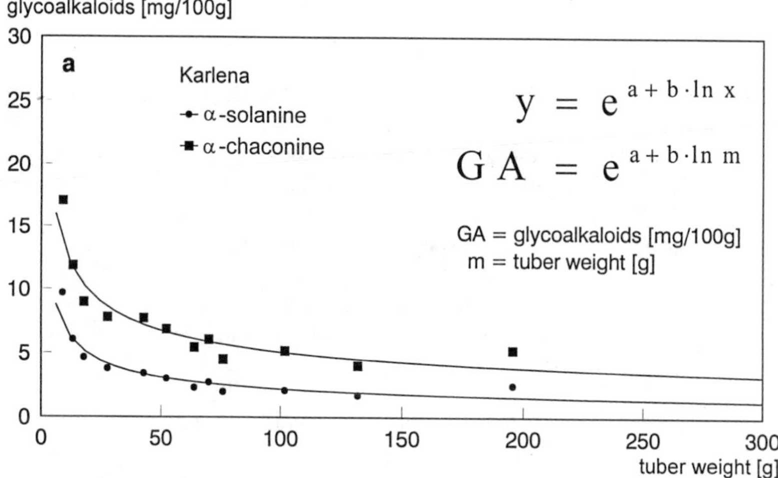

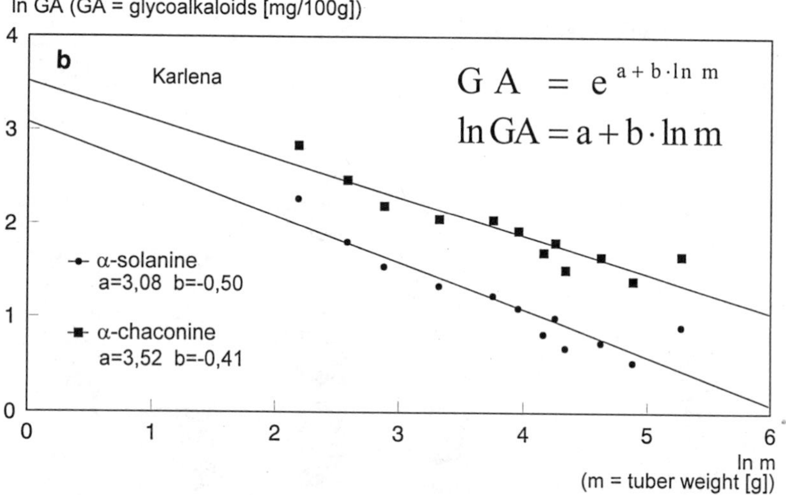

Figure 2. Mathematical correlation between glycoalkaloid content and potato tuber size: upper (a), lower (b).

In order to compare analytical data, a certain reference weight, e.g. 100 g, can be selected and glycoalkaloid concentrations (GA_x) determined for any other tuber size (m_x) can be normalized to this reference weight as follows:

$$GA_{100} = GA_x \cdot e^{b (\ln 100 - \ln m_x)} \quad [3]$$

The application of this concept to the assessment of various conventionally bred potato varieties is shown in Figure 3. On the basis of the data normalized according to equation [3], the average glycoalkaloid content including standard deviation for the reference weight can be determined. This procedure allows the comparison of different cultivars in terms of alkaloid content independent from the tuber sizes analyzed. This is of special importance in the beginning of new breeding programs, when only small tubers are available (*23*). On the basis of these results, the investigation of glycoalkaloid contents in genetically engineered potatoes has been approached. Two types of potatoes modified via recombinant DNA techniques have been investigated.

Genetically modified (anti-GBSS) potato. Potato starch consists of two components: amylose (20%) and amylopectin (80%). Granule-bound starch synthase (GBSS) is the major enzyme involved in the formation of amylose.

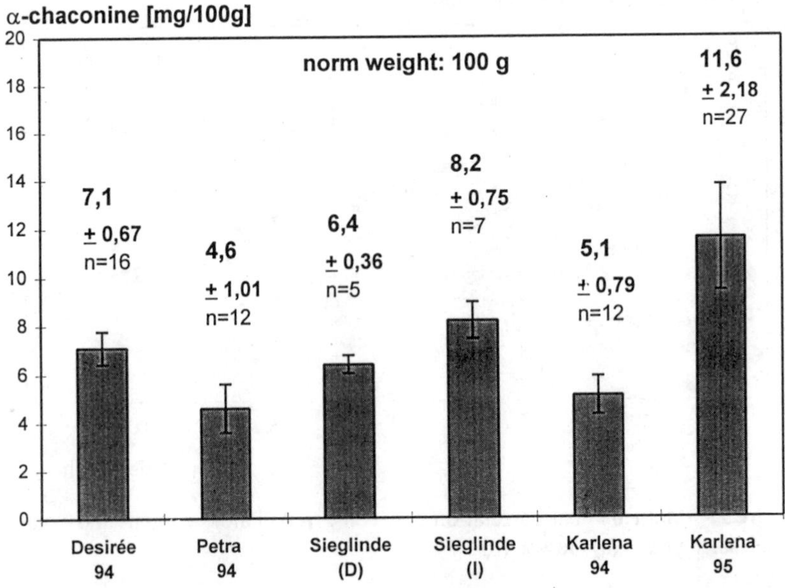

Figure 3. Glycoalkaloid contents in potato varieties normalized to a reference tuber weight of 100 g.

Expression of an anti-sense RNA has been applied to inhibit the biosynthesis of this enzyme, thus leading to potato starch consisting of more than 95% amylopectin (*24*). The potato variety Desirée has been transformed using *Agrobacterium tumefaciens*. Similar experiments to develop transgenic amylose-free potatoes have been described (*25*). In contrast to that approach, in the present study tuber specific expression has been achieved by using the patatin B33 promotor from *Solanum tuberosum.*

The combined analytical/mathematical approach described above has been used to assess the glycoalkaloid content in genetically modified (GM) anti-GBSS potatoes. For example, data on material obtained from a field trial performed in Germany in 1994 are presented in Figure 4. For both, α-chaconine and α-solanine, quantitative determinations over a range of tuber sizes from 8 g to more than 250 g have been carried out (*27*). The curves plotted in Figure 4 show that there is no statistically significant difference in the concentrations of these toxicants between the parental line and the genetically modified potato.

In order to quantitate the data, the normalization according to equation [3] has been carried out. As example, Figure 6 summarizes the results obtained for α-chaconine from field trials performed in 1993. On the basis of 11 parental and 13 anti-GBSS tuber sizes, respectively, average concentrations for a normalized weight of 100 g and the corresponding standard deviations have been calculated. There is no statistically significant difference in the concentrations of α-chaconine between the parental line and the potatoes modified by inhibition of the amylose biosynthesis. According to investigations of glycoalkaloids in amylose-free transgenic potatoes obtained by the use of a plant virus promoter, glycoalkaloid concentrations in transgenic clones were even lower than in the parental line (*26*). However, tuber sizes investigated have not been specified and only total glycoalkaloid concentrations (without differentiation of α-solanine and α-chaconine) have been determined in that study.

Genetically modified (invertase) potato. In the second example investigated, the modification obtained by genetic engineering leads to the biosynthesis of a yeast invertase in potatoes. The corresponding gene from *Saccharomyces cerevisiae* has been fused to the coding region of the N-terminal signal peptide of the proteinase inhibitor II from *Solanum tuberosum,* thus targeting the enzyme to the cell wall (*28*). Tuber specific expression has been achieved by using the patatin B33 promotor. The transformation has been carried out using *Agrobacterium tumefaciens*. Due to the activity of the inserted invertase, sucrose is cleaved into glucose and fructose. As sucrose is the major transport form for photosynthesis products in the plant, this causes changes in the source (leaves) / sink (tubers) interactions, which eventually lead to increased yield and/or increased tuber size (*29*).

The investigation of GM (invertase) potatoes grown under greenhouse conditions indicated a decrease of the glycoalkaloid concentrations in the genetically modified line compared to the parental control (Figure 5a). In order to verify these results, genetically modified material obtained from field trials has been investigated. Glycoalkaloid amounts have been determined in four independently transformed lines. Data determined for α-chaconine in one of these lines obtained

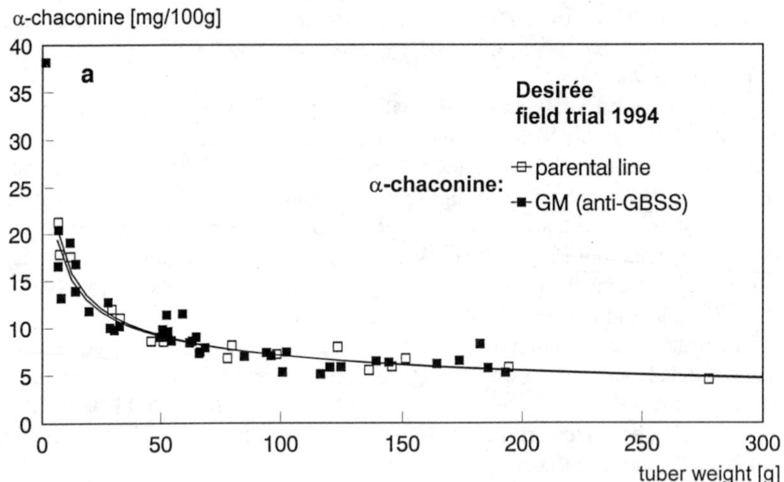

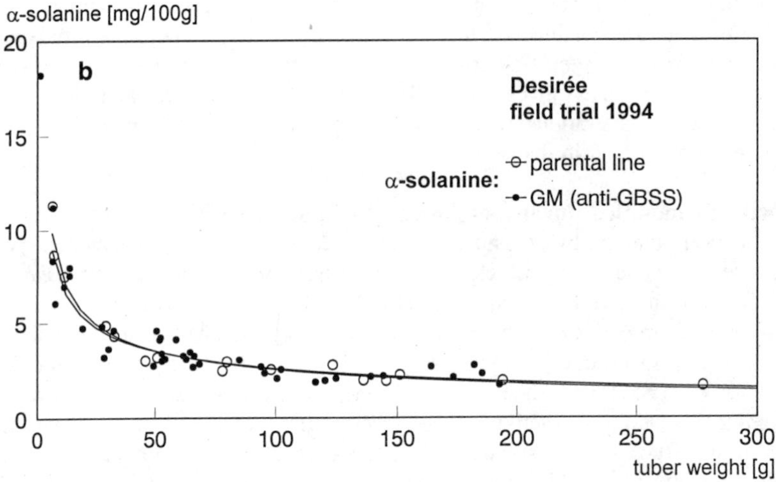

Figure 4. Glycoalkaloid content in genetically modified anti-GBSS potatoes (inhibition of amylose biosynthesis) and the corresponding untransformed control lines: upper (a), lower (b).

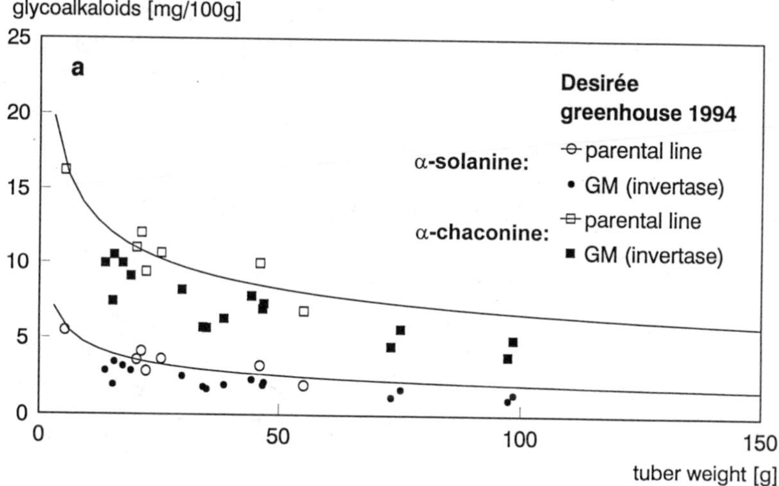

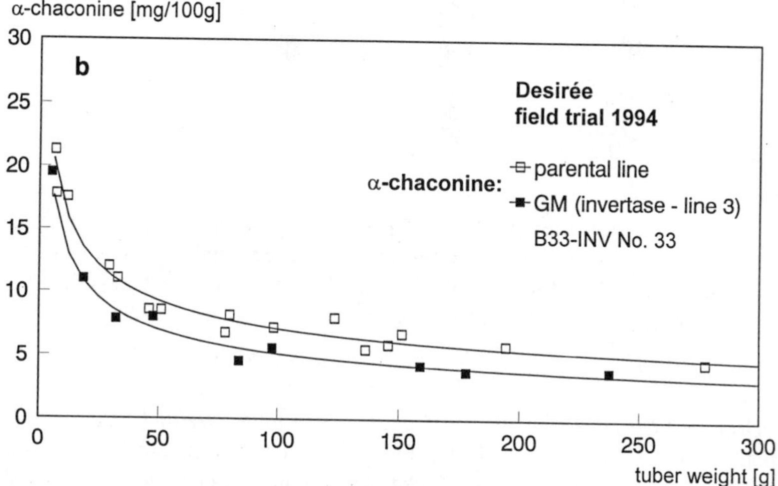

Figure 5. Glycoalkaloid content determined in genetically modified potatoes expressing yeast derived invertase, grown in the greenhouse, upper (5a), and obtained from field trials, lower (5b).

from the field trial in 1994 are shown in Figure 5b. The corresponding graphs show the significant difference between the parental and the genetically modified line.

Quantitation of the data revealed that amounts of α-chaconine calculated for a normalized weight of 100 g in the GM (invertase) potatoes are more than 30% lower than in the parental line (Figure 6). Similar reductions in glycoalkaloid concentrations have been determined for greenhouse and field trial experiments in 1993.

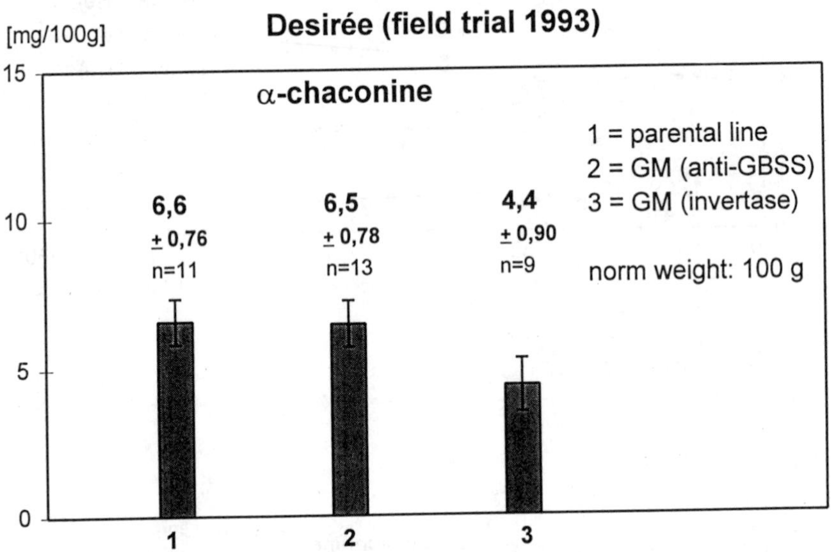

Figure 6. Average concentrations of α-chaconine (normalized to a reference tuber weight of 100 g) determined in untransformed potatoes var. Desirée (1), and in transgenic lines modified by inhibition of amylose biosynthesis (2), and by expression of yeast derived invertase (3), respectively.

Conclusion. A mathematical correlation between tuber size and alkaloid content of potatoes has been established. On the basis of this correlation it is possible to compare potato varieties in terms of alkaloid content by calculation of the data for a normalized weight. This concept can be applied to assess conventionally bred potatoes as well as genetically engineered potatoes.

Amylose inhibition by means of anti-sense RNA expression had no effect on the alkaloid content of the transgenic potatoes. This important result bears heavily on the safety assessment of this crop-gene combination.

Conversely, the expression of a yeast derived invertase in potatoes leads to a reduction of these naturally occurring toxicants. The elucidation of the biochemical reason for this reduction (dilution effect due to the increase of the tubers or metabolic interference with the sugar moieties needed for the biosynthesis of the glycoalkaloids) is part of our ongoing research.

Acknowledgments

The authors thank Dr. A. G. Heyer (Institut für genbiologische Forschung, Berlin, Germany) and Dr. H. Tiemann (Bundesanstalt für Züchtungsforschung in Gross Lüsewitz, Germany) for providing potato samples.

References

1. Willmitzer, L. *New Developments in Plant Biotechnology*; Lecture Publications, Vol. 20; Ernst Schering Research Foundation: Berlin, Germany, 1994.
2. Redenbaugh, K.; Hiatt, W.; Martineau, B.; Emlay, D. In *Genetically Modified Foods: Safety Aspects;* Engel, K.-H., Takeoka, G., Teranishi, R. Eds.; ACS Symposium Series 605; American Chemical Society: Washington, DC, 1995, pp 72-87.
3. Fuchs, R.L.; Se, D.B.; Rogers, S.G.; Hammond, B.G.; Padgette, S.R. In *The Biosafety Results of Field Tests of Genetically Modified Plants and Microorganisms;* Jones, D.D.; Ed.; Proc. of the 3rd International Symposium; The University of California: Oakland, CA, 1995.
4. ACNFP (Advisory Committee on Novel Foods and Processes), Annual Report 1994, MAFF-Publications: London.
5. *The Biosafety Results of Field Tests of Genetically Modified Plants and Microorganisms;* Jones, D.D.; Ed.; Proc. of the 3rd International Symposium; The University of California: Oakland, CA, 1995.
6. *Genetically Modified Foods: Safety Aspects;* Engel, K.-H., Takeoka, G., Teranishi, R., Eds.; ACS Symposium Series 605, American Chemical Society: Washington, DC, 1995, 243 pp.
7. IFBC (International Food Biotechnology Council), *Regulat. Toxicol. Pharmacol.* **1990**, *12*, 1-196.
8. Kessler, D.A.; Taylor, M.R.; Maryanski, J.H.; Flamm, E.L.; Kahl, L.S. *Science*, **1993**, *256*, 1747-1832.
9. *Safety Evaluation of Foods Derived by Modern Biotechnology: Concepts and Principles*, OECD: Paris, 1993.
10. *Application of the Principles of Substantial Equivalence to the Safety Evaluation of Foods or Food Components from Plants Derived by Modern Biotechnology,* Report of a WHO Workshop, WHO: Geneva, 1995.
11. van Gelder, W.M.J. In *Poisonous Plant Contamination of Edible Plants*; Rizk, A.F.M., Ed.; CRC Press: Boca Raton, FL, 1991; pp 117-156.

12. Vayda, M.E.; Belknap, W.R. *Transgenic Res.* **1992**, *1*, 149-163.
13. Lavrik, P.B.; Bartnicki, D.E.; Feldman, J.; Hammond, B.G.; Keck, P.J.; Love, S.L.; Taylor, M.W.; Rogan, G.J.; Sims, S.R.; Fuchs, R.L. In *Genetically Modified Foods: Safety Aspects;* Engel, K.-H., Takeoka, G., Teranishi, R., Eds.; ACS Symposium Series 605; American Chemical Society: Washington, DC, 1995, pp 148-158.
14. Monro, J.A.; James, K.A.C.; Conner, A.J. *Food Info Report No.6*; New Zealand Institute for Crop & Food Research: Palmerston North, New Zealand, 1993, 1-19.
15. Maga, J.A. *Food Rev. Internat.* **1994**, *10*, 385-418.
16. Bushway, R.J.; Bureau, J.L.; King, J. *J. Agric. Food Chem.* **1986**, *34*, 277-279.
17. Carman, Jr., A.S.; Kuan, S.S.; Ware, GM.; Octave, Jr., J.P.; Kirschenheuter, G.P. *J. Agric. Food Chem.* **1986**, *344*, 279-282.
18. Blaas, W.K.; Gabriel, B.; Beckmann, M.; Engel, K.-H. *J. Agric. Food Chem.*, submitted.
19. Wolf, M.J.; Duggar, B.M.J. *Agric. Res.* **1946**, *73*, 1-32.
20. Verbist, J.F.; Monnet, R. *Potato Res.* **1979**, *22*, 239-244.
21. Kozukue, N.; Kozukue, E.; Mizuno, S. *Hort. Science*, **1987**, *22*, 294-296.
22. Johnsson, H.; Hellenas, K.E. *Var Foda* **1983**, *35*, 299-314.
23. Blaas, W.K.; Gabriel, B.; Beckmann, M.N.; Tiemann, H.; Engel, K.-H. *Potato Res.*, submitted.
24. Kossmann, J.; Abel, G.; Buttcher, V.; Duwenig, E.; Emmermann, M.; Lorberth, R.; Springer, F.; Virgin, I.; Welsh, T.; Willmitzer, L. In *Carbohydrate Bioengineering;* Petersen, S.B.; Svensson, B.; Pedersen, S., Eds.; Elsevier Science B.V.: Amsterdam, The Netherlands, 1995, 271.
25. Kuipers, A.G.J. Ph.D. Thesis, Agricultural University of Wageningen: Wageningen, The Netherlands, 1994.
26. Heeres, P.; van Swaaij, A.C.; Bruinenberg, P M.; Kuipers, A.G.J.; Visser, R.G.F.; Jacobsen, E. In *The Biosafety Results of Field Tests of Genetically Modified Plants and Microorganisms;* Jones, D.D.; Ed.; Proc. of the 3rd International Symposium; The University of California: Oakland, CA, 1995.
27. Blaas, W.K.; Gabriel, B.; Beckmann, M.; Heyer, A.G.; Willmitzer, L.; Engel, K.H. *Bio/Technology.*, submitted.
28. Rocha-Sosa, H.; Sonnerwald, U.; Frommer, W.; Stratmann, M.; Schell, J.; Willmitzer, L., *EMBO J 8*, **1989**, 23-29.
29. Sonnewald, U.; Lerchl, J.; Zoenner, R.; Frommer, W. *Plant Cell Environment* **1994**, *17*, 649-658.

Chapter 25

Application of Atmospheric Pressure Ionization Liquid Chromatography–Tandem Mass Spectrometry for the Analysis of Flavor Precursors

Markus Herderich, René Roscher, and Peter Schreier

Lehrstuhl für Lebensmittelchemie der Universität Würzburg, Am Hubland, 97074 Würzburg, Germany

The development of techniques utilizing atmospheric pressure ionization, namely atmospheric pressure chemical ionization (*APCI*) and electrospray ionization (*ESI*), has pioneered the coupling of liquid chromatography (*HPLC*) with mass spectrometry (*MS*) in recent years. Both *ESI* and *APCI* generate ions from polar and labile biomaterials with remarkable ease and efficiency. Particularly the use of liquid chromatography together with tandem mass spectrometry (*MS/MS*) opens further dimensions in the field of bioorganic analysis. Thus, *HPLC-MS/MS* provides the tools to elicit structure and variety of flavor precursor compounds directly in complex matrices. In order to develop efficient and straightforward strategies for the analysis of flavor precursor systems this chapter outlines the potential and limitations of those hyphenated analytical techniques. Finally, current applications demonstrate the successful analysis of glycoconjugates by atmospheric pressure ionization *HPLC-MS/MS*.

In the past the efficient on-line characterization of flavor progenitors by mass spectrometry has been limited by the polar and labile character of these compounds. Combining HPLC with mass spectrometry could have been the method of choice for the analysis of glycoconjugates. But in order to transform molecules from solution to ions in the gas-phase one had to deal with three major incompatibilities (*1*):
(i) the apparent problem of maintaining the high vacuum of the mass spectrometer after coupling with HPLC systems delivering flow rates up to 1 ml/min;
(ii) the "soft" ionization of non-volatile and/or thermally labile compounds in order to obtain molecular weight information and
(iii) the efficient transfer of the analytes as ions into the mass analyzer required for the analysis of trace compounds.

For achieving the above mentioned goals researchers have developed a variety of methods like direct liquid introduction, thermospray ionization and the particle

0097–6156/96/0637–0261$15.00/0

beam interface. It was only recently that the rediscovery of techniques utilizing ionization at atmospheric pressure, namely atmospheric pressure chemical ionization (*APCI*) and electrospray ionization (*ESI*), has directed the application of *HPLC-MS* to an ever-growing role in analytical chemistry.

Electrospray ionization *(ESI)*

Based on earlier experiments of Dole and co-workers (*2*) and the work of Iribarne and Thompson (*3, 4*), Fenn and colleagues developed the technology of electrospray ionization in the mid-eighties (*5, 6, 7*). In an electrospray interface the column effluent is nebulized into the atmospheric pressure region as a result of a strong electric field resulting from the potential difference between the narrow-bore spray capillary and a counter electrode. The field at the capillary tip charges the surface of the emerging liquid mainly by electrophoretic processes and not by electron transfer (*8*). As a result a *Taylor cone* is formed by the interaction of surface tension and coulombic forces from which a fine spray of charged droplets disperses. The droplets evaporate neutral solvent molecules until their surface charge density reaches the *Rayleigh limit*. Then the electrostatic forces overpower the surface tension resulting in a "*Coulomb explosion*" that produces an array of charged microdroplets which also evaporate until they "explode" themselves. As the last solvent molecules evaporate the charge would be retained by the analyte molecule to produce a free ion (*7*). Whether this "*charged residue*" mechanism actually applies is still under debate. An alternative explanation for the formation of single ions represents the "*ion evaporation model*" proposed by Iribarne and Thompson which describes the direct emission of desolvated ions from microdroplets (*4*).

While pure electrospray nebulization is only capable of flow rates up to 20 µl/min, the development of pneumatically assisted electrospray is compatible with flow rates exceeding 200 µl/min (*9*). In order to successfully ionize a compound by the electrospray process the analyte should dissociate in solution to form solvated ions prior to nebulization. Ion formation from species that are not ions themselves always requires the presence of a polar atom or functional group to which solute cations or anions can be attached by ion-dipole forces. Biopolymers carrying multiple functional groups will form multiple charged ions, an observation that has had an outstanding impact on the analysis of peptides, proteins and oligonucleotides (*7*).

Atmospheric pressure chemical ionization *(APCI)*

In an *APCI* interface the column effluent enters a heated nebulizer where the pneumatically assisted desolvation process is almost completed. While still in the spray chamber, ionization of analytes is initiated by corona discharge. The ionization mechanisms in *APCI* are almost identical to those in conventional medium pressure chemical ionization (*1*). Positive ion formation can be achieved by proton transfer, adduct formation or charge exchange reactions, while in the negative mode ions are formed due to proton abstraction, anion attachment and electron capture reactions. The *APCI* interface is compatible with flow rates exceeding 1 ml/min and will generate molecular ions beside some thermal degradation products. Thus, *APCI* possesses a wider range of application for structure elucidation of smaller molecules, while *ESI* is the method of choice for the analysis of biopolymers, in particular, if the amount of sample is limited (Table I).

Table I. Comparison of *APCI* and *ESI*

	APCI	*ESI*
analyte:	no restriction, „active" ionization	chargeable functional groups and preionization in solvent required
upper MW limit:	> 1000 u	> 100000 u
side reactions	low to moderate thermal stress, avoid low pH of solvent	neglible after optimization of ionization conditions
HPLC solvent:	aqueous solvents	aqueous solvents as well as polar organic solvents
max. flow rate:	1 ml/min (4 mm columns)	0.2 ml/min (2 mm columns)

Application of tandem mass spectrometry for the analysis of flavor precursors

Both *ESI* and *APCI* generate molecular ions from polar and labile biomaterials with remarkable ease and efficiency. But the amount of structural information that can be deduced from *ESI* spectra, in particular, is rather limited. Thus, only the use of tandem mass spectrometry (*MS/MS*) together with liquid chromatography opens further dimensions in the field of bio-organic analysis (Fig. 1). Beside retention data and UV spectra one can obtain both molecular mass and substructure specific information of any analyte out of a complex matrix.

Precursor and neutral loss scanning can be used to identify molecular masses of compounds with specific substructures. Subsequent structure elucidation usually involves generation of product ion spectra which reveals the fragmentation patterns of the various molecular ions. Selective reaction monitoring by tandem mass spectrometry is the equivalent to single ion monitoring in conventional GC-MS and LC-MS analysis for the detection of known analytes in complex matrices. A compound is only positively identified by SRM, if it exhibits the correct molecular mass together with the characteristic fragment ion at a specific retention time. Thus, the two-dimensional spectral filter in SRM experiments combines highest structural selectivity with sensitivity of detection for the analysis of minor constituents and allows drastically simplified sample preparation procedures.

Current examples of the successful application of *HPLC-APCI/ESI-MS/MS* during our studies on the generation of flavor compounds include the analysis of

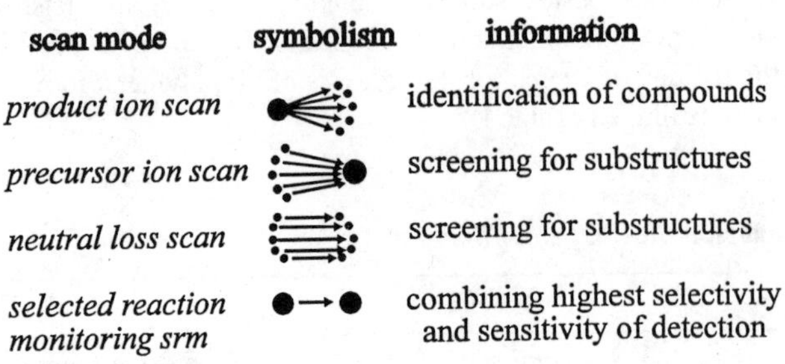

scan mode
symbolism
information
product ion scan
identification of compounds
precursor ion scan
screening for substructures
neutral loss scan
screening for substructures
selected reaction monitoring srm
combining highest selectivity and sensitivity of detection

Figure 1. *MS/MS* experiments.

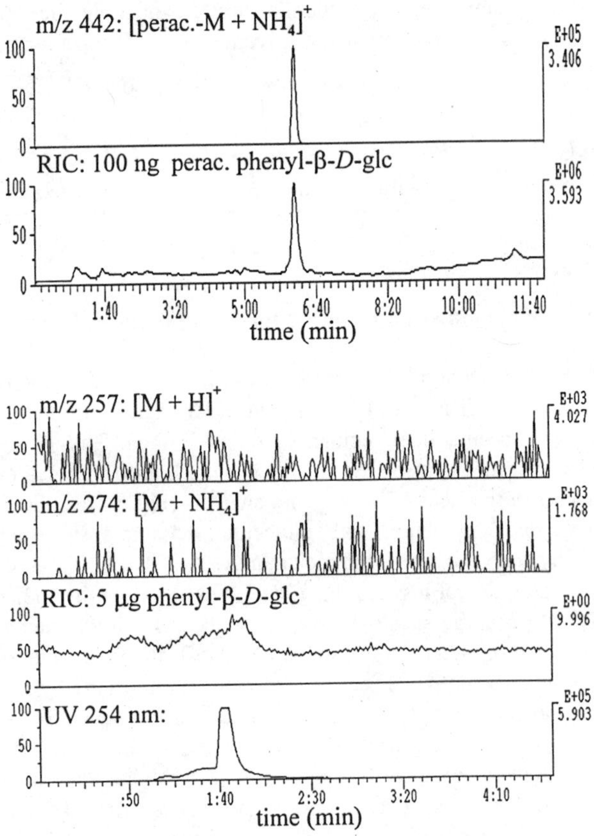

Figure 2. Influence of peracetylation on ionization response. Sample: free and peracetylated phenyl β-*D*-glucopyranoside; *ESI* positive; HPLC: RP-18; 200 µl/min MeOH-H_2O-5 mM NH_4Ac.

terpene phosphates in green tea leaves (*10*) as well as studies on the formation of fatty acid hydroperoxides and their related secondary metabolites catalyzed by lipoxygenase and allene oxide synthase (*11*). As to glycoconjugates, structure elucidation of *O*-glycosides (*12*), *N*-glycosides and esterified glycoconjugates (*13*, *14*) has been performed.

Analysis of glycosylated flavor precursors by *APCI/ESI-MS/MS*

The application of both ionization techniques together with tandem mass spectrometry yields complementary information that allows localization and structure elucidation of glycosylated flavor precursors. Fig. 2 outlines the influence of introducing functional groups on the ionization response during the analysis of glycoconjugates. While free phenyl β-*D*-glucopyranoside is not ionized by *ESI* at all, its tetraacetate exclusively forms the ammonium adduct as molecular ion. For the analysis of acetylated glycoconjugates low detection limits in the sub-pmol range were established, thus demonstrating the excellent sensitivity of *ESI*.

In contrast β-*D*-glucopyranosyl-anthranilate isolated from the tropical pinuela fruit is instantaneously ionized by *ESI* without derivatization. Collision induced dissociation of the protonated molecular ion *m/z* 300 produces a product ion spectrum which is dominated by ions like *m/z* 138 characterizing the nitrogen containing aglycone (Fig. 3). Thus, a screening method yielding molecular mass information of glycoconjugates based on anthranilic acid can be developed by the subsequent analysis of the corresponding precursor ions.

Acetylation of β-*D*-glucopyranosyl-anthranilate did not improve the ionization response and the corresponding product ion spectrum is less informative as most ions are formed by the consecutive loss of acetic acid (-60 u) and ketene (-42 u), respectively. Neutral loss of the aglycone results in an ion *m/z* 331 typical for peracetylated hexose moieties. The fragment ion *m/z* 162 appears to be solely derived from the acetylated aglycone (Fig. 3). In order to gain a better insight in the structure of more complex glycans, permethylation combined with *ESI-MS/MS* of adduct ions obtained by addition of metal ions to the solvent appears to be favorable (*15*).

In the course of our studies on glycoconjugates derived from alcohols we observed that upon fragmentation induced by low energy collision activation the substructure of interest (i.e. the aglycone) did not carry the charge. In that case the neutral loss scan can be used to screen for flavor progenitors derived from compounds like phenylethanol and geraniol. Furthermore, underivatized glycoconjugates like the hexose derivative in Fig. 3 are characterized by the neutral loss of the carbohydrate moiety, thus opening a way for the specific detection of glycosides based on the sugar attached to the aglycone.

Profiling flavor progenitors

Substructure profiling by *APCI/ESI-MS/MS* provides structure specific information useful for analyzing the heterogeneity of glycoconjugated flavor precursors and allows selective localization of hitherto unknown compounds. For example, in

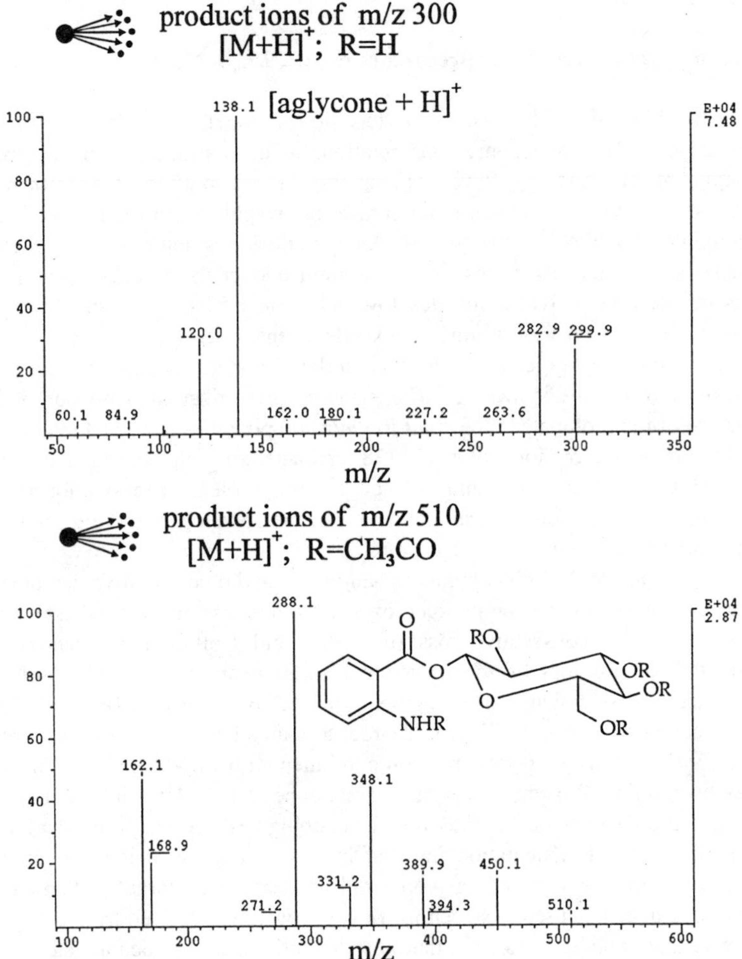

Figure 3. Influence of peracetylation on product ion spectra. Sample: free and peracetylated β-*D*-glucopyranosyl-anthranilate; *ESI* positive; CID: 1.8 mTorr Ar, -15eV; HPLC: RP-18 2 x 100 mm, 200 μl/min MeCN-H_2O-5 mM TFA.

strawberry fruit the glucopyranoside of furaneol (2,5-dimethyl-4-hydroxy-3[*2H*]furanone) accounts for less than 50% of totally bound furaneol (*14*). As the product ion spectrum of underivatized furaneol β-*D*-glucopyranoside exclusively exhibits an ion of the protonated aglycone (Fig. 4), we performed the corresponding precursor ion experiment in order to find additional progenitors of furaneol. Due to the excellent sensitivity of *ESI* we only had to work up one strawberry fruit (20 g of fruit) by adsorption on XAD-2 and subsequent elution of the glycoside fraction with methanol (*16*). *HPLC-MS/MS* analysis of the raw extract instantaneously revealed the presence of furaneol β-*D*-glucopyranoside (*m/z* 291: $[M+H]^+$) and traces of free furaneol (*m/z* 170: $[M+MeCN+H]^+$). However, it was a striking observation that two more peaks at 6:40 min and 8:0 min occurred, featuring both the product ion of protonated furaneol as well as its characteristic UV maximum at 275 nm (Fig. 5). Thus, substructure profiling by *ESI-MS/MS* demonstrated the presence of additional furaneol precursors in strawberry fruit.

Structure elucidation of furaneol precursor by APCI/ESI-MS/MS

The molecular mass of both new furaneol progenitors was determined to be 376 u as the *ESI* mass spectrum revealed the protonated molecular ion *m/z* 377 $[M+H]^+$ together with a characteristic pattern of adduct ions *m/z* 394 $[M+NH_4]^+$ and *m/z* 399 $[M+Na]^+$. Unfortunately, the corresponding product ion spectrum of *m/z* 377 almost exclusively showed the fragment ion *m/z* 129. Thus, it revealed the presence of a furaneol moiety, but did not provide any additional information about the conjugate.

Prior to further structural analysis we tried to derivatize the labile furaneol precursor, but acetylation of pure compound led to a complex mixture of reaction products. Another important observation was the acidic character of the conjugate, as acidifying the solvent significantly improved its chromatographic separation. Consequently, we applied *ESI* in the *negative mode* for the analysis of the deprotonated furaneol precursor (Fig.6).

The mass spectrum of the polar furaneol precursor revealed the deprotonated molecular ion *m/z* 375, while loss of carbon dioxide led to *m/z* 331 (*m/z* 375 - CO_2). The dominating ion *m/z* 391 could be assigned by the corresponding precursor ion experiment as decarboxylation product of the adduct ion *m/z* 435, formed due to the presence of acetate anions. Thus, *negative ESI* confirmed the molecular mass of the furaneol precursor and demonstrated the presence of a carboxylate group beside the furaneol moiety. But again structural assignment was not possible on the basis of *negative ESI* data.

Finally, we analyzed the unknown furaneol progenitors by means of *HPLC-APCI-MS* under optimized ionization conditions (Fig. 7). After adjusting the heated nebulizer to 300°C and reducing the temperature of the heated capillary to 150°C we obtained the protonated molecular ion *m/z* 377 together with the corresponding decarboxylation product *m/z* 333. Obviously, the protonated furaneol ion *m/z* 129 was the result of degradation reactions in the interface. The crucial missing link between the furaneol moiety and the terminal carboxylate group was provided by the fragment *m/z* 291, which revealed the presence of a furaneol hexoside. Thus, the *APCI*

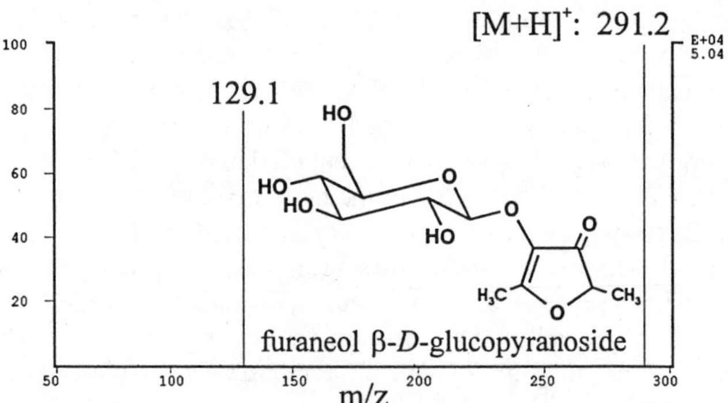

Figure 4. Product ion spectrum of protonated furaneol β-*D*-glucopyranoside. *ESI* positive; CID: 1.8 mTorr Ar, -15eV.

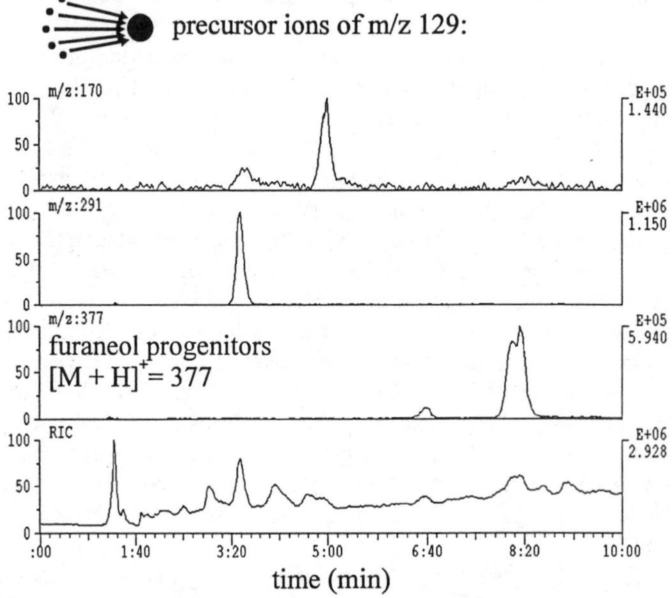

Figure 5. Profiling furaneol precursors by *HPLC-ESI-MS/MS*. Sample: glycosidic raw extract obtained from 20 g strawberry fruit; *ESI* positive; CID: 1.8 mTorr Ar, -15eV;
HPLC: RP-18 2 x 100 mm, 200 µl/min MeCN-H_2O-5 mM TFA.

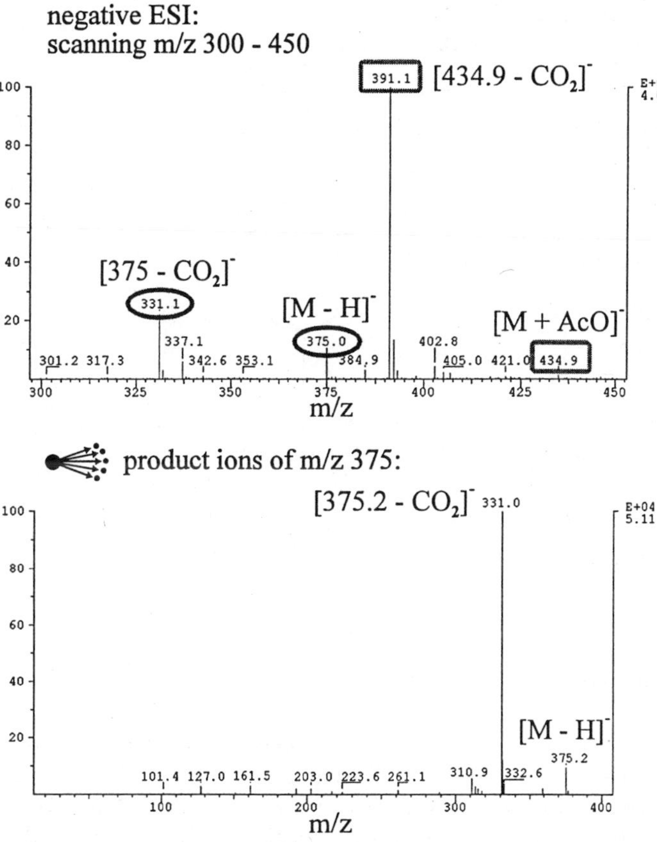

Figure 6. Structure elucidation of furaneol precursor by negative *ESI-MS/MS.* CID: 1.8 mTorr Ar, +12 eV; heated capillary 200°C; HPLC: RP-18 2 x 100 mm, 200 μl/min MeOH-H_2O-5 mM NH_4Ac.

spectrum showed presumably the well-known furaneol glucopyranoside as core structure to which a terminal malonate unit was attached. At last the structure of the major new furaneol precursor (retention time = 8:0 min in Fig. 5) in strawberry was identified as furaneol-(6-*O*-malonyl-β-*D*-glucopyranoside) by comparison with the synthesized reference glycoside, while the minor furaneol progenitor detected in Fig. 5 at 6:40 min appears to be a positional isomer (*14*). In addition, furaneol-(6-*O*-malonyl-β-*D*-glucopyranoside) was identified by *HPLC-MS/MS* amongst the glycoconjugates isolated from mango, pineapple and tomato fruits, thus demonstrating the relevance of this new class of polar flavor progenitors.

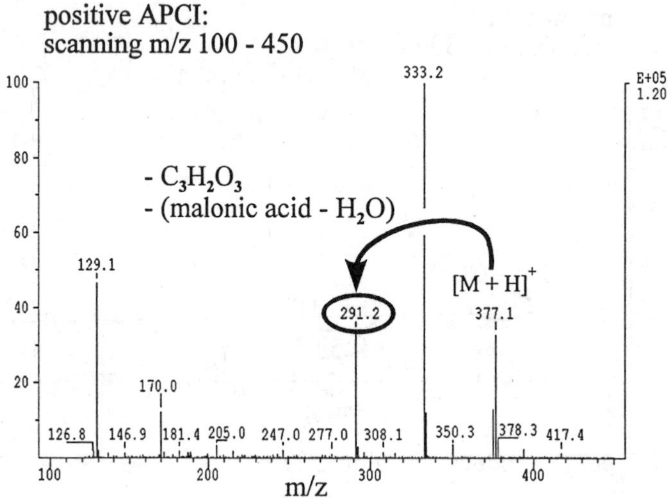

Figure 7. Positive *APCI* spectrum of furaneol precursor. Nebulizer 300°C; heated capillary 150°C.

Acknowledgment

The continuing efforts of B. Kirsch, M. Kleinschnitz and E. Richling within the HPLC-MS lab as well as the contributions of D. Krajewski are gratefully acknowledged. The presented work was granted by the EU (AIR3-CT94-2193) and the Fonds der Chemischen Industrie, Frankfurt. M. H. thanks the Deutsche Forschungsgemeinschaft DFG, Bonn, for generous support (He 2599/1-2).

Literature Cited

1. Niessen, W. M. A.; Tinke, A. P. *J. Chromatogr. A* **1995**, *703*, 37.
2. Dole, M.; Mack, L. L.; Hines, R. L.; Mobley, R. C.; Ferguson, L. D.; Alice, M. B. *J. Chem. Phys.* **1968**, *49*, 2240.
3. Iribarne, J. V.; Thompson, B. A. *J. Chem. Phys.* **1976**, *64*, 2287.
4. Thompson, B. A. J.; Iribarne, J. V. *J. Chem. Phys.* **1979**, *71*, 4451.
5. Yamashita, M.; Fenn, J. B. *J. Phys. Chem.* **1984**, *88*, 4451.
6. Whitehouse, C. M.; Dreyer, R. N.; Yamashita, M.; Fenn, J. B. *Anal. Chem.* **1985**, *57*, 675.
7. Fenn, J. B.; Mann, M.; Meng, C. K.; Wong, S. F.; Whitehouse, C. M. *Mass Spectrom. Rev.* **1990**, *9*, 37.
8. Kebarle, P.; Tang, L. *Anal. Chem.* **1993**, *65*, 972A.
9. Bruins, A. P.; Covey, T. R.; Henion, J. D. *Anal. Chem.* **1987**, *59*, 2642.

10. Ney, I.; Jäger, E.; Herderich, M.; Schreier, P.; Schwab, W. *Phytochem. Anal.* **1996**, in press.
11. Schneider, C.; Schreier, P.; Herderich, M. *Anal. Biochem.* **1996**, submitted for publication.
12. Parada, F.; Krajewski, D.; Herderich, M.; Duque, C.; Schreier, P. *Nat. Prod. Lett.* **1995**, *7*, 69.
13. Parada, F.; Krajewski, D.; Duque, C.; Jäger, E.; Herderich, M.; Schreier, P. *Phytochem.* **1996**, in press.
14. Roscher, R.; Herderich, M.; Steffen, J. P.; Schreier, P.; Schwab, W. *Phytochem.* **1996**, in press.
15. Reinhold, V. N.; Reinhold, B. B.; Costello, C. E. *Anal. Chem.* **1995**, *67*, 1772.
16. Günata, Y. Z.; Bayonove, C. L.; Baumes, R. L.; Cordonnier, R. E. *J. Chromatogr.* **1985**, *331*, 83.

Chapter 26

Application of New Microwave Reactors for Food and Flavor Research

Christopher R. Strauss and Robert W. Trainor

Division of Chemicals and Polymers, CSIRO, Private Bag 10, Rosebank MDC, Clayton, Victoria 3169, Australia

Continuous and batchwise laboratory-scale microwave reactors have been developed for controlled heating under pressure. Reactions can be conducted safely and conveniently for lengthy periods when necessary, and in the presence of volatile organic solvents if required. The advantages of the microwave technology include the capability for rapid heating and quenching; minimal temperature gradients within the sample; and elimination of wall effects. The processes of interest in food and flavor chemistry include hydrolysis of polysaccharides (*e.g.* cellulose) and esters, esterification and transesterification, rearrangement, acetalization, isomerization, and synthesis of aminoreductones. High-temperature water facilitated biomimetic processes, involving linalool, nerol, geraniol, and β-ionone. Indole-2-carboxylic acid and its ethyl ester underwent selective degradation in water or dilute aqueous base at temperatures of 200 °C or greater.

The potential of microwave technology for expediting routine laboratory chemistry, and large-scale chemical processing, prompted our entry into the field in 1988. We believed that if we could resolve technical problems, particularly those associated with the development of microwave equipment for heating organic reactions (*1*), greater efficiency and new methodologies could result. Therefore our aim was to produce microwave reactors which would be capable of reliable and safe operation with volatile organic solvents at elevated temperatures and pressures. Such dedicated systems have now been developed for continuous (*2*) and batchwise (*3*) operations in the laboratory. Their scope has been explored through hundreds of reactions which have been carried out under widely ranging conditions. This report briefly discusses some applications of the CSIRO reactors in food and flavor research.

0097–6156/96/0637–0272$15.00/0

DISCUSSION

The CSIRO Microwave Batch Reactor (MBR). The laboratory-scale MBR (*3*), is illustrated schematically below (Figure 1). With this system, reactions could be performed under carefully controlled heating, at temperatures up to 260 °C and/or pressures up to 10 MPa (100 atmospheres), and then rapidly cooled.

Figure 1. **Features of the MBR**

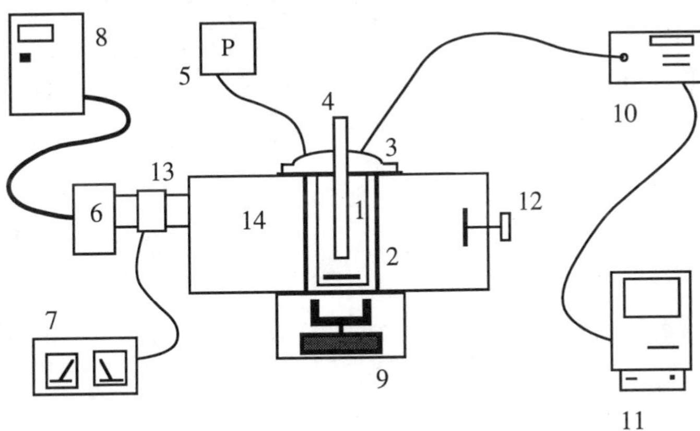

Legend: 1, *Reaction vessel;* 2, *retaining cylinder;* 3, *top flange;* 4, *cold-finger;* 5, *pressure meter;* 6, *magnetron;* 7, *microwave forward/reverse power meters;* 8, *variable power supply to magnetron;* 9, *stirrer unit;* 10, *optic fiber thermometer;* 11, *computer;* 12, *load matching device;* 13, *waveguide;* 14, *microwave cavity (applicator).*

Microwaves were directed into the plate steel chamber containing the microwave-transparent, but chemically inert reaction vessel, of 20-100 mL operating capacity. The vessel was retained by microwave-transparent fittings, and was equipped with a magnetic stirrer bar, an optic fiber thermometer, and cold-finger cooler. A pressure transducer, relief valve, and sample withdrawal tube/inert gas inlet were all attached to a stainless steel flange which formed a gas-tight seal with the vessel. Wettable surfaces were fabricated from polytetrafluoroethylene (PTFE), polyether ether ketone (PEEK) or quartz. Safety aspects have been discussed (*1,3*).

The CSIRO Continuous Microwave Reactor (CMR). The CMR (Figure 2) (*2*) comprised a microwave cavity fitted with a tubular reaction coil (of microwave-transparent, inert material), which was attached to a metering pump and pressure gauge at the inlet end and a heat exchanger and pressure regulating valve at the effluent end. The product stream could be rapidly cooled, under pressure, immediately after it exited the irradiation zone. Temperature was monitored immediately before and after cooling. Variables such as coil length, flow rate, and control of the applied microwave power, allowed flexible operation of the system. Feed-back microprocessor control allowed the operator to set pump rates and temperatures for heating and cooling of reactions. Changes made to the programmed reaction conditions while the unit was operating, would take effect immediately. With a reaction coil of perfluoroalkoxy

(PFA) Teflon (3 mm i.d. and 6 mm o.d.), the unit could be safely operated at temperatures up to 200 °C and pressures up to 1400 kPa.

Figure 2. **Features of the CMR**

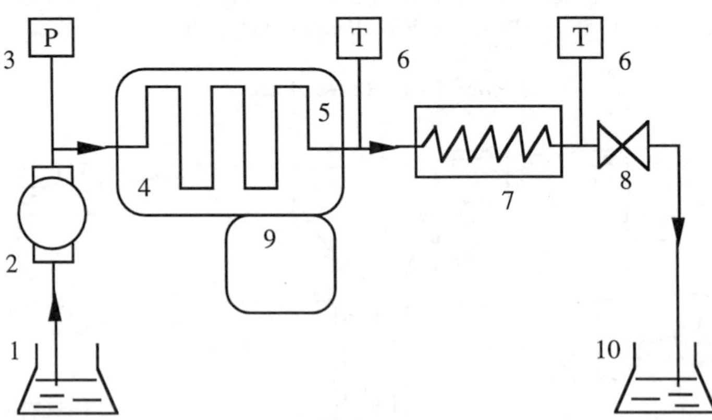

Legend: 1, *Reactants for processing;* 2, *metering pump;* 3, *pressure transducer;* 4, *microwave cavity;* 5, *reaction coil;* 6, *temperature sensor;* 7, *heat exchanger;* 8, *pressure regulator;* 9, *microprocessor controller;* 10, *product vessel.*

Advantages of Microwave Heating. Compounds with high dielectric constants (*e.g.* water and ethanol) tend to heat readily under microwave irradiation, while less polar substances (such as benzene or decane) or compounds with no nett dipole moment (*e.g.* carbon dioxide) and highly ordered materials (such as ice) are poorly absorbing. This circumstance arises because energy transfer is not *primarily* by conduction or convection, as with conventional heating, but by dielectric loss (*1*). The difference in mechanism of heating offers the following advantages for microwave-assisted chemistry with the MBR and CMR:

(i) Rapid response. The sample absorbs the energy directly, so the times required for heating up reaction mixtures are short, allowing selectivity in some reactions and enabling operators to efficiently manage their work.
(ii) Because the reaction mixture is heated directly, the irradiation chamber does not heat up substantially. When the power is turned off, heat input ceases immediately, an important safety consideration.
(iii) There are minimal temperature gradients across the reaction vessel (in the case of the CMR, the vessel is a tube), so the temperature of the material on the wall is not significantly different from that in the body of the reaction mixture. Pyrolysis on the inner wall of the vessel is thus minimized.
(iv) Reactions which are known to require high temperatures and consequently high boiling solvents can be carried out under pressure at these temperatures, but in lower boiling solvents, thereby facilitating work-up.
(v) Low boiling reactants can be heated to high temperatures under the applied pressure and then cooled before removal from the pressurized zone. Losses of volatiles are thus minimized.

(vi) Moderate to high temperature reactions can be carried out in a vessel (or tubing) which is fabricated from an inert material such as PFA Teflon, PTFE, or quartz. This is beneficial where reactants or products are incompatible with metals or borosilicate glass.
(vii) Sampling can be carried out whilst material is being processed. In both systems, the conditions can be readily changed during the run. With the CMR, if necessary, reaction mixtures can be subjected to multiple passes.
(viii) Both reactors allow unstable products to be quickly cooled at the completion of the reaction.
(ix) If solvents of differing polarities are used in a two-phase system, it is possible to differentially heat each phase, or to selectively heat one phase (*1,3*).
(x) The reactors often can be employed complementarily, with small-scale reactions optimized in the MBR, and then readily adapted for continuous processing.

Examples. Either or both microwave reactors have been useful for processes such as esterification, amidation, transesterification, rearrangement, acetalization, nucleophilic substitution, hydrolysis of esters and amides, isomerization, decarboxylation, oxidation, elimination, etherification, and formation of aminoreductones. Examples of such reactions have been tabulated (*2,3*).

Solvents. An array of organic solvents has been employed, including acetone, chloroform, methanol, ethyl acetate, ethanol, butanone, acetonitrile, *i*-propanol, toluene, pyridine, acetic acid, chlorobenzene, propionic acid, DMF, DMSO and hexamethylphosphorous triamide (HMPTA). In some cases, temperatures have been more than 150 °C higher than the boiling point of the solvent at atmospheric pressure, and the heating has been maintained within a narrow temperature range for minutes or hours (in the MBR) as required. By comparison with literature preparations carried out conventionally at lower temperatures under atmospheric pressure, reaction times have been reduced by up to three orders of magnitude. Some procedures which had required many hours at reflux, or days at ambient temperature, have been completed within minutes, and with high selectivity, under microwave heating (*1*).

REACTIONS IN HIGH-TEMPERATURE WATER

Interest is developing in the use of aqueous media as an alternative solvent system for organic reactions. Temperatures of 100 °C and below have been employed for synthesis (*4-7*), often to exploit the hydrophobic effect (*8*). Conversely, at temperatures approaching supercritical (T_C = 374 °C) and beyond, water has found degradative applications (*9*).

An objective of our work is to study biomimetic reactions in an aqueous environment at temperatures between 100 and 300 °C. Water has a dielectric constant which decreases from 78 at 25 °C, to 20 at 300 °C (*1,9*), this latter value being comparable with that of solvents such as acetone at ambient temperature. Thus water can behave as a pseudo-organic solvent, dissolving otherwise insoluble organic compounds at elevated temperatures. This property has allowed us to study the behavior of naturally occurring organic compounds in water directly, without the requirement that such molecules first be derivatized with characteristic water-solubilizing groups such as phosphate esters or glycosidic units. In addition, isolation of products is normally facilitated by the decrease in the solubility of the organic material, which occurs upon post-reaction cooling.

Significantly, the ionic product (dissociation constant) of water also is greatly influenced by temperature, increasing by three orders of magnitude from $10^{13.99}$ at 25 °C to $10^{11.30}$ at 300 °C (*9*). Water is therefore a much stronger acid and base at elevated temperatures than at ambient. These properties of water can be exploited with the microwave reactors, so biomimetic reactions that would normally be acid-catalyzed, can be carried out in the absence of added acidulant.

Treatment of Monoterpene Alcohols. In one such series of experiments, geraniol(1), nerol (2) and linalool (3) were each heated in water at 220 °C for 10 min in the MBR. These three terpenols are practically insoluble in water at ambient temperature (*10*). Additionally, although they are acid labile (*11-14*), they do not readily react in water at moderate temperature and neutral pH.

In unacidified water at 220 °C in the MBR however, the terpenols (1), (2) and (3) rapidly decomposed. The major products and their relative proportions (Table 1), were consistent with those previously reported by others for the carbocationic rearrangement chemistry of derivatives of linalool, nerol, and geraniol under acidic conditions (*13,15-16*).

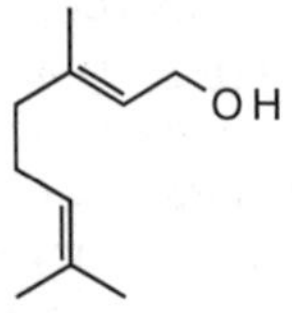

OH OH OH OH

(1) 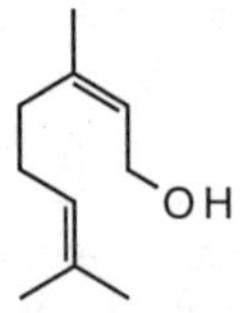 (2) 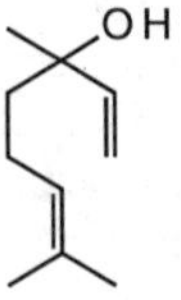 (3) 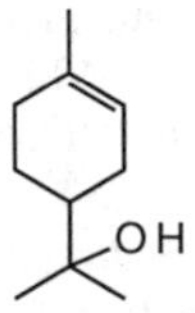 (4)

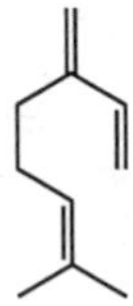

(5) 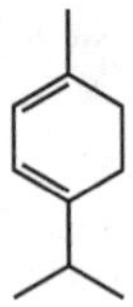 (6) 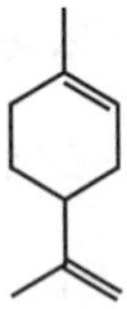 (7) 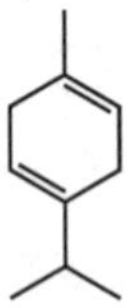 (8)

O

(9) 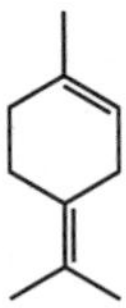 (10)  (11) 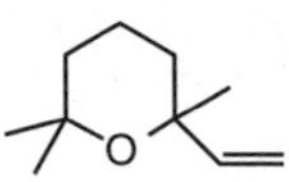 (12)

In the present work, geraniol (1) predominantly afforded the monoterpene alcohols α-terpineol (4) and linalool (3), along with lesser amounts of each of the hydrocarbons, myrcene (5), α-terpinene (6), limonene (7), γ-terpinene (8), the ocimenes (9), α-terpinolene (10) and alloocimenes (11). The cyclic oxide, 2,6,6-trimethyl-2-vinyltetrahydropyran (12) was also present as a trace component. Nerol (2) and linalool (3) both underwent considerably more elimination, and hence afforded substantially less of the isomeric monoterpene alcohols (of MW 154), and a greater proportion of the above array of hydrocarbons (of MW 136) than did geraniol (1).

The product distribution from the reaction of geraniol (1) was also similar to that which is known to result from acid-catalyzed degradation of naturally occurring linalyl and geranyl glycosides in fruit juices and wines (*17*).

***Table 1.* Products from Treatment of Alcohols (1), (2), (3) in Water at 220 °C**

		PRODUCT DISTRIBUTIONS (%)*		
COMPOUND	#	*Geraniol (1)*	*Nerol (2)*	*Linalool (3)*
2,2,6-trimethyl-2-vinyltetrahydropyran	(12)	3	1	–
Myrcene	(5)	6	6	3
α-Terpinene	(6)	3	12	10
Limonene	(7)	11	28	21
Ocimenes	(9)	9	6	7
γ-Terpinene	(8)	2	9	7
α-Terpinolene	(10)	10	24	35
Linalool	(3)	16	1	11
Alloocimenes	(11)	4	2	3
α-Terpineol	(4)	18	3	2
Nerol	(2)	3	2	–
Geraniol	(1)	9	–	–

*Product percentages from GCMS analysis. Products listed in order of increasing retention time.

Treatment of β-Ionone (13) in Water. Traditional methods for the dehydration of α- and β-ionones to give ionene (14) have involved catalysis by hydriodic acid along with small amounts of phosphorus, or distillative heating in the presence of 0.5% iodine (*18*). This latter procedure is now known to co-produce about 10% 1,1,6-trimethyl-1,2-dihydronaphthalene (*19*). However, we have found that the cyclization could be effected by merely heating β-ionone in water at 250 °C for twenty minutes in the MBR. The microwave-assisted reaction was not optimized, and proceeded with only moderate conversion. Unlike in the literature method (*18*), however, the work-up did not require exhaustive washing procedures.

(13) —Water/250°/20 min, *30% conversion*→ (14)

Applicability of High-Temperature Water. At first glance, an aqueous environment would not seem a useful medium for conducting dehydrations, such as those observed here for the terpenols and β-ionone. However, dehydration reactions in high-temperature water are not novel, with the conversion of cyclohexanol to cyclohexene having been recently investigated by Kuhlmann *et al.* (*20*). Acid catalyzed dehydrations of alcohols usually proceed by E1 mechanisms, *via* carbocation intermediates (*24-26*), so the product distributions obtained here from the monoterpenoids in high temperature water were not unexpected.

The results of the thermolyses demonstrate the potential of high-temperature water for the study of biomimetic transformations. Furthermore, they confirm that for certain reactions, elevated temperatures under neutral pH conditions can offer advantages over the use of more acidic (or basic) reagents at lower temperatures. Some microwave-assisted reactions proceed rapidly, in the presence of: (i) less catalyst, (ii) a milder catalyst or, (iii) no added catalyst, as found here.

Reactions under Aqueous Conditions in the Presence of Acid. The rapid heating and cooling capabilities of the microwave reactors have also allowed us to conduct high-temperature microwave-assisted reactions in acidic or basic media, to obtain thermally labile products. An example is the acid-catalyzed depolymerization of polysaccharides such as cellulose (15): the generated monosaccharides and oligosaccharides are susceptible to degradation under the conditions of high temperatures in aqueous acid that were conventionally required to hydrolyze the glycosidic bonds at an acceptable rate. Although several sets of satisfactory conditions were obtained with the MBR, the most successful method involved heating a suspension of cellulose in 1% aqueous sulfuric acid, from ambient to 215 °C within 2 minutes, maintaining this temperature briefly, and then rapidly cooling. The entire operation was completed within 4 minutes and glucose (16) was obtained in good yield (*3,27*).

(15) —— 1% aq. H_2SO_4, MBR/215°/30 sec ——> (16) 39%

Reactions under Aqueous Conditions in the Presence of Base. Many indole derivatives are potent odorants, and others possess highly significant, diverse biological properties which directly influence the metabolism of plants and animals (*28-29*). Accordingly, reactions involving the indole nucleus are of importance to food and flavor chemistry.

In this context, the degradation of 2-carboethoxyindole (17) and indole 2-carboxylic acid (18) were investigated. The study confirmed that highly selective reactions could be effected with microwave heating in the MBR. For example, ester (17) was readily hydrolyzed with 0.2 M NaOH, at 200 °C within 10 minutes to afford indole-2-carboxylic acid (18). However, at 255 °C in 0.2 M NaOH for 20 minutes, ester (17) gave indole (19) in excellent yield.

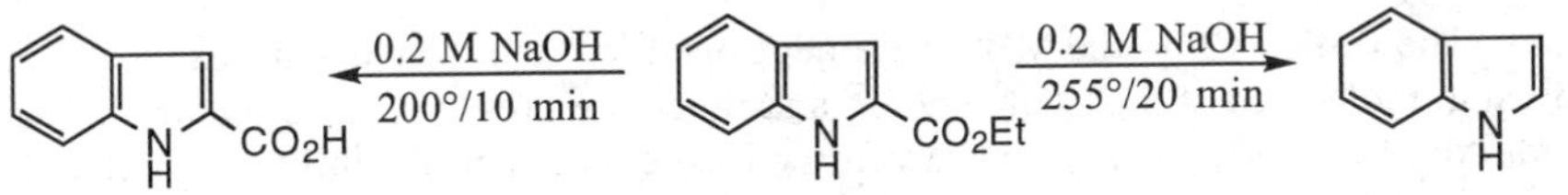

(18) $\xleftarrow[200°/10\ min]{0.2\ M\ NaOH}$ (17) $\xrightarrow[255°/20\ min]{0.2\ M\ NaOH}$ (19)

94% 93%

In neutral water, at 255 °C and 4.2 MPa (*c.* 42 atmospheres), decarboxylation of indole-2-carboxylic acid (18) was quantitative within 20 minutes, but the corresponding ester (17) underwent only about 20% conversion to indole (19) under these conditions. The presence of only traces of the acid (18) in the product mixture indicated that the hydrolytic step was rate-limiting.

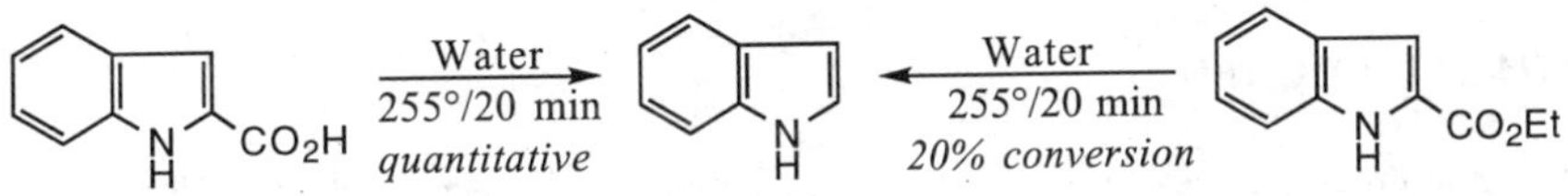

(18) $\xrightarrow[255°/20\ min]{Water}$ *quantitative* (19) $\xleftarrow[255°/20\ min]{Water}$ *20% conversion* (17)

Ease of Optimization. In high-temperature water, the thermal range available for synthetic reactions and controlled degradation studies can extend to about 300 °C. For such reactions, optimal conditions may be found to occur only over a narrow range of time and temperature. Yet in other instances, several sets of optima can be established for a given reaction, if longer times are accepted in return for a lower operating temperature or the use of less catalyst (*1*).

WORK BY OTHERS

Although the preceding sections have focussed predominantly upon the work of our group, others have recently entered this field and made significant contributions. Among them, Tong *et al.* (*30-31*) produced continuous and batchwise microwave reactors for the study of food chemistry, and applied these to thiamin degradation, and Maillard reactions. The latter reactions have also been the subject of separate investigations by Parliment (*32*), Shibamoto and Yeo (*33-35*) as well as Hidalgo and Zamora (*36-37*). Studies on isomerization of eugenol (*38*), syntheses of dihydroactinidiolide (*39*), isoflavones (*40*), and the catalytic transfer hydrogenation of soybean oil (*41*), highlight the diversity of applications for microwave technology in food and flavor chemistry.

CONCLUSION

The directness of microwave heating offers several advantages in flavor research. The reaction vessels are usually fabricated from microwave-transparent inert materials such as PTFE, thereby benefiting product stability and isolation. With microwave heating, thermal gradients within the sample can be minimized, and pyrolytic wall effects are eliminated. The heating and subsequent cooling of samples are rapid, so thermally labile products can be readily isolated.

The CSIRO CMR and MBR were developed for safe, convenient reactions with organic molecules at high temperature and moderate pressure, under controlled and monitored operation. They have afforded excellent reproducibility, essential for process optimization and kinetics studies (*42-43*). So, despite earlier misgivings by others (*44*), it has now been demonstrated that microwave-assisted organic reactions can be conducted advantageously without incident, at elevated temperatures, and even for lengthy periods with volatile solvents. Without the development of such reactors, microwave-assisted organic chemistry would have been largely restricted to niche applications involving either open-vessel reactions, or small-scale operations in closed systems (*1*). Since this technology is new there have been few opportunities so far for its exploitation in the food, flavor and fragrance areas. Hence it offers considerable promise for highly novel research and development activities.

LITERATURE CITED

1. For a recent review, see Strauss, C. R.; Trainor, R. W. *Aust. J. Chem.* **1995**, *48*, 1665 and references therein.
2. Cablewski, T.; Faux, A. F.; Strauss, C. R. *J. Org. Chem.* **1994**, *59*, 3408.
3. Raner, K. D.; Strauss, C. R.; Trainor, R. W.; Thorn, J. S. *J. Org. Chem.* **1995**, *60*, 2456.
4. Breslow, R. *Acc. Chem. Res.* **1991**, *24*, 159.
5. Grieco, P. A. *Aldrichimica Acta* **1991**, *24*, 59.
6. Lubineau, A.; Auge, J.; Queneau, Y. *Synthesis* **1994**, 741.
7. Li, C.-J. *Chem. Rev.* **1993**, *93*, 2023.
8. Blokzijl, W.; Engberts, J. B. F. N. *Angew. Chem. Int. Ed. Engl.* **1993**, *32*, 1545.
9. Siskin, M.; Katritzky, A. R. *Science* **1991**, 231.
10. *The Merck Index;* Budavari, S; Ed.; Merck & Co.: Rahway, N. J., 1989; Eleventh Edition.
11. Baxter, R. L.; Laurie, W. A.; McHale, D. *Tetrahedron* **1978**, *34*, 2195.
12. Pickett, J. A.; Coates, J.; Sharpe, F. R. *Chem. & Ind.* **1975**, 571.
13. Stevens, K. L.; Jurd, L.; Manners, G. *Tetrahedron* **1972**, *28*, 1939.
14. Valenzuela, P.; Cori, O. *Tetrahedron Lett.* **1967**, *32*, 3089.
15. Bunton, C. A.; Cori, O.; Hachey, D.; Leresche, J.-P. *J. Org. Chem.* **1979**, *44*, 3238.
16. Cramer, F.; Rittersdorf, W. *Tetrahedron*, **1967**, *23*, 3015
17. Williams, P. J.; Strauss, C. R.; Wilson, B.; Massy-Westropp, R. A. *J. Agric. Food Chem.* **1982**, *30*, 1219.
18. Bogert, M. T.; Fourman, V. G. *J. Am. Chem. Soc.* **1933**, *55*, 4670.
19. Strauss, C. R.; Williams, P. J. *J. Inst. Brew.* **1978**, *84*, 148 and references therein.
20. Kuhlmann, B.; Arnett, E. M.; Siskin, M. *J. Org. Chem.* **1994**, *59*, 3098.
24. Saunders, W. H. Jr. *The Chemistry of Alkenes*, Patai S., Ed.; Interscience: New York, 1964; Ch. 2, p 149.
25. Cram, D. J. *J. Am. Chem. Soc.* **1952**, *74*, 2137.
26. Manassen, J.; Klein, F. S. *J. Chem. Soc.* **1960**, 4203.
27. Radojevic, I.; Trainor, R. W.; Strauss, C. R.; Kyle, W. S. A.; Britz, M. L. Presented at the 11th Australian Biotechnology Conference, Perth, Western Australia, September, 1993.

28. Hughes, D. L. *Org. Prep. Proced. Int.* **1993**, *25*, 609.
29. Hugel, H. M.; Kennaway, D. J. *Org. Prep. Proced. Int.* **1995**, *27*, 3.
30. Welt, B. A.; Steet, J. A.; Tong, C. H.; Rossen, J. L.; Lund, D. B. *Biotechnol. Prog.* **1993**, *9*, 481.
31. Peterson, B. I.; Tong, C. H.; Ho, C. T.; Welt, B. A. *J. Agric. Food Chem.* **1994**, *42*, 1884.
32. Parliment, T. H. In *Food Flavors, Ingredients and Composition* Charalambous, G. Ed.; Elsevier: Amsterdam, 1993, p. 657.
33. Yeo, H. C. H.; Shibamoto, T. *Trends in Food Sci. Technol.* **1991**, 329
34. Yeo, H. C. H.; Shibamoto, T. *J. Agric. Food Chem.* **1991**, *39*, 948.
35. Yeo, H. C. H.; Shibamoto, T. *J. Agric. Food Chem.* **1991**, *39*, 370.
36. Hidalgo, F. J.; Zamora, R. *J. Agric. Food Chem.* **1995**, *43*, 1023.
37. Zamora, R.; Hidalgo, F. J. *J. Agric. Food Chem.* **1995**, *43*, 1029.
38. Loupy, A.; Thach, L. N. *Synthetic Commun.* **1993**, *23*, 2571.
39. Subbaraju, G. V.; Manhas, M. S.; Bose, A. K. *Tetrahedron Lett.* **1991**, *32*, 4871.
40. Chang, Y. C.; Nair, M. G.; Santell, R. C.; Helferich, W. G. *J. Agric. Food Chem.* **1994**, *42*, 1869.
41. Leskovsek, S.; Smidovnik, A.; Koloini, T. *J. Org. Chem.* **1994**, *59*, 7433.
42. Raner, K. D.; Strauss C. R. *J. Org. Chem.* **1992**, *57*, 6231.
43. Raner, K. D.; Strauss, C. R.; Vyskoc, F.; Mokbel, L. *J. Org. Chem.* **1993**, *58*, 950.
44. Gutierrez, E.; Loupy, A.; Bram, G.; Ruiz-Hitzky, E. *Tetrahedron Lett.* **1989**, *30*, 945.

Chapter 27

Characterization of Citrus Aroma Quality by Odor Threshold Values

H. Tamura, Y. Fukuda, and A. Padrayuttawat

Department of Bioresource Science, Kagawa University, Miki-cho, Kagawa, 761–07 Japan

Limited odor unit (Lo) can be used to characterize aroma quality and can be expressed by an odor unit like equation. The Lo value is an index of the significance of each aroma component at the concentration of volatile oils (ppm) diluted below the threshold level. The aroma intensity (Lo) of each aroma component is expressed by dividing the concentration of individual components at the threshold level of the volatile oils (Cr), by its detection threshold or recognition threshold (Tc). For this equation, we examined which was the most suitable, the detection threshold or the recognition threshold as the Tc value. This olfactory measurement is proposed as an objective evaluation method for food aroma quality. It was found that limonene contributes to the characteristic aroma as the base note. Limonene, linalool, and octanal were shown to be essential for the characteristic aroma. The usefulness of Lo values for the characterization of the aroma of *Citrus sinensis* OSBECK, cv. Shiroyanagi, was shown by the fact that a mixture of 11 compounds selected by this method (using the detection threshold as Tc) - limonene, linalool, octanal, decanal, dodecanal, geranial, neral, myrcene, α- and β-sinensal, and citronellal -duplicated the aroma of navel oranges very closely. Thus, Lo values using detection thresholds were found to be useful for sensory evaluation of the character impact compounds in navel oranges.

Characteristic aroma components in foods and off-flavor substances in processed foods are called character impact compounds. It would be desirable to develop methods for finding such compounds with sensory methods because such information is useful in the food industry. A compilation of odor and taste threshold values was edited by Fazzalari (*1*). Olfactory-trigeminal response to odorants was measured using rabbits (2). However, aroma quality can be evaluated only by human sense. In spite of this fact, olfactory judgment by humans can not give constant data like

0097–6156/96/0637–0282$15.00/0

instrumental results for aroma quality. So, it is very difficult to obtain reliable, reproducible aroma quality and intensity data constant from olfaction experiments. Detection of food aroma is a complex reaction to many parameters. An expert perfumer can evaluate the aroma quality of a complex essential oil, but the evaluation is unique to that expert perfumer. The meaning of words used to describe a certain aroma quality vary considerably with individuals depending on their past experiences with such aromas. Words used by an individual for describing certain aromas are ambiguous and are not suitable for qualitative and quantitative descriptions.

Expert flavorists and perfumers have identified subjectively some single compounds as character impact compounds for some foods. To reduce the ambiguity of descriptions, Acree et al. (*3*) designed a gas chromatography (GC)-sniff test. This method was developed to measure human responses and to find character impact compounds. Aroma Extraction Dilution Analysis (*4*) and Charm Analysis (*5*) can be used to determine dilution values of each effluent. We have demonstrated the value of a GC-sniff test using a flavor wheel to determine the significant compounds in *Ulva pertusa* and citrus oils (*6, 7*). In the GC-sniff test, intensity and quality of each aroma component were more easily described by a simple word which was selected from a flavor wheel. The characteristic compounds in the oxygenated fraction of the navel orange was reported by Yang et al. (*7*) to be composed of nine compounds: octanal, nonanal, decanal, dodecanal, linalool, neral, geranial, and α- and β-sinensal.

The "odor-unit" value, which was proposed by Guadagni et al. (*8*) and Rothe and Thomas (*9*), was introduced to evaluate the importance of individual aroma constituents in food volatiles. This value is defined in the following equation [1]:

$$\text{Odor Unit (Uo)} = \frac{\text{Fc}}{\text{Tc}} \qquad [1]$$

where: Fc is the concentration of the volatiles in food (ppm) and Tc is the concentration of the individual components (ppm).

Compounds having greater than 1 odor unit value are assumed to contribute to the chararacteristic aroma in the food. Compounds having less than 1 odor unit value are assumed not to contribute to the characteristic aroma or are less important. Guadagni et al. and Tamura et al. applied this method to the volatiles in apple essence (*8*) and in *Ulva pertusa* (*10*), respectively.

Recently Buttery et al. (*11*) showed the importance of logarithmic values of odor units (log [Uo]) in the evaluation of tomato flavor. The logarithmic factor is correlated to Weber-Fechner's law (*12*) in regard to odor intensity. Weber-Fechner's law is defined in the following equation:

$$S = k \log [I] \qquad [2]$$

where: S is the intensity judgement of the human sense
k is a constant
I is the intensity of the odor stimulus

This equation suggests that the human olfaction sense responds linearly to the logarithmic value of the intensity of the odor stimulus. Sugisawa et al. (*13*, *14*) applied this equation to the evaluation of the aroma of navel oranges and acid citrus fruits. In both cases, compounds with large positive values of log [Uo] proved to be important in contributing to the characteristic sensory properties. Multivariate statistical analysis using logarithmic values of odor units of each component identified in 6 acid citrus fruits was found to be effective for characterizing the difference of mutual aroma quality of each citrus (*15*).

We have applied a modified odor unit equation for evaluating aroma quality of the volatiles of *Citrus sinensis* OSBECK, ev. Shiroyanagi. Although the concept of odor units in flavor research was proposed by Rothe et al. (*9*) as a objective index of aroma quality, the concentration of individual components in a food (Fc in equation [1]) depends on the extraction efficiency of the essential oils. If the test sample is a solid, we can not calculate the exact concentration. Because the aroma oils, for example, may exist in different cells in the peels of citrus, we cannot take out only specified cells. It does not give a homogeneous concentration. Therefore, the odor units of individual aroma components in a food do not always give a constant value. Equation [1] should be applied to beverages such as apple juice, citrus juice, coffee, milk and so forth. The modified odor unit equation (*15*) for liquid and solid samples is shown as follows:

$$\text{limited odor unit (Lo)} = \frac{Cr}{Tc} \qquad [3]$$

where: Cr is the concentration of the individual components at the recognition threshold of volatile oils (ppm), and
Tc is the threshold of the individual components (ppm).
Tcr: recognition threshold, Tcd: detection threshold

Tamura et al. proposed this equation and then applied it to the volatile oils in *Citrus lemon* and *Citrus sudachi* (*15*).

Because the aroma quality of the peel oils of *Citrus sinensis* has been investigated (*7*, *13*) and the important aroma compounds have been determined, we chose to use navel oranges to test the applicability of the new equation in detail.

Isolation of volatile oils from *Citrus sinensis* OSBECK, ev. Shiroyanagi

Reagents. Guaranteed chemicals were purchased from Wako-pure chemicals in Japan and Aldrich Chemical Co. in the U.S. The purity of all chemicals was checked by GC before measuring odor threshold values. Purity of all chemicals used in this experiment was more than 99.2%.

Material. Shiroyanagi navel oranges (*Citrus sinensis* OSBECK, cv. Shiroyanagi) is a cultivated variety of navel oranges, grown in the Ehime Prefecture, Japan. *Citrus iyo* is also grown in the Ehime Prefecture.

Isolation and separation of volatile oils. The peels (605 g) were chopped into small pieces. The chopped pieces were mixed with de-ionized water (2.2 L) and homogenized at 0°C for 15 minutes. Volatiles were extracted from the homogenized mixture (300 g aliquot) with two portions (30 mL) of ethyl ether by simultaneous distillation extraction (SDE) for 60 minutes. The extracts were combined, and the solvent was removed by evaporation under reduced pressure and under nitrogen atmosphere for 5 hours. The amount of volatile oils was 5.7 g (0.94% yield). The essential oil (4.0 g) was fractioned by silica gel column chromatography (30 cm X 3 cm id) with pentane (1000 mL) and diethyl ether (1000 mL) as eluants. The solvents were partially removed by evaporation. Each fraction was concentrated at room temperature under purified nitrogen. The yields were: pentane fraction, 98.18%, and ether fraction, 1.82%.

Citrus iyo peels (600 g) were treated similarly as above. The amount of volatile oils was 5.2 g (0.87% yield). The essential oil (4.0 g) was separated into a hydrocarbon fraction (3.4 g) and oxygenated fraction (0.24 g) by silica gel column chromatography.

Identification. The components in the fractionated oils mentioned above were analyzed by GC and GC/MS resulting in the identification of 12 hydrocarbon terpenes and 72 oxygenated compounds. Gas chromatographic conditions were: OV-101 column, 50 m X 0.25 mm i.d., 80°C to 200°C at 2 °C/min. GC/MS was conducted with a JEOL JMS-SX102AQQ (EI mode, 70 eV) equipped with a JEOL MP-7010 data processor unit. The analytical conditions were almost the same as the GC conditions: OV-101, 50 m X 0.25 mm i.d.; temperature program, 80-200°C at 2 °C/min; He, 0.64 mL/min; split injection (1:50). NIST public data library of MS spectra (data number: 49,496) which was supplied from JEOL LTD was used for matching of mass spectral pattern.

Concentration of individual components. Concentration of individual components was determined by using the peak area of the FID detector. Peak area % of individual components integrated by Shimadzu C-R6A are listed in Table I. The individual concentrations in the oils were calculated on the basis of the oil weight and peak area % of each component.

Detection Thresholds of Aroma Components in *Citrus sinensis*

The odor threshold (Tc) was classified by the nomenclatures listed in Table II. The odor-detection threshold (Tcd) of each component was determined by the same method previously described (*15*). First, volatile compounds were dissolved in a small amount of methanol. The solution was diluted by deionized water until the solution was judged as odorless. Because the detection threshold of methanol is very high (more than 100 ppm), it does not interfere with determinations of the thresholds of the aroma components in *Citrus sinensis* OSBECK, cv Shiroyanagi. Each aqueous solution of volatiles was diluted by a factor of ten. Weber-Fechner's law predicts that the human nose can clearly distinguish only ten fold differences of concentration. We made three series (A, B, C) of diluted solutions to test detection thresholds (A: 90 ppm, 9 ppm, 0.9 ppm, and 0.09 ppm. B: 60 ppm, 6 ppm, 0.6 ppm and 0.06 ppm, C: 30 ppm, 3 ppm, 0.3 ppm and 0.03 ppm). Independently, panel members were

Table I. Composition of the Essential Oil in Navel Oranges

Compounds	Content (X 10^2 %)
limonene	9260.6
myrcene	175.21
sabinene	74.89
linalool	49.99
α-pinene	46.50
octanal	28.40
decanal	23.65
valencene	18.83
terpinen-4-ol	18.32
farnesol	18.23
Δ^3-carene	18.14
γ-terpinene	16.66
geranial	15.72
α-terpineol	14.84
neral	13.12
α- and β-sinensal	7.71
n-butyl n-butyrate	5.39
dodecanal	3.95
1-dodecanol	3.95
citronellal	3.57
nonanal	2.28
perillaldehyde	1.75
tetradecanal	0.90

Table II. Nomenclatures of the Odor Thresholds

Test sample	Threshold	
	Recognition	Detection
Essential Oil	T_{or}	T_{od}
Chemicals	T_{cr}	T_{cd}

subjected to the three sets of samples. Two cups out of five contained aqueous solutions of the volatile compounds; the other three contained pure water. Six panel members were asked to determine which two cups contained the volatile compounds. The panel consisted of 5 males and 1 female who were students and I in my laboratory. They had experience at training detection thresholds of chemicals for more than 3 months. Their ages ranged from 21 to 36 years with a mean of 24.5 years. Threshold values determined by the 6 panel members were converted to logarithmic values. The mean values of the logarithmic values does not mean the threshold values. Therefore, the mean values were converted back to ordinal numbers. The same panel members contributed to all samples tested. Tcd values of the volatile components of navel oranges are listed in Table III.

Research Progress for Determining Recognition Thresholds. Leonardos et al. (*16*) reported odor recognition thresholds (Tcr) of 53 chemicals. In this report, odorous chemicals (0.1 mL) were injected into the test room (13,200 liters) and Tcr was determined to be the concentration at which all four panel members recognized the odors. Hellman et al. (*17*) also determined odor thresholds of 101 petrochemicals using sensory methods. They defined Tcr to be the concentration at which 100% of the panel members recognized the odor being representative of the odorant being studied.

Odor thresholds of some organic compounds in foods and beverages were reported by Guadagni et al. (*18*). The odor thresholds were measured in aqueous solutions. They applied this method for measuring odor value, which is expressed as the odor concentration divided by the odor threshold.

Recognition Thresholds of Volatile Components of *Citrus sinensis*

Another Tc value was determined using recognition thresholds rather than detection thresholds. Recognition thresholds (Tcr) were measured by sensorially recognizing two kinds of odor differences, the concentration and the odor quality, of individual chemicals presented to panel members.

Two kinds of aroma chemicals were presented to panel members, the test sample and the standard sample. Both were diluted to the concentration of detection threshold levels. The diluted test samples were presented to at least 15 panel members (usually 15-21) who had experience at testing detection thresholds of chemicals for more than 3 months. The panel consisted of 9-13 males and 6-8 females. Their ages ranged from 20 to 36 years with a mean of 23 years. Concentration difference of each series was ten-fold. Four cups of test sample solutions having different concentrations were presented with three cups of standard solution at different concentrations and one cup of water. Three series of diluted test samples were presented to the panel members as shown in Fig. 1. Each glass (68 mm high, 3.0 mm i.d., volume approximately 45 mL) contained 20 mL of solution. The panel members had to select and arrange both series of the 8 samples in order from the highest concentration to the lowest. The minimum concentration which could be arranged in the correct order to the logarithmic value. The means of logarithmic values were converted back to ordinal numbers as the recognition threshold (Tcr). As a standard sample, the adaptability of octanal, limonene, and linalool was

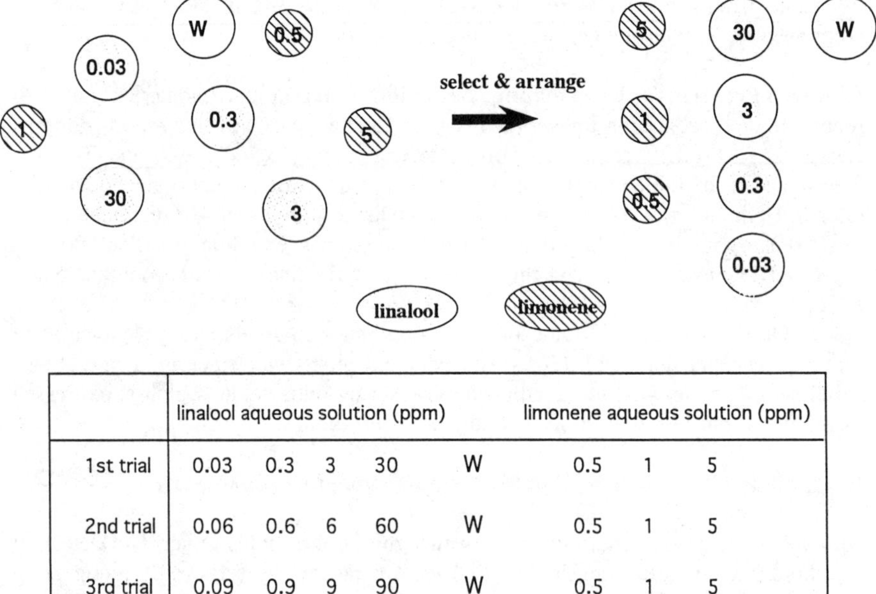

	linalool aqueous solution (ppm)					limonene aqueous solution (ppm)		
1st trial	0.03	0.3	3	30	W	0.5	1	5
2nd trial	0.06	0.6	6	60	W	0.5	1	5
3rd trial	0.09	0.9	9	90	W	0.5	1	5

Panel members have to select and arrange both series of the 8 samples in order from higher concentration to lower. W: Water

Figure 1. Method for determining the Tcr value of linalool using limonene as the standard.

examined by the t-test. From the result of 15 trials, there was no statistically significant difference among these three chemicals as the standard sample. Any chemical can be used as a standard sample. Because limonene contributes to a base note of citrus and because it is a main constituent (more than 90% of the oil) of the oil, we chose limonene as the standard. The recognition threshold values of the compounds found in the volatile oils of *Citrus sinensis* OSBECK, cv. Shiroyanagi, were determined using limonene as the standard for sensory evaluations. Tcr values of the individual components in the volatile oils are listed in Table III with detection threshold values.

As expected, Tcr values of all compounds were higher than the Tcd values. The difference between recognition and detection threshold values was not constant with respect to individual compounds. Large differences were observed with linalool (21-fold), citronellol (22.9-fold), eugenol (35-fold), nerol (49-fold), 1-hexanol (47-fold), nerolidol (21-fold), and (E)-2-hexenol (24-fold). Detection and recognition threshold values were similar with myrcene (1.37-fold), (E)-2-hexenal (1.33-fold), α-pinene (1.0-fold), β-pinene (1.77-fold), and 1-octanol (1.64-fold).

Recognition Thresholds of Volatile Oils of *Citrus sinensis* from That of *Citrus Iyo*

To determine the Lo values, the Cr values and Tc values were determined by the method described above. The Cr values are the concentration of the individual components not at the detection threshold of the volatile oils (Tod), but at the recognition threshold (Tor) because at the Tod level of the oil, only a few compounds can be sensorially detected and these contribute to the aroma only as the base notes.

The recognition threshold value of a complex mixture (Oils), generally speaking, probably varies with the standard sample. When limonene was chosen as the standard sample, the recognition threshold value of the essential oils of *Citrus sinensis* was 5.26 ppm. The odor quality under the recognition threshold concentration was recognized as an orange-like aroma but could not be specified as navel orange aroma. *Citrus iyo*, which is one of the sweet oranges produced only in Matsuyama Prefecture in Japan, was selected as the standard sample for recognition threshold studies because its odor is similar to that of the navel orange. The characteristics of *Citrus iyo* aroma was reported by Yamanishi et al. (*19*). When *Citrus iyo* was chosen as the standard sample, the navel orange was distinguished from the *Citrus iyo* aroma at the 8.32 ppm level.

Evaluation of Lo Values Using Detection Thresholds and Recognition Thresholds

To determine the Lo value, the Tc values are also needed. Using the two types of thresholds, recognition (Tcr) and detection thresholds (Tcd), the Tc value was measured. Both Tc values were compared to each other for several chemical compounds. The evaluation of Lo values using recognition threshold and detection threshold values are shown in Table IV. The only compound having an Lo value more than one (Lo>1) in both cases was limonene. Even though limonene is essential for the characteristic arom aof navel oranges, it does not evoke the characteristic

Table III. Odor Threshold Value of Volatiles in Navel Oranges[a]

compounds	Threshold (ppm)	
	T_{cr}	T_{cd}
limonene	3.500	1.000
linalool	1.082	0.050
octanal	0.248	0.030
decanal	0.245	0.036
citronellal	0.100	0.006
geranial	0.410	0.037
dodecanal	0.063	0.014
myrcene	0.915	0.670
α- and β-sinensal	0.310	0.040
neral	0.888	0.100
nonanal	0.260	0.040
terpinen-4-ol	4.370	0.590
perillaldehyde	0.290	0.056
1-dodecanol	2.470	0.158
Δ^3-carene	2.030	0.770
α-pinene	2.082	2.080
tetradecanal		0.053
γ-terpinene	2.890	1.000
n-butyl n-butyrate	1.089	0.400

a: these are a part of volatile components in navel oranges

Table IV. Limited Odor Unit Value of Volatiles in Navel Oranges[a]

Compounds	L_o value	
	using T_{cr} value	using T_{cd} value
limonene	2.201	7.705
linalool	0.043	0.885
octanal	0.095	0.788 (1.67)
decanal	0.080 (0.22)[b]	0.547
citronellal	0.030	0.496
geranial	0.032	0.353 (3.07)
dodecanal	0.052	0.218
myrcene	0.159 (0.49)	0.218
α- and β-sinensal	0.021	0.160
neral	0.012 (0.52)	0.109 (3.77)
nonanal	0.007	0.048
terpinen-4-ol	0.004	0.026
perillaldehyde	0.005	0.026
1-dodecanol	0.001	0.021
Δ^3-carene	0.007	0.020
α-pinene	0.019	0.014
tetradecanal		0.015
γ-terpinene	0.005	0.014
n-butyl n-butyrate	0.004	0.011

a: these are a part of volatile components in navel oranges.
b: Parentheses are the accumulated number of Lo values except that of limonene.

Table V. Comparison of Uo Value and Lo Value of Volatiles in Navel Oranges[a]

compounds	U_o	L_o[b]
limonene	8689.2	7.705
linalool	938.1	0.885
octanal	888.3	0.788
decanal	616.4	0.547
citronellal	558.8	0.496
geranial	398.5	0.353
dodecanal	265.0	0.218
myrcene	245.4	0.218
α- and β-sinensal	180.8	0.160
neral	123.1	0.109
nonanal	53.5	0.048
carvacrol	43.9	0.007
perillaldehyde	29.3	0.026
terpinen-4-ol	29.1	0.026
nerol	25.2	0.022
citronellol	25.0	0.022
1-dodecanol	23.5	0.021
Δ^3-carene	22.1	0.020
γ-terpinene	15.6	0.014
α-pinene	15.6	0.014
n-butyl n-butyrate	12.6	0.011

a:these are a part of volatile components in navel oranges.
b:these data are based on T_{cd} value.

navel aroma by itself. As Guadagni et al. (*20*) had found an additive effect of sub-threshold concentrations, we considered the additive effect of several compounds having Lo values less than 1. When recognition thresholds are used for Tc values, even by accumulating additive effects of all compounds, without limonene (Lo=2.2), a meaningful Lo value (Lo>1) cannot be attained. However, when detection thresholds are used for Tc values, the additive effect of 10 compounds without limonene (Lo= 7.7) became a meaningful Lo value (Lo>3.5). Table V compares data of Uo and Lo values. There are many compounds which have Uo values more than 1. Many compounds must be considered to be important compounds by the definition of Uo. It seems that the large Uo values of many chemicals may be due to the overestimation of Fc values in equation [1] because of the concentration in a heterogeneous solution (or in a solid food).

To determine how many chemicals are necessary for producing the aroma of navel oranges, 11 compounds having large Lo were selected from Table IV. These 11 compounds selected by this method are similar to the compounds previously determined as the characteristic compounds (*10*). Eleven compounds were divided into 4 groups in order of the size of the Lo values shown in Fig. 2. Group 1 consisted of limonene. Group 2 was composed of linalool and octanal. Group 3 was

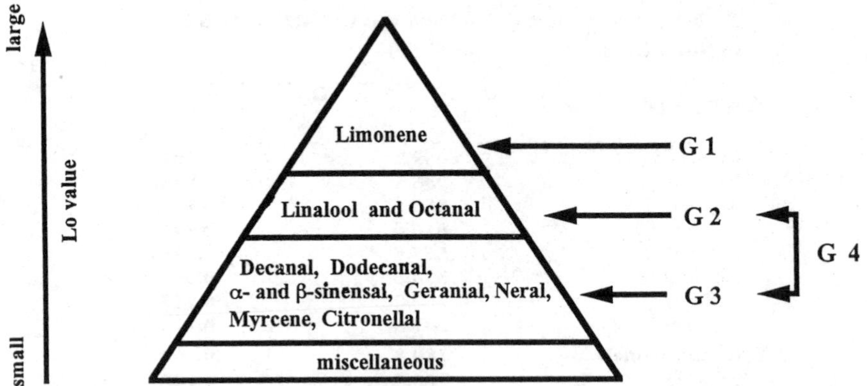

G 1: limonene, G 2: linalool and Octanal, G 3: Decanal, Dodecanal, Geranial, Neral, Myrcene, α- and β-Sinensal and Citronellal, G 4: G 2 plus G 3

Figure 2. Sensory evaluation by 11 compounds selected by L_{od} value.

composed of decanal, geranial, neral, dodecanal, myrcene, α-, and β- sinsensal. Group 4 was composed of compounds belonging to Group 2 plus Group 3. The amount of component was based on the concentration found in the peel oils of *Citrus sinensis* (Table I). The essential oil and the mixed oils of the 6 groups prepared were diluted to 20 ppm in order to sensorially evaluate as shown in Figures 3 and 4. The panel consisted of 11 males and 7 females. Their ages ranged from 20 to 24 years with a mean of 22 years. α- and β-Sinensals were used as the mixtures. The isomeric ratio of α- and β-sinensal mixtures was almost one to one. In Fig. 3, the aroma quality of the six mixtures of the groups with essential oil of navel orange were judged whether each mixture was similar to navel orange or not without a reference sample. The panel members selected the solutions of group 1 plus group 2 and group 1 plus group 4 as the odor of navel orange. When the essential oil of navel orange (Fig. 4) was presented as a standard to the panel members, the solutions of group 1 plus group 2 and group 1 plus group 4 were judged by half the members of the panel to be like navel orange oil. These results might indicate that limonene contributes to the base note and that limonene, linalool and octanal are essential for the characterisitic navel orange aroma. Moreover, the mixture of Group 1 plus Group 4, which contains limonene, linalool, octanal, decanal, dodecanal, geranial, neral, myrcene, α- and β-sinensal, and citronellal formed a very close approximation of the navel orange oil aroma. This Lo value method using detection thresholds was concluded to be useful for sensory evaluation of characte impact compounds in navel orange aroma.

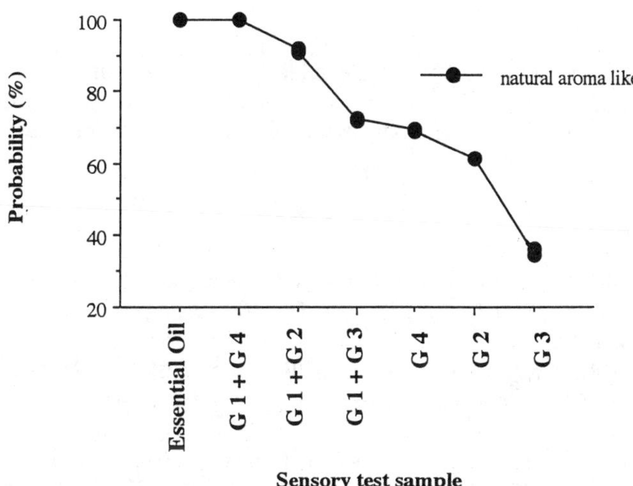

Figure 3. Probability of being the natural aroma of navel oranges without a reference sample.
G1, G2, G3 and G4 show the same meaning as the abbreviation in Fig. 2.

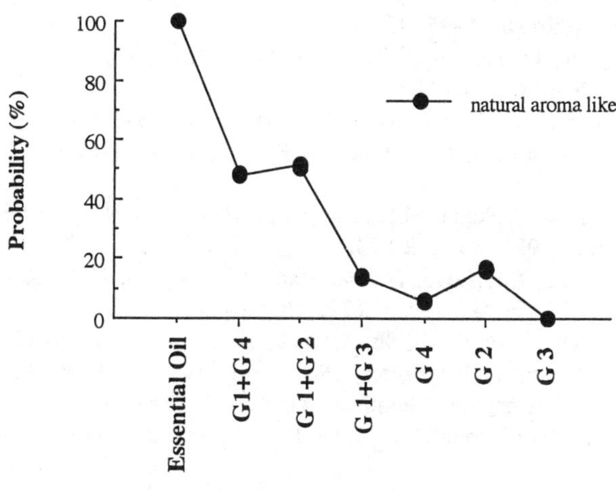

Figure 4. Probability of being the natural aroma of navel oranges with a reference citrus oil.
G1, G2, G3 and G4 show the same meaning as the abbreviation in Fig. 2.

Acknowlegements

The authors are very grateful to Dr. Nobuo Takagi, of the Fruit Tree Experimental Station, Ehime, Japan, for supplying samples of *Citrus sinensis* OSBECK, cv. Shiroyanagi (navel). We thank Ms. Yuri Izumiya and Ms. Yoko Inoue for their technical assistance.

Literature Cited

1. Fazzalari, F.A. In Composition of Odor and Taste Threshold Values Data: Fazzalari, F.A., ed., American Society for Testing Materials: Philadelphia, 1978, 1-212.
2. Stone, H.; Carregal, E.J.; Williams, B. *Life Sci.* **1966**, *5*, 2195-2201.
3. Acree, T.E.; Butts, R.M.; Nelson, R.R.; Lee, C.Y. *Anal. Chem.* **1976**, *48*, 1821-1822.
4. Ullrich, F.; Grosch. W. *Z. Lebensm. Unters. Forsch.* **1987**, *184*, 277-282.
5. Acree, T.E.; Barnard, J.; Cunningham, D.G. *Food Chem.* **1984**, *14*, 273-286.
6. Sugisawa, H.; Nakahara, K; Tamura, H. *Food Rev. Int.* **1990**, *6*, 573-589.
7. Yang, R.-H.; Otsuki. H.; Tamura, H.; Sugisawa, H. *Nippon Nogeikagaku kaishi,* **1987**, *61,* 1435-1439.
8. Guadagni, D.G.; Okano, S.; Buttery, R.G.; Burr, H.K. *Food Technol.* **1966**, *20*, 166-169.
9. Rothe, M.; Thomas, B. *Z. Lebensm. Unters. Forsh.* **1963**, *119*, 302-310.
10. Tamura, H.; Nakamoto, H.; Yang, R.-H.; Sugisawa, H. *Nippon Shokuhin Kogyo Kakkaishi,* **1995**, *42*, 887-891.
11. Buttery, R.G.; Teranishi, R.; Ling, L.C.; Turnbaugh, J.G. *J. Agric. Food Chem.* **1990**, *38*, 336-340.
12. Stevens, S.S. In *Handbook of Sensory Physiology: Principles of Receptor Physiology,* Vol. 1, Loewensen, W.R., Ed.; Springer-Verlag: Berlin, 1971; p 226.
13. Sugisawa, H.; Takeda, M.; Yang, R.-H., Takagi, N. *Nippon Shokuhin Kogyo Gakkaishi,* **1991**, *38*, 668-674.
14. Yang, R.-H.; Sugisawa, H.; Nakatani, K.; Tamura, H.; Takagi, N.; *Nippon Shokuhin Kogyo* Gakkaishi, **1992**, *39*, 16-24.
15. Tamura, H.; Yang, R.-H.; Sugisawa, H. In *Bioactive Volatile Compounds from Plants;* Teranishi, R.; Buttery, R.G.; Sugisawa, H., Eds.; ACS Symposium Series 525, American Chemical Society; Washington, DC: 1993; 121-136.
16. Leonardos, G.; Kendall, D.; Banard, N. *J. Air Pollution Control Assoc.* **1969**, *19*, 91-95.
17. Hellman, T.; Small, F.H. *J. Air Pollution Control Assoc.* **1974**, *24*, 979-982.
18. Guadagni, D.G.; Buttery, R.G., Okano, S. *J. Sci. Food Agric.* **1963**, *14*, 761-765.
19. Yamanishi, T.; Fukawa, S.; Takei, Y. *Nippon Nogeikagaku Kaishi*, **1980**, *54*, 21-25.
20. Guadagni, D.G.; Buttery, R.G.; Okano, S.; Burr, H.K. *Nature* **1963**, *200*, 1288-1289.

Chapter 28

Carotenoid-Derived Aroma Compounds: Biogenetic and Biotechnological Aspects

Peter Winterhalter

Institut für Pharmazie und Lebensmittelchemie, Universität Erlangen-Nürnberg, Schuhstrasse 19, D–91052 Erlangen, Germany

Due to their attractive aroma properties often combined with low odor thresholds carotenoid-derived aroma compounds (*synonyms*: norisoprenoids/norterpenoids) are highly esteemed by the flavor and fragrance industry. Despite considerable efforts in the understanding of the formation of carotenoid-derived flavor compounds from non-volatile precursors including the stereodifferentiation of an increasing number of aroma-relevant candidates, still little is known about the initial steps of carotenoid biodegradation. This chapter gives a survey of the present knowledge in the biogenesis of norisoprenoid compounds and outlines possible ways for their biotechnological production.

A vast number of apparently carotenoid-derived substances has been identified in plant extracts (*1-4*), but knowledge of the biochemistry of carotenoid catabolism is still extremely limited. The *in vivo* cleavage of the carotenoid chain is generally considered to be catalyzed by dioxygenase systems (*5-8*). Although all the in-chain double-bonds seem to be vulnerable to enzymatic attack, thus resulting in the formation of major fragment classes with 10, 13, 15 or 20 carbon atoms (cf. Fig.1), in fruit tissues a bio-oxidative cleavage of the 9,10 (9',10') double bond seems to be the most preferred (*5*). Since this last cleavage reaction creates the majority of aroma-relevant products, only the formation of carotenoid metabolites with a 13 carbon skeleton, i.e. C_{13}-norisoprenoids, will be discussed here.

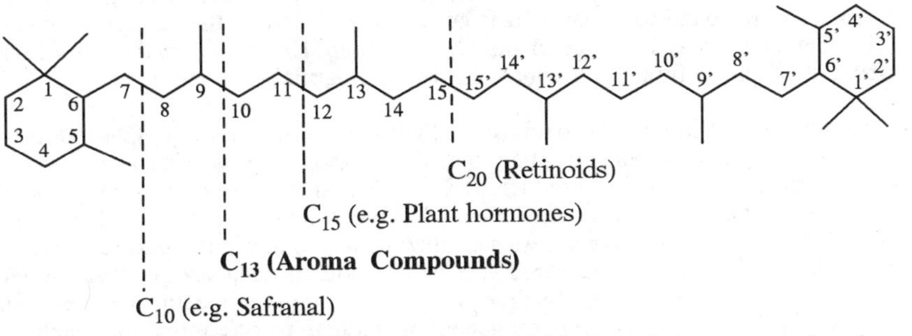

C_{20} (Retinoids)

C_{15} (e.g. Plant hormones)

C_{13} (Aroma Compounds)

C_{10} (e.g. Safranal)

Fig. 1: Major fragment classes of carotenoid biodegradation assumed to be formed by regioselective attack of dioxygenases.

0097–6156/96/0637–0295$15.00/0

Biogeneration of C_{13}-Norisoprenoid Aroma Compounds

Evidence for an Apo-carotenoid Pathway. Whereas only in rare cases, e.g. for the cyanobacterium *Microcystis*, has the carotenoid degrading system been partially characterized (*9*), little is still known about the presumed dioxygenases in higher plants. In *Microcystis*, a membrane bound, cofactor-independent, iron-containing dioxygenase has been identified which cleaves carotenoids specifically at the positions 7,8 (7',8'). Although oxygen is essential for the reaction and a high incorporation rate of ^{18}O (86%) into the carotenoid fragments was observed, the enzyme itself was reported to be extremely unstable in the presence of air. The observed instability of the dioxygenase systems is likely to be one of the reasons why knowledge about carotenoid degrading enzymes in higher plants is still missing. Thus, many fundamental aspects of carotenoid biodegradation in higher plants are rather speculative.

In the ongoing discussion whether C_{13}-norisoprenoid compounds are carotenoid breakdown products or not, three important facts are supportive for an apo-carotenoid pathway. (i) *Stereochemistry of the fragments:* In all cases examined (*10*, and refs. cited), the stereochemistry of the C_{13}-fragment was identical with the stereochemistry found in the endgroup of the presumed parent carotenoid. Since the stereospecific functionalization of the carotenoid endgroup (i.e. hydroxylation, epoxidation) is the final step in carotenoid biosynthesis, it can be concluded that the dioxygenase attack occurs after the carotenoid chain has been completely synthesized. Hence, the formation of C_{13}-norisoprenoids is likely to be a catabolic process. (ii) *Change in carotenoid concentration upon ripening or processing:* For tea (*11*), tobacco (*12*), and grapes (*13*), it was found that an increase in the amount of C_{13}-norisoprenoids correlates with a decrease in carotenoid concentration. (iii) *Identification of additional carotenoid fragments:* In addition to the C_{13}-endgroups several C_{27}-apocarotenoids as well as a C_{14}-diapocarotenoid are known to occur in fruit tissues (*14*). Based on these findings, a two-step mechanism has been proposed for the enzymatic degradation of carotenoids (*14*). In the first step, a regioselective cleavage catalyzed by a 9,10 (9',10')-dioxygenase gives rise to a C_{13}-endgroup as well as a C_{27}-fragment. The C_{27}-fragment can then undergo a second dioxygenase cleavage, thus generating an additional C_{13}-endgroup and a C_{14}-fragment, which is derived from the central portion of the carotenoid chain.

Similarily, bio-oxidative cleavage reactions involving 7,8 (7',8')- and 11,12 (11',12')-dioxygenases are plausible for the generation of C_{10}- and C_{15}-fragments. Such fragments are known constituents of saffron (*15*), quince (*16*) and starfruit (*17*). In the case of the C_{15}-isoprenoids, it is noteworthy that attractive aroma compounds were found to be derived from the central part of the carotenoid chain which remains after the cleavage of the C_{15}-endgroup. Examples are the key flavor compounds of quince fruit, i.e. marmelo oxides and marmelo lactones (*16*).

Conversion of Primary Degradation Products into Aroma Compounds. Carotenoid fragments obtained after the initial dioxygenase cleavage are the so-called primary degradation products (cf. Fig. 2). Among these frequently found plant constituents, only the ionones **1** and **2** have a low odor threshold. For the oxygenated representatives **3-6**, a series of enzymatic and acid catalyzed reactions is required to finally yield the desired aroma compound, such as the well-known ß-damascenone, theaspiranes, vitispiranes, etc. The suggested pathways for their formation involve oxidations, dehydrogenations, reductions or elimination reactions as summarized earlier (*2,18,19*). An example illustrating the above mentioned reactions is the formation of the intensely odorous C_{13}-compound ß-damascenone **7** (odor threshold: 0.002 ppb). The primary oxidative cleavage product of neoxanthin, grasshopper ketone **6**, has to undergo an enzymatic reduction (cf. Fig. 3) before finally being acid-catalyzed converted into ketone **7** (*20,21*).

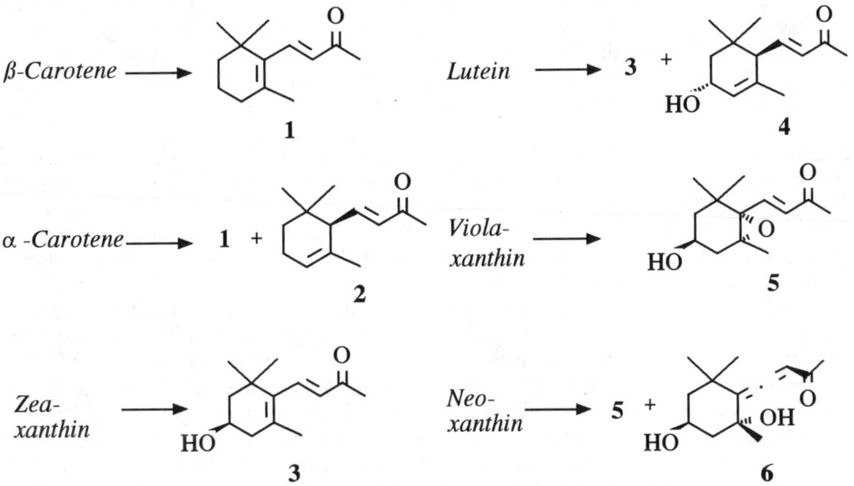

Fig. 2: Structures of primary cleavage products of common carotenoids (ß-ionone **1**, α-ionone **2**, 3-hydroxy-ß-ionone **3**, 3-hydroxy-α-ionone **4**, 3-hydroxy-5,6-epoxy-ß-ionone **5**, and grasshopper ketone **6**).

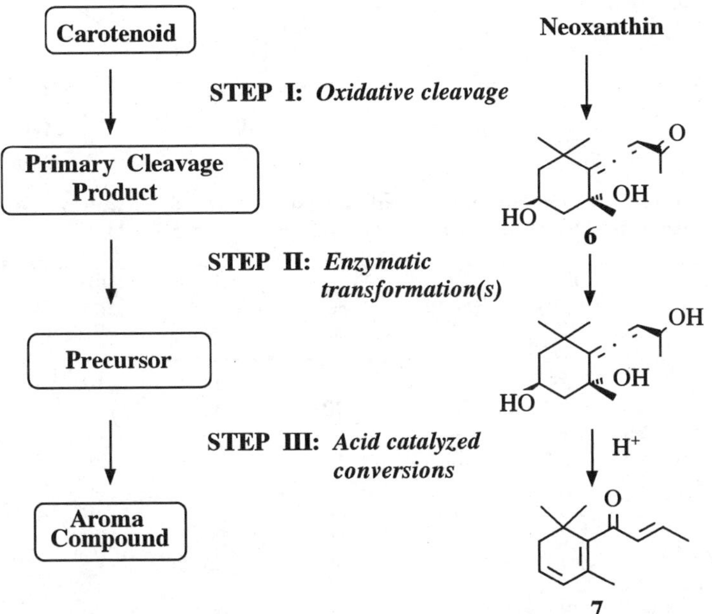

Fig. 3: General steps for the conversion of carotenoids into flavor compounds (left) and as example: ß-damascenone **7** formation from neoxanthin (right).

The three essential steps for the formation of C_{13}-norisoprenoid aroma compounds can therefore be considered to consist of (i) the initial dioxygenase cleavage, (ii) the subsequent enzymatic transformation(s) of the primary degradation product in natural tissues, and (iii) the final acid-catalyzed conversion of a flavorless polyol into an aroma compound under processing conditions.

Enantioselectivity of Biogenetic Routes. The enantiomeric composition of natural aroma compounds is known to reflect the enantioselectivity of their biogenesis (22). Thus, by elaborating the chiral composition of C_{13}-norisoprenoids in fruits, we expected to obtain some information about step II in the formation of C_{13}-norisoprenoids, i.e. the enzymatic transformation(s) of the primary carotenoid degradation products into labile aroma precursors (cf. Fig. 3).

The tremendous progress in chirospecific analysis of natural flavors in recent years was mainly due to the commercial introduction of modified cyclodextrin columns as well as multidimensional gas chromatographic (MDGC) instrumentation. An equally important prerequisite for the accurate determination of the chiral composition of fruit flavors, however, is the availability of optically pure reference samples. Since for most chiral C_{13}-volatiles the absolute configuration was not known, research has been initiated in this direction. The optically pure material can subsequently be used to determine the order of elution of different stereoisomers by enantioselective MDGC. In this way, information about the chiral composition of, e.g., theaspiranes **8**, theaspirones **9**, vitispiranes **10**, and edulans **11** in different natural substrates has been obtained (*23-26*). As examples, the enantiomeric composition of theaspiranes **8** and vitispiranes **10** in various natural sources are summarized in Table I.

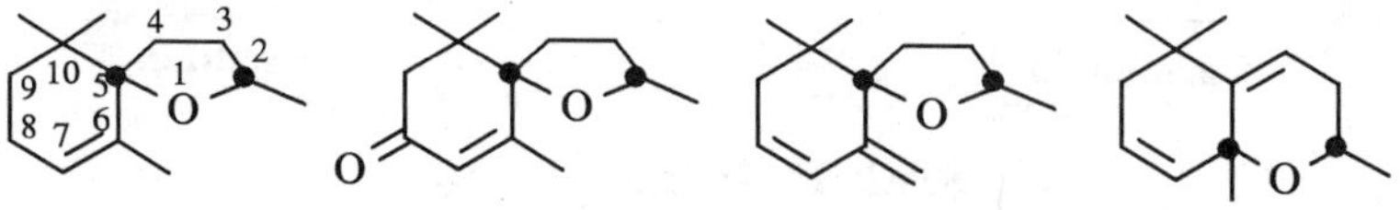

8 **9** **10** **11**

Table I. Enantiomeric composition of theaspiranes 8a-d as well as vitispirane isomers 10a-d in various natural sources (adapted from refs. *23,25*)

	8a (2R,5R) %	8c (2S,5S) %	ee %	8d (2S,5R) %	8b (2R,5S) %	ee %
guava	1	99	98	99	1	98
quince	5	95	90	94	6	88
Osmanthus	16	84	68	88	12	76
gooseberry	29	71	42	74	26	48
raspberry						
(France)	37	63	26	60	40	20
(France)	75	25	50	20	80	60
(Yugoslavia)	90	10	80	85	15	70
blackberry						
leaves*	35	65	30	64	36	28
passionfruit*	14	86	72	85	15	70
green tea*	14	86	72	84	16	68

Table I (cont.)

	10a (2R,5R) %	10b (2S,5S) %	ee %	10c (2R,5S) %	10d (2S,5R) %	ee %
vanilla	n.d.	100	100	n.d.	100	100
quince	8	92	84	18	82	64
passionfruit	24	76	52	24	76	52
gooseberry	29	71	42	74	26	48
mate tea	69	31	38	81	19	62
grape juice	62	38	24	88	12	76
leaves from						
gooseberry*	35	65	30	18	82	64
blackberry*	75	25	50	57	43	14
whitebeam*	96	4	92	99	1	98

(*obtained from bound precursors; ee enantiomeric excess; n.d. not detected)

The data collected so far (*23-28*) has shown that the norisoprenoids **8-11** are present in natural substrates in well-defined mixtures, thus enabling *inter alia* authenticity control of natural flavoring material. (*Note:* The only exception are raspberries, which showed considerable variations in their composition of isomeric theaspiranes depending on their origin). The data also shows that obviously different enzymes are operative in plants, catalyzing the formation of different optical antipodes of C_{13}-volatiles. This is evident from the fact that, e.g., vanilla beans were found to contain optically pure (2*S*)-isomers of vitispiranes **10b/d**, whereas whitebeam (*Sorbus aria*) leaves contain mainly the (2*R*)-configured isomers **10a/c** (*25*).

With regard to the metabolism of primary carotenoid degradation products, recent results obtained by Tang and Suga (*31*) have indicated a two-step mechanism for the conversion of the primary carotenoid metabolite ß-ionone **1** to **14** in *Nicotiana tabacum* plant cells (cf. Fig. 4). These results revealed, that first of all, the side-chain double bond is hydrogenated by the action of *carvone reductase* (co-factor: NADH), before the carbonyl function is enzymatically reduced in the second step. Importantly, the enantioselectivity of the latter reductase is decisive for the enantiomeric composition in fruits as outlined for theaspirane formation from the labile precursor diol **12** (cf. Fig. 5). The formation of vitispiranes **10**, edulans **11** and related C_{13}-norisoprenoids (*29,30*) is known to proceed via similar mechanisms.

NADH H$^+$ NAD$^+$ O *Carvone Reductase* **1** O *"Reductase" (not specified)* **13** OH **14**

Fig. 4: Biotransformation of the primary carotenoid metabolite ß-ionone **1** into 7,8-dihydro-ß-ionone **13** and 7,8-dihydro-ß-ionol **14** by immobilized cell cultures of *Nicotiana tabacum* according to Tang and Suga (*31*).

It is well documented that the stereochemistry of a flavor compound can determine its sensory properties as well as its aroma intensity (*28* and refs. cited). Considerable differences in sensory properties have also been found for some of the C_{13}-volatiles under investigation, as shown for theaspiranes in Fig. 5.

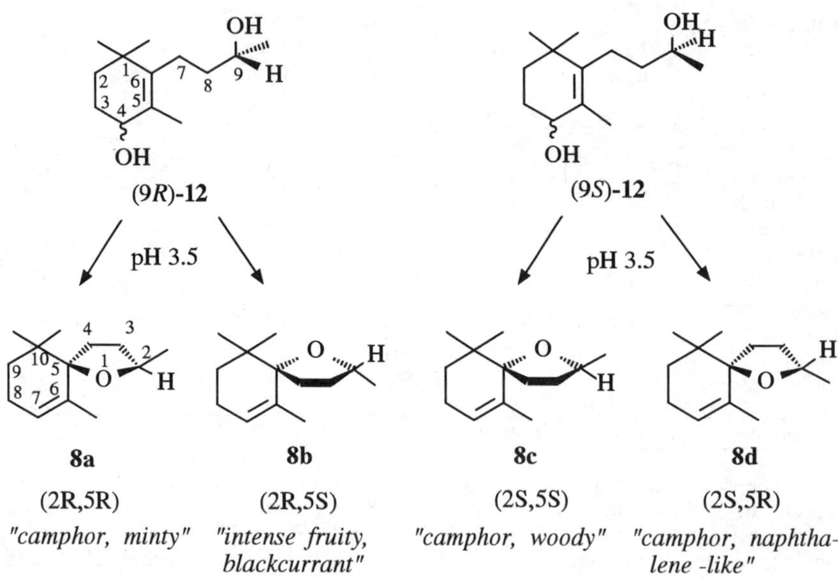

Fig. 5: Formation of isomeric theaspiranes **8a-d** from the reactive 1,6-allyldiol **12** demonstrating the decisive role of the C-9 configuration in precursor **12**.

In this case, only the (2*R*,5*S*)-isomer **8b** was found to exhibit a highly attractive *"fruity, blackcurrant"* note, whereas the (2*S*,5*R*)-configured isomer **8d** showed an unpleasant *"naphthalene-like"* note (*23*). With regard to any envisaged biotechnological production of C_{13}-volatiles it is important to keep these different sensory properties in mind and to steer the production in such a way that - ideally - only the sensory attractive stereoisomers are formed.

Biotechnological Production of C_{13}-Norisoprenoid Aroma Compounds.

In general, three major biotechnological methods are used in the production of natural flavors: use of enzymes, microorganisms, and of plant tissue or cell cultures (*32-34*). In the following, the use of these techniques for the biotechnological production of C_{13}-compounds will be discussed.

Use of Enzymes. The big advantage of enzyme applications is their well-known specificity in reaction pathways. A survey in the biotechnological use of enzymes showed that 65 % of all reported applications fell into the classes of esterolytic reactions (40%) and dehydrogenase reactions (25 %); pig liver esterase, pig pancreatic lipase, the lipase from *Candida cylindracea* and baker's yeast being the most frequently used biocatalysts (*35*). For the generation of C_{13}-aroma volatiles, however, two additional enzyme systems have gained importance, (i) hydrolases and (ii) dioxygenases, with the former converting flavorless glycosidic precursors into aroma compounds and the latter degrading carotenoid precursors into volatile C_{13}-aroma substances.

Enzymatic Cleavage of Plant Glycosides by Hydrolases. In many fruits and other plant tissues oxygenated norisoprenoids (so-called "polyols"), like other secondary metabolites, accumulate as non-volatile flavorless glycosides (*36*). These glycosidic or *"bound aroma"* fractions were found to contain important aroma precursors, some of which are easily converted into a series of attractive aroma substances under technological conditions. This finding and the fact that the glycosidic fraction often exceeds by a factor 5 to 10 the amount of the *"free aroma"* fraction, has stimulated much research on plant glycosides in recent years. The structures, occurrence and precursor function of C_{13}-norisoprenoid glycosides identified before 1994 has been the subject of review (*37*).

Enzymatic cleavage of glycosidic aroma precursors by glycosidase treatment is not only considered as a possibility to enhance the fruit specific aroma of juices or wine, but also as a method for producing natural flavors from otherwise waste material, such as peelings, skins, stems, etc. With regard to the exploitation of the glycosidic pool in plants, three aspects have to be discussed: (i) specificity of glycosidases, (ii) sensory properties of enzymatically released aglycones, and (iii) possible artefact and off-flavor formation due to enzymatic hydrolysis.

(i) ***Specificity of glycosidases.*** Although ß-D-glucosides constitute the vast majority amongst the C_{13}-glycosides isolated thus far (*37*), in some fruits, disaccharidic conjugates prevail. An example is the glycosidic fraction of grape juice or wine, which is known to contain three major disaccharidic sugar moieties, 6-O-α-L-arabinofuranosyl-ß-D-glucopyranose, 6-O-α-L-rhamnopyranosyl-ß-D-glucopyranose (rutinose), and 6-O-α-L-apiofuranosyl-ß-D-glucopyranose (*38,39*). Importantly, the common feature of the glycosidic aroma precursors is the presence of a glucose moiety or substituted glucose moiety. Based on this finding, Williams *et al.* (*40*) developed a method for estimating the glycoside content in natural substrates by using acid hydrolysis of the glycoconjugates. The released glucose is then determined enzymatically. This procedure enables *inter alia* a rapid screening for substrates rich in glycosides.

For the biotechnological production of natural C_{13}-flavoring material, first of all, appropriate substrates with high levels of glycosides are required. Furthermore, enzyme preparations with multiple glycosidase activities should be used to ensure hydrolysis of the majority of the conjugates. Such multiple glycosidase activities are known to be present in fungal derived enzyme preparations, with most of the glycosidases being *exo*-glycosidases, i.e. the intersugar linkage has first to be cleaved by the action of either an α-rhamnosidase, α-arabinosidase or ß-apiosidase before in the second step ß-D-glucosidase activity is able to liberate the aglycone (*41*). So far, only in one case, i.e. for rutinosides, an *endo*-ß-glucosidase from an *Aspergillus niger* strain is available that generates intact rutinose and the aglycone in a one-step process (*42*).

With regard to practical applications, one should also bear in mind, that most of the above mentioned enzyme preparations tolerate only low sugar and ethanol concentrations and some of them exhibit only low activity at pH values of fruit juices and wine. Detailed properties of enzyme preparations used for the hydrolysis of glycosides have been published previously (*43,44* and refs. cited). Finally, attention should be drawn to a patent that claims the production of aroma components from glycosidic precursors by using sequential enzymic hydrolysis (*45*).

(ii) ***Sensory properties of enzymatically released aglycones.*** Contrary to glycosidically bound monoterpenols (e.g. conjugates of linalool) or phenols (e.g. conjugates of raspberry ketone) which after enzymatic hydrolyses liberate an "attractive" aglycone, most of the known C_{13}-glycosides liberate an aglycone (polyol) which is flavorless. In this case, further modifications, e.g., acid-catalyzed conversion at elevated temperatures (*46,47*), are required to finally generate the odor-active form (*19* and refs. cited). Examples are shown in Fig. 6.

(iii) ***Artefact and off-flavor formation.*** In order to liberate the majority of glycosidically bound constituents, high concentrations of fungal glycosidases were routinely used in the past. The detection of aglycones that were clearly unable to form glycosidic linkages prompted more detailed studies into side-activities of the commercial enzyme preparations. Three different fungal enzyme preparations as well as a protease were reported to oxidize some of the labile C_{13}-aglycones (*48,73*). This, however, alters the aroma precursors in such a way, that they are unable to generate the desired aroma component, e.g, ß-damascenone **7** (cf. Fig. 7).

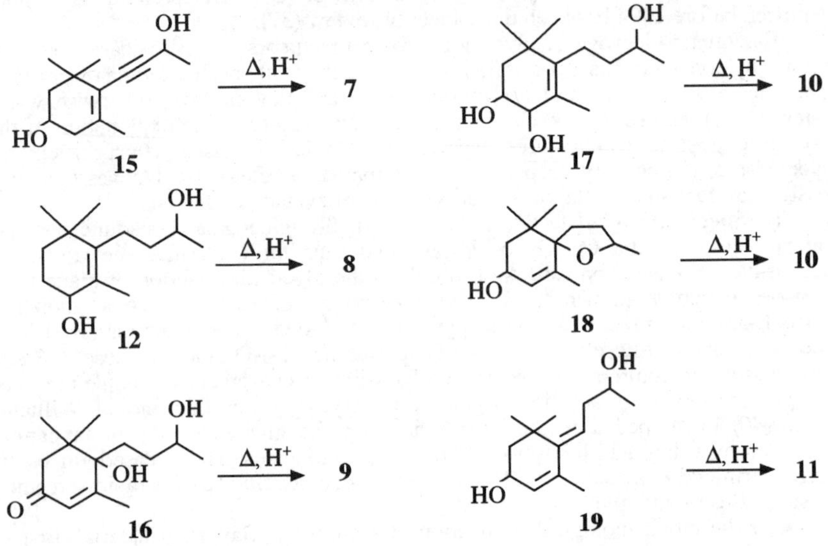

Fig. 6: Structures of glycosidically-bound aroma precursors of ß-damascenone **7**, theaspiranes **8**, theaspirones **9**, vitispiranes **10**, and edulans **11**, i.e. 3-hydroxy-7,8-didehydro-ß-ionol **15**, 4-hydroxy-7,8-dihydro-ß-ionol **12**, 7,8-dihydrovomifoliol **16**, 3,4-dihydroxy-7,8-dihydro-ß-ionol **17**, 8-hydroxytheaspirane **18**, and 3-hydroxy-*retro*-α-ionol **19** (for details cf. ref. *19*).

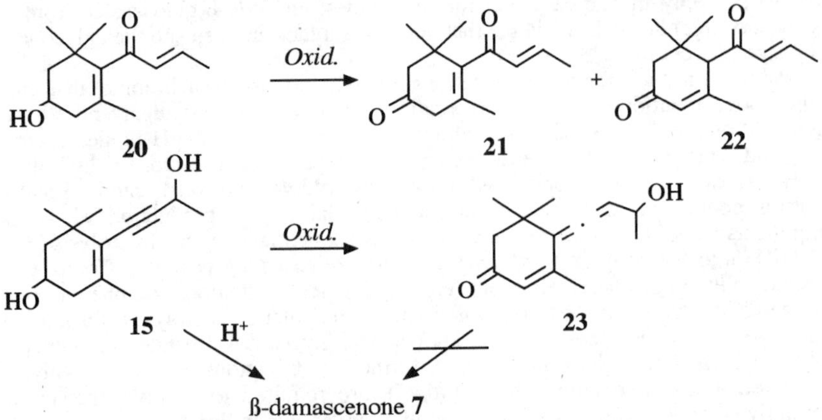

ß-damascenone **7**

Fig. 7: Formation of oxidation products **21-23** from glycosidic forms of aglycones **15** and **20** after enzymatic hydrolysis using fungal enzyme preparations (adapted from ref. *48*).

Due to the fact that the C_{13}-aroma precursors are present in a highly complex mixture together with, e.g., monoterpenols and shikimates, the possible formation of malodorous compounds after enzymic treatment has also to be considered. Although there is still little information available on the structure of enzymatically released unpleasant aroma substances, two examples may demonstrate this problem: (i) Pectinase preparations are known for their potential to liberate glycosidic precursors in wine. After application of a commercial pectinase preparation during winemaking, formation of malodorous vinylphenols was reported, which occurred following enzymatic release of cinnamic acid derivatives and subsequent transformation by decarboxylase activity from wine yeast (*49*). (ii) Among the enzymatically liberated C_{13}-aglycones in wine, polyol structures were identified which yielded the malodorous hydrocarbon 1,1,6-trimethyl-1,2-dihydronaphthalene (*"kerosene"* off-flavor) at elevated temperature (*50*). These examples illustrate that a careful sensory testing of the hydrolysates is recommended for each application, especially in cases when a sequence of glycosidase and acid treatment at elevated temperatures is used.

Co-oxidation of carotenoids. A different enzymatic approach for the biotechnological production of C_{13}-compounds uses the carotenoid fraction of plants as aroma precursors. In this case, the cleavage of the carotenoid chain is carried out by "co-oxidation" reactions using lipoxygenase or other oxidase systems.

In vitro cleavage of carotenoids has been achieved by using different types of enzymes in aqueous solutions, e.g., phenoloxidase (*11*), lactoperoxidase (*51*), and more recently xanthine oxidase (*52*). By far most co-oxidation studies, however, have been carried out with lipoxygenase (LOX) (linoleate : oxygen oxidoreductase, EC 1.13.11.12), a thoroughly studied enzyme that catalyzes the oxidation of polyunsaturated fatty acids containing a *Z,Z*-1,4-pentadiene subunit to a conjugated dienoic hydroperoxide (*53-58*). The well-known capacity of LOX to co-oxidize carotenoids has been ascribed to the fact that a large proportion of the initially formed peroxyl radicals is not directly converted into the hydroperoxides (*55*). The former remain as aggressive radicals being able to attack activated sites of the polyene chain, thus leading *inter alia* to the formation of volatile break-down products. In the case of ß-carotene, formation of ß-ionone, 5,6-epoxy-ß-ionone, and dihydroactinidiolide as major volatile products is observed (*55*). Experimental support for the suggested free radical mechanisms has been obtained in two recent studies, in which no enantioselectivity for LOX catalyzed co-oxidation reactions was observed (*59,60*). Similarily, a radical phenomenon was recently established for xanthine oxidase bleaching of carotenoids (*52*).

Co-oxidation reactions require the enzyme (LOX), polyunsatured fatty acids and the polyene compounds. Since recent results have revealed an almost equal co-oxidative reactivity of LOX isoenzymes under aerobic conditions (*60*), a crude mixture of soybean LOXs can be used. With regard to polyene compounds, natural sources rich in carotenoids, such as palm oil, plant extracts (e.g. carrots) or extracts from the algae *Dunaliella* are suitable starting materials (*61*). Initially, most co-oxidation reactions have been carried out in aqueous solutions by solubilizing linoleate as well as the carotenoid with different detergents (*53*). The limitations of this procedure were due to the fact that in the complex reaction mixture - consisting of fatty acid metabolites, carotenoid breakdown products and detergents - the yield of norisoprenoids was rather low (*56*). Nevertheless, an improved co-oxidation procedure is already used for the industrial production of natural C_{13}-flavoring material. According to a recent patent (*62*) for co-oxidation using a bioreactor, the maximum yields of ß- and α-ionone were 210 mg and 185 mg per kg of multiphase reaction mixture. The latter consisted of carrot juice as carotenoid source, soy flour as source of LOX and vegetable oil as source of unsaturated fatty acids.

The subsequent conversion of the accessible primary carotenoid breakdown products has already been carried out by using (i) plant cell cultures and (ii) microorganisms.

Use of Plant Cell Cultures for the Bioconversion of Primary Carotenoid Degradation Products. Although all attempts to generate significant amounts of volatile flavors using plant cell cultures have failed so far, such cultures have a potential for biotransformation reactions. Again, the regioselectivity and stereospecificity of the conversion of exogenously administered substrates can be considered as the major advantage of plant cell technology. Examples are summarized in ref. *63*. Disadvantages, such as low efficiency and the formation of by-products, have been overcome in part by using immobilized cells (*64*). Immobilized cells of *Nicotiana tabacum* were *inter alia* found to convert ß-ionone **1** 12-times faster than the suspension cells (*31*). Similar results have been obtained with α-ionone **2**. The structure of conversion products as well as reaction yields are outlined in Fig. 8.

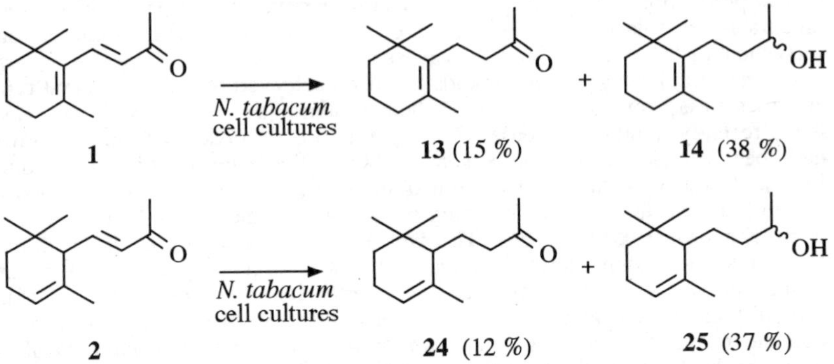

O OH + *N. tabacum* cell cultures **1** **13** (15 %) **14** (38 %)

O OH + *N. tabacum* cell cultures **2** **24** (12 %) **25** (37 %)

Fig. 8: Structures of biotransformation products of ß- and α-ionones **1** and **2** by using immobilized cells of *N. tabacum* (adapted from ref. *31*).

As mentioned earlier, in a first step the carbon-carbon double bond adjacent to the carbonyl group in structures **1** and **2** is reduced by *carvone reductase*, thus yielding ketones **13** and **24**. The production of hydrogenated products **13** and **24** was maximised by conducting the biotransformation in a medium with a pH value close to the optimal pH of *carvone reductase*. In the medium usually employed for cell culturing (pH 5.2) a subsequent reduction of the keto-function took place, thus generating C_{13}-alcohols **14** and **25** as major reaction products. Contrary to previous results reported for comparable biotransformation reactions carried out with carvone and pulegone (*65*), a relatively poor stereoselectivity in the enzymatic conversion of ionones was observed. The enantiomeric excess values for alcohols **14** and **25** were found to be only 7 % and 14 % (*S*-configured isomer), respectively (*31*). Nevertheless, these examples demonstrate the ability of plant cell technology to convert primary carotenoid degradation products, i.e. α,ß-unsaturated ketones, into saturated alcohols. Importantly, such a reaction sequence is essential in the biogeneration of the immediate precursors **12**, **16**, and **17** of the important C_{13}-volatiles theaspiranes **8**, theaspirones **9**, and vitispiranes **10** (cf. Fig. 6).

Use of Microorganisms for the Bioconversion of Primary Carotenoid Degradation Products. Microbial systems are frequently used for biotransformation reactions, especially in cases where a functionalization (hydroxylation, epoxidation and multienzyme controlled oxidation) of the administered substrate has to be carried out (*34* and refs. cited). Such conversion reactions are still restricted to the use of intact organisms in the form of growing, resting or immobilized cultures. The limitation to intact cells has been reported to be due to the instability of the oxidizing enzymes and their dependence on cofactor systems (*66*). Whereas particular attention has been directed to the microbial conversion of monoterpenes, only a few reports of microbial transformations of C_{13}-norisoprenoids have been reported so far (*67-72*).

Mikami *et al.* (*67,68*) have first investigated the microbial transformation of ß-ionone **1** using a strain of the fungus *Aspergillus niger*. In the complex reaction mixture obtained, which exhibited an odor similar to tobacco, two major products were identified, i.e. (*R*)-4-hydroxy-ß-ionone **26** and (*S*)-2-hydroxy-ß-ionone **27**. Krasnobajew and Helmlinger (*69*) extended this research by performing a screening of appropriate microorganisms for ß-ionone bioconversions. Among the tested microorganisms the closely related genera of *Botryodiplodia*, *Botryoshaeria* and *Lasiodiplodia* showed a relatively high activity in biotransforming ß-ionone. The most suitable fungus was reported to be *Lasiodiplodia theobromae* (synonym: *Diplodia gossipina*). Fermentation of ß-ionone with pre-grown cultures of this species yielded a complex mixture of metabolites (*69*). Again a pleasant, tobacco-like odor was observed. Importantly, at the beginning of the fermentation hydroxylases were found to be predominantly active, thus generating oxygenated C_{13}-structures **26**, **28** and **29** as major products. In the later phase of the fermentation, however, *Lasiodiplodia* was found to degrade the C_{13}-structures in a *Baeyer-Villiger* type of oxidation to C_{11}-compounds **30-32** by shortening the side-chain by a C_2-unit (Fig. 9). Further biotransformation reactions using the fungus *Botrytis cinerea* were carried out with isomeric damascones as substrates (*70*).

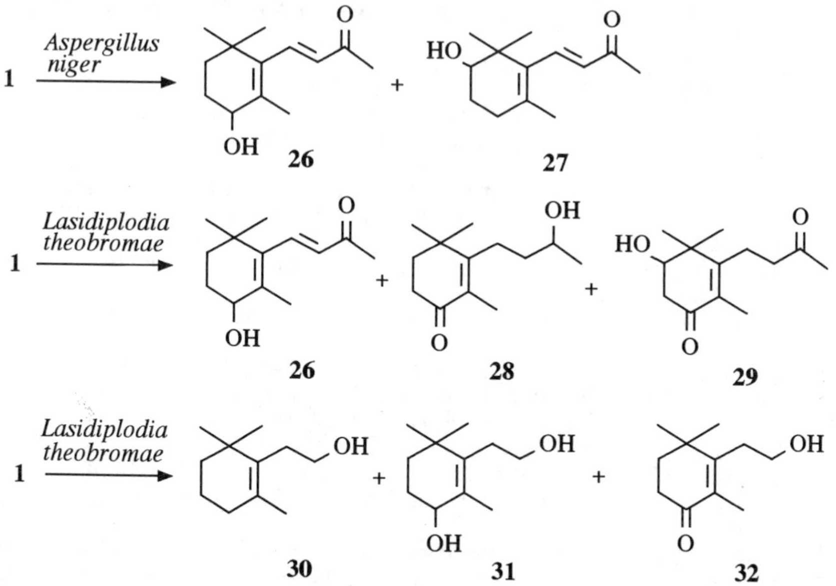

Fig. 9: Structures of major microbial transformation products of ß-ionone **1** according to refs. *67-69*.

Conclusions

Although there are still many unanswered questions with regard to the enzymatic pathways of carotenoid biodegradation, the first approaches in using the potential of carotenoids as flavor precursors appear to be promising. It has been demonstrated that standard biotechnological procedures already allow a production of certain C_{13}-norisoprenoid aroma volatiles from carotenoids as precursor substances. For a major breakthrough, however, more detailed knowledge about the carotenoid cleavage enzymes is required. Progress in this area will finally open the avenue for a biotechnological production of C_{13}-aroma substances.

Acknowledgement

Dr. G. Skouroumounis is thanked for his helpful comments on this paper. The Deutsche Forschungsgemeinschaft, Bonn, is thanked for funding the research.

Literature Cited

1. Ohloff, G.; Flament, I.; Pickenhagen, W. *Food Rev. Int.* **1985**, *1*, 99-148.
2. Wahlberg, I.; Enzell, C.R. *Nat. Prod. Rep.* **1987**, *4*, 237-276.
3. Williams, P.J.; Sefton, M.A.; Wilson, B. In *Flavor Chemistry - Trends and Developments*; Teranishi, R.; Buttery, R.G.; Shahidi, F., Eds.; ACS Symp. Ser. 388; American Chemical Society: Washington, DC, 1989, pp. 35-48.
4. Winterhalter, P.; Schreier, P. In *Bioflavour '87*; Schreier, P., Ed.; W. de Gruyter: Berlin, 1988, pp. 255-273.
5. Enzell, C.R. *Pure & Appl. Chem.* **1985**, *57*, 693-700.
6. Rock, C.D.; Heath, T.G.; Gage, D.A.; Zeevart, J.A.D. *Plant Physiol.* **1991**, *97*, 670-676.
7. Erdman, J. *Clin. Nutr.* **1988**, *7*, 101-106.
8. Winterhalter, P.; Schreier, P. *Food Rev. Int.* **1995**, *11*, 237-254.
9. Jüttner, F.; Höflacher, B. *Arch. Microbiol.* **1985**, *141*, 337-343.
10. Enzell, C.R.; Wahlberg, I.; Aasen, A.J. *Progress Chem. Org. Natural Products* **1977**, *34*, 1-79.
11. Sanderson, G.W.; Co, H.; Gonzales, J.G. *J. Food Sci.* **1971**, *36*, 231-236.
12. Enzell, C.R. In *Flavor '81*; Schreier, P., Ed.; W. de Gruyter: Berlin, 1981, pp. 449-478.
13. Razungles, A.; Bayonove, C.L.; Cordonnier, R.E.; Sapis, J.C. *Am. J. Enol. Vitic.* **1988**, *39*, 44-48.
14. Eugster, C.H.; Märki-Fischer, E. *Angew. Chem. Int. Ed. Engl.* **1991**, *30*, 654-672.
15. Pfander, H.; Schurtenberger, H. *Phytochemistry* **1982**, *21*, 1039-1042.
16. Lutz, A.; Winterhalter, P. *J. Agric. Food Chem.* **1992**, *40*, 1116-1120.
17. Lutz, A.; Winterhalter, P. *Phytochemistry* **1994**, *36*, 811-812.
18. Williams, P.J.; Sefton, M.A.; Francis, I.L. In *Flavor Precursors - Thermal and Enzymatic Conversions*; Teranishi, R.; Takeoka, G.R.; Güntert, M., Eds.; ACS Symp. Ser. 490; American Chemical Society: Washington, DC, 1992, pp. 74-86.
19. Winterhalter, P. In *Flavor Precursors - Thermal and Enzymatic Conversions*; Teranishi, R.; Takeoka, G.R.; Güntert, M., Eds.; ACS Symp. Ser. 490; American Chemical Society: Washington, DC, 1992, pp. 98-115.

20. Ohloff, G.; Rautenstrauch, V.; Schulte-Elte, K.H. *Helv. Chim. Acta* **1973**, *56*, 1503-1513.
21. Isoe, S.; Katsumura, S.; Sakan, T. *Helv. Chim. Acta* **1973**, *56*, 1514-1516.
22. Engel, K.H.; Albrecht, W.; Tressl, R. In *Aroma - Perception, Formation, Evaluation*; Rothe, M.; Kruse, H.P., Eds.; Deutsches Institut für Ernährungsforschung: Potsdam-Rehbrücke, 1995, pp. 229-246.
23. Schmidt, G.; Full, G.; Winterhalter, P.; Schreier, P. *J. Agric. Food Chem.* **1992**, *40*, 1188-1191.
24. Full, G.; Winterhalter, P.; Schmidt, G.; Herion, P.; Schreier, P. *HRC* **1993**, *16*, 642-644.
25. Herion, P.; Full, G.; Winterhalter, P.; Schreier, P.; Bicchi, C. *Phytochem. Anal.* **1993**, *4*, 235-239.
26. Schmidt, G.; Full, G.; Winterhalter, P.; Schreier, P. *J. Agric. Food Chem.* **1995**, *43*, 185-188.
27. Werkhoff, P.; Brennecke, S.; Bretschneider, W. *Chem. Mikrobiol. Technol. Lebensm.* **1991**, *13*, 129-152.
28. Werkhoff, P.; Brennecke, S.; Bretschneider, Güntert, M.; Hopp, R.; Surburg, H. *Z. Lebensm. Unters. Forsch.* **1993**, *196*, 307-328.
29. Dollmann, B.; Full, G.; Schreier, P.; Winterhalter, P.; Güntert, M.; Sommer, H. *Phytochem. Anal.* **1995**, *6*, 106-111.
30. Schmidt, G.; Neugebauer, W.; Winterhalter, P.; Schreier, P. *J. Agric. Food Chem.* **1995**, *43*, 1898-1902.
31. Tang, Y.X.; Suga, T. *Phytochemistry* **1994**, *37*, 737-740.
32. Drawert, F. In *Bioflavour '87*; Schreier, P., Ed.; W. de Gruyter: Berlin, 1988, pp. 3-32.
33. Winterhalter, P.; Schreier, P. In *Flavor Science - Sensible Principles and Techniques;* Acree, T.E.; Teranishi, R., Eds.; ACS Professional Reference Book, American Chemical Society: Washington, DC, 1993, pp. 225-258.
34. Berger, R.G. *Aroma Biotechnology*, Springer: Berlin, Heidelberg, 1995.
35. Crout, D.H.G.; Christen, M. In *Modern Synthetic Methods 1989;* Scheffold, C.R. Ed.; Springer: Berlin, Heidelberg, 1989, pp. 1-114.
36. Williams, P.J.; Sefton, M.A.; Marinos, V.A. In *Recent Developments in Flavor and Fragrance Chemistry;* Hopp, R.; Mori, K.; Eds.; VCH Verlagsgesellschaft: Weinheim, Germany, 1993, pp. 283-290.
37. Winterhalter, P.; Schreier, P. *Flav. Frag. J.* **1994**, *9*, 281-287.
38. Williams, P.J.; Strauss, C.R.; Wilson, B.; Massy-Westropp, R.A. *Phytochemistry* **1982**, *21*, 2013-2020.
39. Voirin, S.; Baumes, R.L.; Bitteur, S.; Günata, Z.Y.; Bayonove, C.L. *J. Agric. Food Chem.* **1990**, *38*, 1373-1378.
40. Williams, P.J.; Cynkar; W.; Francis, L.; Gray, J.D.; Iland, P.G.; Coombe, B.G. *J. Agric. Food Chem.* **1995**, *43*, 121-128.
41. Günata, Y.Z.; Bitteur, S.; Brillouet, J.M.; Bayonove, C.L.; Cordonnier, R.E. *Carbohydr. Res.* **1988**, *134*, 139-149.
42. Shoseyov, O.; Bravdo, B.A.; Ikan, R.; Chet, I. *Phytochemistry* **1988**, *27*, 1973-1976.
43. Günata, Z.; Dugelay, I.; Sapis, J.C.; Baumes, R.; Bayonove, C. In *Progress in Flavour Precursor Studies - Analysis, Generation, Biotechnology;* Schreier, P.; Winterhalter, P., Eds.; Allured Publ.: Carol Stream, IL, 1993, pp. 219-234.
44. Williams, P.J. In *Flavor Science - Sensible Principles and Techniques;* Acree, T.E.; Teranishi, R., Eds.; ACS Professional Reference Book, American Chemical Society: Washington, DC, 1993, pp. 287-308.
45. Gunata, Z.; Bitteur, S.; Baumes, R.; Brillouet, J-M.; Tapiero, C.; Bayonove, C.; Cordonnier, R. *European Patent Appl. EP O332 281 A1* (1989).

46. Francis, I.L.; Sefton, M.A.; Williams, P.J. *J. Sci. Food Agric.* **1992**, *59*, 511-520.
47. Francis, I.L.; Sefton, M.A.; Williams, P.J. *Am. J. Enol. Vitic.* **1994**, *45*, 243-251.
48. Sefton, M.; Williams, P.J. *J. Agric. Food Chem.* **1991**, *39*, 1994-1997.
49. Dugelay, I.; Günata, Z.; Bitteur, S.; Sapis, J.C.; Baumes, R.; Bayonove, C. In *Progress in Flavour Precursor Studies - Analysis, Generation, Biotechnology;* Schreier, P.; Winterhalter, P., Eds.; Allured Publ.: Carol Stream, IL, 1993, pp. 189-193.
50. Winterhalter, P. *J. Agric. Food Chem.* **1991**, *39*, 1825-1829.
51. Ekstrand, B.; Björck, L. *J. Agric. Food Chem.* **1986**, *34*, 412-415.
52. Bosser, A.; Belin, J.M. *Biotechnol. Prog.* **1994**, *10*, 129-133.
53. Ben Aziz, A.; Grossman, S.; Ascarelli, I.; Budowski, P. *Phytochemistry* **1971**, *10*, 1445-1452.
54. Grosch, W.; Laskawy, G.; Weber, F. *J. Agric. Food Chem.* **1976**, *24*, 456-459.
55. Weber, F.; Grosch, W. *Z. Lebensm. Unters. Forsch.* **1976**, *161*, 223-230.
56. Hohler, A. *Thesis*, Technische Universität München, Germany, 1986.
57. Grosch, W. *Lebensmittelchem. Gerichtl. Chem.* **1976**, *30*, 1-8.
58. Barimalaa, I.S.; Gordon, M.H. *J. Agric. Food Chem.* **1988**, *36*, 685-687.
59. Matsui, K.; Kajiwara, T.; Hatanaka, A.; Waldmann, D.; Schreier, P. *Biosci. Biotech. Biochem.* **1994**, *58*, 140-145.
60. Waldmann, D.; Schreier, P. *J. Agric. Food Chem.* **1995**, *43*, 626-630.
61. Eisenbrand, G.; Schreier, P., Eds.; *Römpp Lexikon Lebensmittelchemie*, Thieme: Stuttgart, 1995, p. 161.
62. BFA Laboratoires, *International Patent Appl. WO 94/08028* (1994)
63. Suga, T.; Hirata, T. *Phytochemistry* **1990**, *29*, 2393-2406.
64. Knorr, D. In *Biocatalysis in Agricultural Biotechnology*; Whitaker, J.R.; Sonnet, P.E., Eds.; ACS Symp. Ser. 389; American Chemical Society: Washington, DC, 1988, pp. 65-80.
65. Suga, T.; Tang, Y.-X. *J. Nat. Prod.* **1993**, *56*, 1406-1409.
66. Abraham, W.-R.; Arfmann, H.-A.; Stumpf, B.; Washausen, P.; Kieslich, K. In *Bioflavour '87*; Schreier, P., Ed.; W. de Gruyter: Berlin, 1988, pp. 399-414.
67. Mikami, Y.; Watanabe, E.; Fukunaga, Y.; Kisaki, T. *Agric. Biol. Chem.* **1978**, *42*, 1075-1077.
68. Mikami, Y.; Fukunaga, Y.; Arita, M.; Kisaki, T. *Appl. Environ. Microbiol.* **1981**, *41*, 610-617.
69. Krasnobajew, V.; Helmlinger, D. *Helv. Chim. Acta* **1982**, *65*, 1590-1601.
70. Schoch, E.; Benda, I.; Schreier, P. *Appl. Environ. Mikrobiol.* **1991**, *57*, 15-18.
71. Yamazaki, Y.; Hayashi, Y.; Arita, M.; Hieda, T.; Mikami, Y. *Appl. Environ. Microbiol.* **1988**, *54*, 2354-2360.
72. Hartman, D.A.; Pontones, M.E.; Kloss, V.F.; Curley, R.W.; Robertson, L.W. *J. Nat. Prod.* **1988**, *51*, 947-953.
73. Nakanishi, T.; Konishi, M.; Murata, H.; Inada, A.; Fujii, A.; Tanaka, N.; Fujiwara, T. *Chem. Pharm. Bull.* **1990**, *38*, 830-832.

Chapter 29

Biotechnology and the Development of Functional Foods: New Opportunities

Alvin L. Young and Daniel D. Jones

Office of Agricultural Biotechnology, Cooperative State Research, Education, and Extension Service, U.S. Department of Agriculture, 14th Street and Independence Avenue, Southwest, Washington, DC 20250–0904

The ability to add or delete specific genes and change the levels of enzyme activities in transgenic organisms has given scientists a new way of manipulating and exploring metabolism. Discoveries made through these biotechnological approaches are changing the way that plants and animals are grown, boosting their value to growers, processors, and consumers alike. Foods produced in new ways that meet consumer demands for safety, quality, nutrition, and taste are now approaching commercialization. The greatest benefit to the public and to public health, however, may be the development of compositionally enhanced food products, sometimes called functional foods, that have positive health outcomes beyond the satisfaction of basic nutritional requirements. Growing scientific evidence suggests that specific food components can promote human health by reducing disease risk and that these components can be incorporated into food products by the newer methods of molecular genetics.

Food and nutrition play a pivotal role in optimizing personal health, reducing the risk of diet-related disease and in maintaining societal productivity. Dietary factors profoundly affect growth, development, and the risks of such chronic diseases as diabetes, heart disease, cancer, and osteoporosis. For the majority of people who do not smoke, do not drink excessive amounts of alcohol, and are not exposed to environmental hazards on a continuing basis, diet is the most significant controlling factor determining their long-term health. Inadequate nutrition underlies many chronic conditions, but the relationship between the genetic make-up of an individual, the diet, and disease is only now becoming the focus of government research funding. The Committee on Health, Safety, and Food of the National Science and Technology Council *(1)*, recently noted that the challenge for the 21st century will be to apply the knowledge generated by the tools of molecular biology to the understanding of nutrient-gene interactions, to produce healthier foods, and to detect and prevent disease. The Committee proposed a research agenda to meet this challenge.

Phytochemicals

In addition to nutrients needed to satisfy basic nutritional requirements of humans, many foods contain components, sometimes called *phytochemicals*, that appear to exhibit positive effects in disease prevention. Phytochemicals are substances found in edible fruits and vegetables that may be ingested by humans and that exhibit a potential for modulating human metabolism or the metabolism of cancerous cells, in a manner favorable for cancer prevention *(2)*, or more broadly, for disease prevention in general.

The National Cancer Institute, through its Experimental Food Program *(3)*, has identified many phytochemicals that can interfere with and potentially block the biochemical pathways that lead to malignancy in animals. These phytochemicals fall into approximately 14 classes of substances that are present in common foods. They include sulfides, phytates, flavonoids, glucarates, carotenoids, coumarins, monoterpenes, triterpenes, lignans, phenolic acids, indoles, isothiocyanates, and polyacetylenes.

Examples of phytochemicals and their biological effects include isothiocyanates and sulforaphane which are found in vegetables such as broccoli and have been shown to trigger enzyme systems that block or suppress cellular DNA damage and reduce tumor size in animal studies *(4)* Allylic sulfides, found in onions and garlic, can enhance immune function, increase the production of enzymes that help to excrete carcinogens, decrease the proliferation of tumor cells, and reduce serum cholesterol levels *(5)*. Isoflavonoids in soy have also been shown to reduce serum cholesterol levels in humans *(6)*.

Our supermarkets today provide an abundance in food choices, yet most Americans do not follow the research based advice on good nutrition summarized in _The Dietary Guidelines for Americans_. Instead, people plan and select a diet of traditional foods that they believe will meet their needs. In the best of cases, this can result in a balanced, healthful diet that makes a positive contribution to the quality of life. In the worst of cases, it can result in over-consumption of some nutrients, under-consumption of other nutrients, and a generally unbalanced diet that results in a general or specific decrement in health. Our goal must be to consume the variety of foods necessary to obtain in the appropriate amounts (i.e., the levels associated with the lowest prevalence of disease and longest life), all the nutrients and protective compounds found in food *(1)*.

Functional Foods

"Functional food" may be broadly defined as any modified food or food ingredient that may provide a health benefit to humans beyond the traditional nutrients it contains *(7)*. Another term that is sometimes used is "designer food" which usually denotes a processed food that is supplemented with food ingredients naturally rich in recognized disease-preventing substances *(7)*. A third term sometimes used is "nutraceutical" which is any substance that may be considered a food or part of a food and provides medical or health benefits, including the prevention of disease *(8)*. These functional food concepts can be viewed as an outgrowth of the medical food industry which has served hospitalized and other patients for decades with hundreds of specialized medical food products *(7)*.

There are several ways of providing functional foods for health in a diet *(9)*. They include:

- o selection by the consumer of an appropriate diet of traditional foods;
- o traditional plant breeding;
- o selective processing;
- o addition of chemical substances to food, and;
- o biotechnology or genetic engineering.

Biotechnology is thus only one method of producing functional foods, but it is one that may find favor among food developers in the future because of its molecular precision, genetic specificity, and relative speed. A brief survey of field test permits issued by USDA over the past few years provides a snapshot in time of the types of genetic modifications currently being practiced in crop plants.

From 1987 to the end of the 1994 growing season, USDA issued over 1200 permits or notification receipts for field tests of genetically modified plants. As shown in Table I, the largest percentage of these, 30%, was for herbicide tolerance with changes in product quality in second place, accounting for about 24%. Insect resistance, viral resistance, fungal resistance, and others accounted for the remainder. Thus, for this group of crops about three quarters of the field tests were undertaken for the purpose of agronomic improvement rather than for changes in product quality.

Among the genetic modifications for product quality, shown in Table II, the largest percentage was for delayed ripening at 32%. Modified oil profile, modified metabolic enzymes, modified storage protein, increased solids, and others accounted for the remainder. The "other" category under product quality includes such modifications as antisense manipulation, increased fatty acids, increased starch, modified nutritional value, modified color, and chalone suppression.

Some of the changes shown in Table II might be useful for developing new functional foods. In time there will probably be many other desirable goals for which genetic modification of plants is undertaken including insertion of genes for specific phytochemical substances. The scientific and technical options will be vastly multiplied as further research identifies specific genes of food and agricultural importance, elucidates the genome maps of agricultural plants, animals, and microbes, and develops new transformation techniques.

Biotechnology Research at USDA

USDA currently funds $212 million dollars in biotechnology programs. Basic research accounts for $162 million of this, the largest part of which is agricultural research, with smaller expenditures on agriculturally related energy, environmental, manufacturing/bioprocessing, and biotechnology risk assessment research. USDA also

Table I. USDA Permits/Notifications for Field Tests Of Genetically Modified Plants and Microorganisms: Distribution by Genetic Trait[1]

Trait	Number	Percent
Herbicide tolerance	381	30.3
Product quality	301	23.9
Insect Resistance	265	21.1
Viral Resistance	175	13.9
Other	100	8.0
Fungal Resistance	35	2.8
Totals	1257	100.1

[1]From 1987 to end of growing season.

funds biotechnology infrastructure, regulatory and information services programs. The 1996 Fiscal Year Budget for biotechnology programs at USDA continues the investment in this technology.

Genome Research

Just as improvements in mapmaking facilitated the age of global exploration in the 15th and 16th centuries, so will genome mapping in crops and animals make significant contributions to improvements in the breeding of plants and animals and to the design of foods prepared from them. The Department of Agriculture at the Federal level and States through the State Experiment Stations support genome research in both plants and animals. Areas of emphasis in the plant genome program include:

- o Construction of genomic maps of appropriate resolution for different species
- o Detailed mapping and sequencing of specific regions of the plant genome
- o Development of new mapping, cloning, and sequencing technologies

The USDA Plant Genome Program provided 76 awards in 1991, and 104 awards in 1994. Dollar amounts were about $10.5 million in 1991, $11.7 million in 1994, with over $12 million in awards in 1992 and 1993. A partial list of the plant species funded in the Plant Genome Program (Table III) shows that many of them have also been identified as significant sources of phytochemicals that may play a role in maintaining human health.

Some functional foods may come from animal sources and USDA also funds an animal genome program focusing on identifying genetic mechanisms and gene mapping. Current areas of emphasis include:

- o gene mapping and identification of genes;
- o identification and mapping of DNA segregation markers;
- o interactions between nuclear and cytoplasmic genes;
- o development and application methods to modify the animal genome;
- o genome organization, including that coding for the immune system;
- o identification of genetic diversity; and
- o genetic localization of economically important traits.

Biotechnology and Nutrition

A food biotechnology workshop, co-sponsored by USDA and the National Agricultural Biotechnology Council (NABC) in 1990, addressed several issues in the improvement of nutritional quality of foods through biotechnology *(10)*. The participants at the workshop agreed on several research and development objectives regarding "antinutrients" and "protective substances"[1] in food. One was to decrease antinutrient content and increase phytogenic substances of health significance in foods. Advantages of this approach enumerated by the workshop participants included:

- o Increased levels of certain protective substances may reduce cancer risk;
- o Reduction of certain antinutrients may improve bioavailability of nutrients;
- o Reduction of antinutrients may make previously inedible foods available for food utilization; and
- o Removal of antinutrients by biotechnology may reduce current processing costs.

Another objective the participants at the NABC workshop identified was to explore development of specific foods for specific subpopulations. Young and Lewis (*11*), for example, have discussed specific designer foods and nutraceuticals for children. Genes for the synthesis of desirable or therapeutic compounds, such as the potential cancer-preventing genes for β-carotene in cauliflower, could be added

[1] "Antinutrients" are substances in food such as goitrogens, trypsin inhibitor, oxalic acid, and gossypol. "Protective substances" include many of the classes of phytochemicals referred to earlier in the text.

Table II. USDA Permits/Notifications for Field Tests of Plants Genetically Modified for Product Quality

Intended Technical Effect	Percent
Delayed ripening	32
Modified oil profile	17
Modified metabolic enzymes	14
Modified storage protein	14
Increased solids	12
Other[1]	11
Total	100

[1]Includes antisense suppression, increased fatty acids, increased starch, modified nutritional value, modified color, and chalone suppression.

Table III. Selected Plant Species Included in USDA Plant Genome Awards

Alfalfa	Citrus	Pea
Apple	Cucumber	Peach
Arabidopsis	Flax	Pepper
Asparagus	Grape	*Prunus*
Barley	Legumes	Rice
Bean	Lettuce	Rye
Blueberry	Maize	Soybean
Brassica	Mungbean	Spinach
Cabbage	Oats	Strawberry
Carrot	Onion	Wheat

Source: Datko, A.; Kaleikau, E.; Heller, S.; Miksche, J.; Smith, G.; Bigwood, D. "USDA Plant Genome Research Program Progress Report, *Probe* **1994,** *4,* 1-6.

through genetic engineering of foods popular with children and adolescents. Another approach would be to produce fruit and vegetable drinks by liquefying fruits and vegetables with pectinolytic enzyme mixtures such as pectinase and macerase. Such drinks could provide nutrients, help to lower blood cholesterol levels, and help to fight obesity and tooth decay *(11).*

The participants at the NABC nutrition workshop concluded by recommending that the research community identify active phytogenic components and their mechanisms of action in plant cells, and determine the safe and adequate intake ranges of phytogenic substances in foods consumed by humans *(10).*

The American Dietetic Association *(12)* has also taken a position on functional foods. "It is the position of The American Dietetic Association (ADA) that specific substances in foods (eg, phytochemicals as naturally occurring components and functional food components) may have a beneficial role in health as part of a varied diet. The Association supports research regarding the health benefits and risks of these substances. Dietetics professionals will continue to work with the food industry and government to ensure that the public has accurate scientific information in this emerging field" *(12).*

Conclusion

The development of functional foods for human health represents an important opportunity to improve the health and well-being of vast numbers of people worldwide. Through processed foods that appeal to a variety of consumers, functional foods have the potential to improve the nutrition and health of people who may not have the knowledge, time, or inclination to seek out specific health-promoting foods on their own. In addition, functional foods offer the possibility of extending current uses and finding new uses of agricultural products. There is a need for continued progress in identifying active phytogenic substances and determining their safe and adequate intake ranges for humans. It will also be important for producers, consumers, and government to stay in communication with each other regarding these and other exciting new biotechnological developments in food and agricultural science and technology.

Literature Cited

1. Meeting The Challenge: A Research Agenda for America's Health, Safety and Food. Committee on Health, Safety and Food, National Science and Technology Council, Executive Office of the President, Washington, DC, 1996, 54 p.
2. Jenkins, M.L.Y. *Food Technology* **1993,** *47,* 76, 78-79.
3. Caragay, A.B. *Food Technology* **1992,** *46,* 65-68.
4. Wattenberg, L.; Lipkin, M.; Boone, C.W.; Kelloff, G.J.; eds. *Cancer Chemoprevention,* 1992, CRC Press, Boca Raton, Florida.
5. Nishino, H.; Iwashima, A.; Itakura, Y.; Matsuura, H.; Fuwa, T. *Oncology* **1989,** *46,* 277-280.

6. Carroll, K. *J. Am. Diet. Assoc.* **1991,** *91,* 820-827.
7. Committee on Opportunities in the Nutrition and Food Sciences, Food and Nutrition Board, Institute of Medicine; *Opportunities in the Nutrition and Food Sciences, Research Challenges and the Next Generation of Investigators,* 1994, Thomas, P.R. and Earl, R. eds., National Academy Press, Washington, D.C.
8. Anonymous, *Food Technology* **1992,** *46,* 77-79.
9. Fitch-Haumann, B. *INFORM* **1993,** *4,* 344-346, 348-350, 352.
10 National Agricultural Biotechnology Council, *Agricultural Biotechnology, Food Safety and Nutritional Quality for the Consumer,* NABC Report 2, 1990, J.F. MacDonald, ed., National Agricultural Biotechnology Council, Ithaca, NY.
11. Young, A.L. and Lewis, C.G. *Pediatric Clin. N. Amer.* **1995,** *42,* 917-930.
12 Bloch, A.; Thomson, C.A. *J. Am. Diet. Assoc.* **1995,** *95,* 493-496.

INDEXES

Author Index

Affiliation Index

Subject Index

D

G

R

S